NUMBER THEORY

NUMBER THEORY

W. Narkiewicz
Wroclaw University, Poland

Translated by
S. Kanemitsu
Kyushu University, Japan

World Scientific

World Scientific Publishing Co Pte Ltd
P O Box 128
Farrer Road
Singapore 9128

ISBN 9971-950-13-8
     9971-950-26-X pbk

Printed in Singapore by Singapore National Printers (Pte) Ltd.

FOREWORD

This book aims to present the fundamentals of number theory, one of the oldest mathematical disciplines. An exhaustive treatment of this large independent field of mathematics is obviously impossible in a book of reasonable size. We have thus confined ourselves in this book to some selected results in number theory, and the various chapters are devoted to the most typical problems of various aspects of the theory. We discuss in Chapter I, after some introductory remarks, the theory of congruences. Particular attention is given to congruences of degree two, and the quadratic reciprocity law is proved. Chapter II discusses classical arithmetic functions (Euler's function, sigma function) and a proof of the theorem of Erdös on the characterization of the logarithm among additive functions is given. Birch's result characterizing the powers among multiplicative functions is also presented. These two chapters are concerned with the so-called elementary number theory, and are presented rather simply as there is a detailed treatment in Professor Wacław Sierpiński's book, *Theory of Numbers.*

In Chapter III, we give the fundamental results of the theory of prime numbers, namely the prime number theorem and Dirichlet's theorem on primes in progressions. These results are proved using the Tauberian theorem of Delange-Ikehara.

Chapter IV discusses the sieve methods developed in recent years. The Eratosthenes' sieve, Selberg's sieve and the large sieve are studied and some applications pointed out. Brun's result on twin primes and Gallagher's theorem on primitive roots are proved.

Chapter V is concerned with geometrical problems. We first give some elementary fact concerning convex sets and lattices, and then prove Minkowski's theorem on convex bodies. Vinogradov's elementary method of finding the number of lattice points in plane regions is also discussed.

In Chapter VI we consider additive number theory. The reader will find in it elements of Schnirelman's density, which we use to prove Schnirelman's theorem on the representation of natural numbers as the sum of prime numbers. We prove also Mann's theorem concerning this density and the theorem of Waring-Hilbert.

Chapter VII gives the elements of probabilistic number theory. We prove three fundamental results of this theory, namely the inequality of Turán-Kubilius, the Erdös-Kac
theorem on the normal decomposition, and the theorem of Erdös on asymptotic distribution of additive functions.

In Chapter VIII we consider Diophantine approximation, i.e. the approximation of
irrational numbers by rational numbers. We discuss continued fractions and prove the
theorem of Hurwitz on the best approximation. Here we also introduce the concept of
uniform distribution, prove Weyl's criterion, and the classical result of Weyl concerning
the uniform distribution of the sequence of values of polynomials.

Chapter IX, the last chapter, is concerned with the generalization of the concept of
integers. We discuss algebraic integers and give an elementary theory of algebraic number
fields and of Dedekind rings. We also introduce p-adic integers, define p-adic fields and
prove their fundamental properties.

As one of the objectives of this book is to show the relationship of number theory
with other fields of mathematics, we do not restrict oursleves to the conventional elementary methods. Methods and techniques in algebra, topology, analysis and probability
theory are used quite liberally, but attempts have been made to keep the prerequisites to
that of the undergraduate level. The reader is expected to have a knowledge of the fundamental concepts of algebra such as groups, rings and fields. In Chapter III, familiarity
with fundamental facts of the theory of analytic function is needed, and in Chapter VII,
elements of probability theory. Some notion of topology in metric spaces and elements
of the theory of extension of fields will be necessary to read Chapter IX.

This book is based on lectures given at Wrocław University in memory of B. Bierut,
and at Bordeaux University I. It was Assistant Professor Marceli Stark who persuaded me
to write it. Without his encouragement, the plan to write this book would never be
realized.

I wish to thank Professor Andrzej Schinzel for many valuable remarks and for pointing
out a series of inaccuracies, and Professor Helmut Koch and Master Jan Śliwa for valuable
simplifications of proofs. I thank Mrs. Teresa Bochynek for typing the manuscripts.

Wrocław, February 1975   *Władysław Narkiewicz*

CONTENTS

NOTATION

We shall give here notation and symbols which will be used in the text without specific explanation. We shall denote by the letter $\mathscr{Z}$ the ring of rational integers, $\mathscr{N}$ will denote the set of natural numbers, where we make a convention that 0 is not an element of $\mathscr{N}$. We denote the set $\mathscr{N} \cup \{0\}$ by $\mathscr{N}_0$. The letter $\mathscr{Q}$ denotes the field of rational numbers, and the letter $\mathscr{R}$ the field of real numbers. We denote the field of complex numbers by $\mathscr{3}$.

By the symbol $[x]$ we denote the integral part of the number x and by the symbol $\{x\}$ its fractional part, i.e. if $x = n + r$, where $n \in \mathscr{Z}$ and $r \in [0, 1)$, then $[x] = n$, $\{x\} = r$.

If $F(x)$, $G(x)$ are real-valued functions defined on some set X, and moreover there exists a positive constant B such that for all $x \in X$ the inequality

$$|F(x)| \leqslant BG(x)$$

holds and $G(x) > 0$, then we write

$$F(x) = O(G(x)) \quad .$$

If the set X is a subset of the real line or the complex plane and for some $x_0 \in X$ we have

$$\lim_{x \to x_0} \frac{F(x)}{G(x)} = 0 \ ,$$

then we write

$$F(x) = o(G(x)) \qquad \text{(as } x \text{ tends to } x_0) \ .$$

We use the same symbol in the case when

$$\lim_{x \to \infty} \frac{F(x)}{G(x)} = 0 \quad .$$

If from the context it is clear what x_0 is, then we simply write

$$F(x) = o(G(x)) \quad .$$

These notations were introduced by E. Landau. Instead of $F(x) = o(G(x))$, we often write $F(x) \prec G(x)$; I. M. Vinogradov introduced this symbolism.

We note that in using the symbols o and O it is necessary to take care because e.g. on the set of natural numbers greater than 1 we have

$$x = O(x^2)$$

and

$$x^{\frac{1}{2}} = O(x^2) \; ,$$

but the equality $x = x^{\frac{1}{2}}$ is false. One should, however, not think that these notations are not precise and could lead to a contradiction. One should simply understand the designation $F = O(G)$ as one which means that the function F belongs to the family of those functions which are bounded when divided by G.

In using Landau's symbol it is worth remembering the following properties whose proof the reader should do it himself if he wishes:

(i)   If $f_1 = O(f_2)$ and $f_2 = O(f_3)$, then $f_1 = O(f_3)$.

(ii) If $f_1, f_2 = O(f)$, then $f_1 \pm f_2 = O(f)$.

(iii) If $f = O(g)$, then $fh = O(g|h|)$.

(iv) If $f_1 = o(f_2)$ and $f_2 = O(f_3)$, then $f_1 = o(f_3)$.

(v)   If $f_1, f_2 = o(f)$, then $f_1 \pm f_2 = o(f)$.

(vi) If $f_1 = o(f_2)$ and g does not vanish, then $f_1 g = o(f_2 g)$.

CHAPTER I

DIVISIBILITY, CONGRUENCES

§1. Divisibility. Prime numbers

1. We say that an integer $m \neq 0$ *divides* an integer a if there exists an integer n such that $mn = a$, that is when the number a/m is an integer. We express this fact by $m \mid a$. If an integer m does not divide an integer a, then we write $m \nmid a$. Replacing the word "an integer" by "an element of a ring R" in this definition, we obtain the notion of divisibility in the ring R. In Chapter IX we shall have an occasion to apply this general notion.

From the definition follow the following properties of the notion of divisibility:

Theorem 1.1

(i) *If $m \mid a$ and $m \mid b$, then $m \mid a+b$ and $m \mid a-b$.*

(ii) *If $m \mid a$ and $a \mid n$, then $m \mid n$.*

(iii) *If $m \mid a$ and $b \in \mathscr{Z}$, then $m \mid ab$.*

(iv) *If $a \mid b$ and $b \neq 0$, then $|a| \leqslant |b|$.*

(v) *If $a \mid b$ and $b \mid a$, then $b = a$ or $b = -a$.*

Proof. If $m \mid a$ and $m \mid b$, then we can write $a = a_1 m$, $b = b_1 m$, where $a_1, b_1 \in \mathscr{Z}$, and this gives $a \pm b = (a_1 \pm b_1)m$, thereby proving (i). To prove (ii), let us write $a = a_1 m$, $n = an_1 (a_1, n_1 \in \mathscr{Z})$ and note that we then have $n = m(a_1 n_1)$. Now (iii) follows in view of $a \mid ab$ and (ii). Next for a proof of (iv), we note that from $a \mid b$ follows the integrality of the number b/a, and so by $b \neq 0$, we must have $|b/a| \geqslant 1$. Finally, (v) follows from (iv) on noting that if a and b are real numbers having the same absolute value, then $b = a$ or $b = -a$. ∎

Remark. From (i) and (iii), it follows that the set of all integers divisible by a given integer m forms an ideal in the ring $\mathscr{Z}$.

Theorem 1.2 (division algorithm). *If a, $b \in \mathscr{Z}$, and $b \neq 0$, then there exists exactly one pair of integers q, r satisfying the conditions:*

$$a = bq + r, \qquad\qquad 0 \leqslant r < |b| \; .$$

Furthermore, $b \mid a$ holds iff $r = 0$.

Proof. In the case $b > 0$, let us take $q = [a/b]$ and $r = a - bq$. Then, in view of $q \leqslant a/b < q+1$ we have $bq \leqslant a < bq + b$, so that $0 \leqslant r < b = |b|$. In the case of a negative b, we take $q = -[a/|b|]$.

If we have $a = bq_1 + r_1 = bq_2 + r_2$, $0 \leqslant r_1, r_2 < |b|$, then $r_2 - r_1 = b(q_1 - q_2)$, hence $b \mid (r_2 - r_1)$. If we have $r_1 \neq r_2$, then by Theorem 1.1 (iv) the inequality $|b| \leqslant |r_2 - r_1| < b$ would hold, which is impossible. Therefore $r_1 = r_2$, and so $q_1 = q_2$. The last part of the theorem now follows from Theorem 1.1 (i). ■

We call the integer r the *residue* of a divided by b and call q the *(incomplete) quotient* of this division.

2.   Every natural number $n > 1$ has at least two natural divisors —— 1 and n. If there are no other divisors, then we say that n is a *prime number*. We denote the set of all prime numbers by $\mathscr{P}$. Integers $n > 1$ which are not prime numbers are called *composite numbers*. Hence the integers 2, 3, 5, 7, 11, 257, 65537 are prime numbers, and the integers 4, 6, 8, 21, 35, 99999 are composite numbers.

The greatest known prime number is $2^{19937} - 1$ with 6002 digits. It was found by B. Tuckerman [1] in 1971[*]. We shall soon note that the set $\mathscr{P}$ is an infinite set, so the discovery of larger and larger prime numbers may verify the development of computational technique but does not contribute significantly to the theory itself.

Theorem 1.3. *Every natural number $n > 1$ can be expressed in the form* $n = p_1 \cdots p_k$, $p_i \in \mathscr{P}$.

Proof. We use induction. For $n = 2$ we take $k = 1$, $p_1 = 2$. Now suppose the validity of the assertion for all $n < N$. If N is a prime number, then we take $k = 1$, $p_1 = N$; and if $N = ab$, $1 < a$, $b < N$, then by the induction hypothesis we have $a = p_1 \cdots p_r$, $b = p_{r+1} \cdots p_s$ with $p_i \in \mathscr{P}$, and we obtain $N = p_1 \cdots p_s$. ■

Corollary 1. *Every natural number $n > 1$ has at least one prime factor.* ■

Corollary 2. *Every natural number $n > 1$ can be expressed in the form* $n = p_1^{\alpha_1} \cdots p_t^{\alpha_t}$, *where $\alpha_i \in \mathscr{N}$, and $p_1, \ldots, p_t$ are prime numbers.* ■

---

[*]Translator's note: This is actually the 24th Mersenne prime M_{19937} (cf. Exercise 6 at the end of this section). Recently, three further Mersenne primes have been found: M_{21701}, M_{23209}, M_{44497}, see e.g. D. Slovinski [1] and L.-K. Hua [3], p. 21.

Added in proof: Meanwhile D. Slovinski has found a much larger prime $2^{86243} - 1$ with 25,962 digits. But it is not known whether it is the 28th Mersenne prime (see Math. Intelligencer 5, No. 1 (1983), p. 60).

We denote the number of distinct prime divisors of an integer n by $\omega(n)$. Additionally we define the value of the function $\omega(n)$ for $n = 1$ by putting $\omega(1) = 0$. We shall be concerned with the investigation of various properties of this function in later chapters.

Denote the number of primes not exceeding x by $\pi(x)$. We have then, $\pi(2) = 1$, $\pi(3) = 2, \ldots, \pi(10) = 4$, and it can be checked that $\pi(100) = 25$, $\pi(1000) = 168$. As early as in Euclid's *Elements*, we can find a proof of the fact that there are infinitely many primes, that is $\pi(x) \to \infty$.

Let us give three proofs of this result:

Theorem 1.4. *The set $\mathscr{P}$ of all primes is infinite.*

Proof I (Euclid). Assume that the set $\mathscr{P}$ is finite, $\mathscr{P} = \{p_1, \ldots, p_n\}$. Then the integer $N = 1 + p_1 \ldots p_n$ is greater than 1 and moreover it gives the residue 1 when divided by $p_1, \ldots, p_n$, which contradicts Corollary 1 to Theorem 1.3. ∎

Proof II (Euler). As in the preceding proof, let us assume that $\{p_1, \ldots, p_n\}$ is the set of all prime numbers. For $x > 1$ we have

$$(1.1) \qquad \prod_{k=1}^{n} \left(1 - \frac{1}{p_k^x}\right)^{-1} = \prod_{k=1}^{n} \sum_{j=0}^{\infty} \frac{1}{p_k^{jx}}$$

$$= \sum_{j_1=0}^{\infty} \cdots \sum_{j_n=0}^{\infty} \frac{1}{(p_1^{j_1} \ldots p_n^{j_n})^x} \geq \sum_{m=1}^{\infty} \frac{1}{m^x},$$

because by Corollary 2 to Theorem 1.3 each natural number can be expressed in the form $m = p_1^{j_1} \ldots p_n^{j_n}$.

Let $\alpha = \prod_{k=1}^{n} \left(1 - \frac{1}{p_k}\right)^{-1}$ and let T be a natural number chosen in such a way that the inequality

$$\sum_{m=1}^{T} \frac{1}{m} > \alpha.$$

holds. Then, using (1.1), we have

$$\alpha = \lim_{x \to 1} \prod_{k=1}^{n} \left(1 - \frac{1}{p_k^x}\right)^{-1} \geq \limsup_{x \to 1} \sum_{m=1}^{\infty} \frac{1}{m^x}$$

$$\geq \limsup_{x \to 1} \sum_{m=1}^{T} \frac{1}{m^x} = \sum_{m=1}^{T} \frac{1}{m} > \alpha.$$

The obtained contradiction proves our contention. ∎

Proof III (G. Pólya and G. Szegö). The numbers $2^{2^n} + 1$ are greater than 1 for $n = 1, 2, \ldots$, hence they have prime divisors. Denote by q_n any one of prime divisors of numbers $2^{2^n}+1$, e.g. the least one. Let us show that the prime numbers $q_1, q_2, \ldots$ are all distinct. In fact, if for some $m < n$ we have $q_m = q_n$, then for some $a \in \mathcal{N}$

$$2^{2^m} = aq_m - 1,$$

therefore

$$2^{2^n}+1 = (2^{2^m})^{2^{n-m}}+1 = (aq_m-1)^{2^{n-m}}+1.$$

Applying Newton's binomial expansion, we see that all the terms except the last one are divisible by q_m, and the last term is equal to $(-1)^{2^{n-m}} = 1$. This gives in turn (for a suitable natural number b) the equality $2^{2^n}+1 = bq_m+2$. Since from the supposition q_m must divide $2^{2^n}+1$, hence q_m divides 2, that is $q_m = 2$, which is impossible because the number $2^{2^m}+1$ is not divisible by 2. ∎

The numbers appearing in the third proof of Theorem 1.4 are called the *Fermat numbers* and denoted by F_n. It is not difficult to check that $F_1 = 5$, $F_2 = 17$, $F_3 = 257$, $F_4 = 65{,}537$ are prime numbers, and Fermat in 1640 raised a conjecture that for each n the number F_n is a prime number. However, it is not true, as Euler noted, $F_5 = 641 \cdot 6700417$ and next it was shown that all Fermat numbers F_n for $n = 5, 6, \ldots, 16$ are composite. (See for example W. Sierpiński [4], pp. 344-345.) We do not know any Fermat prime greater than F_4.

The method used in the second proof of the preceding theorem can also be adopted to prove the following more powerful result:

Theorem 1.5. *If $p_1 < p_2 < \ldots$ is a sequence consisting of all the primes, then the series*

$$\sum_{n=1}^{\infty} \frac{1}{p_n}$$

is divergent.

Proof. For a natural number N we have

$$\prod_{n=1}^{N} \left(1 - \frac{1}{p_n}\right)^{-1} = \prod_{n=1}^{N} \left(1 + \frac{1}{p_n} + \frac{1}{p_n^2} + \ldots\right) \geqslant \sum_{m \leqslant p_N} \frac{1}{m},$$

because each integer $m \leqslant p_N$ is the product of powers of primes not exceeding p_N. But the right-hand side of this inequality for a suitable choice of N can be made arbitrarily large, which, however, proves the divergence of the product

$$\prod_{n=1}^{\infty} \left(1 - \frac{1}{p_n}\right)^{-1}.$$

It remains to note that the convergence of $\sum_{n=1}^{\infty} \frac{1}{p_n}$ would imply the convergence of this product. ∎

Other simple proofs of this theorem can be found in the following papers: R. Bellman [1], P. Erdös [2], L. Moser [1].

To determine whether a given integer N is prime or not is not in general a simple problem. In an obtrusive method relying upon the test whether N is divisible by prime numbers $< N$, one can restrict oneself to primes $\leqslant N^{1/2}$ because each composite number N has a prime divisor $\leqslant N^{1/2}$, but the number of these primes exceed $\dfrac{N^{\frac{1}{2}}}{\log N}$ for large N, (which follows from the so-called prime number theorem to be proved in Chapter III), therefore one should execute at least $\dfrac{N^{\frac{1}{2}}}{\log N}$ operations. The fastest known method is due to J. M. Pollard [1] which requires at most $BN^{1/4}$ operations, where B is some constant. Of course, for integers having a special form the number of operations can be made smaller.

3. One of the fundamental results in elementary number theory is the theorem on unique factorization of natural numbers into prime factors, which we shall now prove.

Theorem 1.6. *Every natural number n greater than 1 can be uniquely expressed in the form*

$$(1.2) \qquad n = p_1 \ldots p_k,$$

where $p_i \in \mathcal{P}$ $(i = 1, \ldots, k)$ and $p_1 \leqslant p_2 \leqslant \ldots \leqslant p_k$. Each prime divisor of the integer n must appear among the primes $p_1, \ldots, p_k$.

Proof. Let us first show that the second part of the assertion of the theorem is a consequence of the first part. In fact, if n has a unique decomposition into prime factors, $n = p_1 \ldots p_k$ and $p|n$, then decomposing $n/p = q_1 \ldots q_l$ into prime factors, we obtain $n = pq_1 \ldots q_l = p_1 \ldots p_k$, hence the prime p must appear among the primes p_i.

We prove the first part of the theorem by two different methods.

Proof I. By Theorem 1.2 the ring $\mathcal{Z}$ of rational integers is a Euclidean domain, hence it must be a unique factorization domain, and this coincides with our assertion. ∎

Proof II (without the use of algebraic concepts). It is obvious that each prime number has only one representation of the form (1.2), therefore our theorem is true for $n = 2$. Suppose that our theorem is false and denote by N the smallest number having at least two different decompositions (1.2), say

$$N = p_1 \ldots p_r = q_1 \ldots q_s,$$

where $p_1 \leqslant p_2 \leqslant \ldots \leqslant p_r$, $q_1 \leqslant q_2 \leqslant \ldots \leqslant q_s$ are prime numbers. Without loss of generality we can suppose that $p_1 \leqslant q_1$. If $p_1 = q_1$, then the number

$$N/p_1 = p_2 \ldots p_r = q_2 \ldots q_s$$

is an integer less than N having two different decompositions which is against the choice of N. Hence $p_1 < q_1$. Therefore we can write

$$q_1 = ap_1 + b \qquad (a \geqslant 0,\ 0 < b < p_1,\ a, b \in \mathcal{Z}),$$

whence

$$N = (ap_1 + b)q_2 \ldots q_s = ap_1 q_2 \ldots q_s + bq_2 \ldots q_s.$$

The integer b is either equal to 1 or has a unique factorization into prime factors. Let $b = Q_1 \ldots Q_t$ be this factorization (if $b = 1$, then we take $t = 0$). As the integer $m = bq_2 \ldots q_s$ is less than N, it also a unique factorization and we see that the prime numbers $Q_1, \ldots, Q_t, q_2, \ldots, q_s$ arranged in increasing order appear as factors in this factorization. But $m = N - ap_1 q_2 \ldots q_s$ is an integer divisible by p_1, so from the remark made in the beginning of the proof it follows that p_1 is one of $Q_1, \ldots, Q_t,$ $q_2, \ldots, q_s,$ hence $p_1 = Q_i$ for some i. Therefore $p_1 | b$, which is contradictory to $0 < b < p_1$. The obtained contradiction proves that no integer exists with two different decompositions. ∎

Corollary.1. *Every natural number $n > 1$ can be uniquely expressed in the form*

$$n = \prod_{i=1}^{r} p_i^{\alpha_i},$$

where $p_1 < p_2 < \ldots < p_r$ are prime numbers, $r = \omega(n)$ and $\alpha_i \in \mathcal{N}$.

Proof. This follows from the theorem by grouping together the identical prime factors. ∎

Corollary 2. *Every $n \in \mathcal{Z}$ different from 0 and ± 1 can be uniquely expressed in the form*

$$n = \operatorname{sgn} n \prod_{i=1}^{r} p_i^{\alpha_i},$$

where $p_1 < \ldots < p_r$ are prime numbers,

$$\operatorname{sgn} n = \begin{cases} 1, & \text{if} \quad n > 0, \\ -1, & \text{if} \quad n < 0 \end{cases}$$

and $\alpha_i \in \mathcal{N}$.

Proof. Let us write $n = |n| \operatorname{sgn} n$ and apply the preceding corollary to $|n|$. ∎

Corollary 3. *Every non-zero integer n can be uniquely written in the form*

$$n = \operatorname{sgn} n \prod_{p \in \mathcal{P}} p^{\alpha_p(n)},$$

where $\alpha_p(n) \in \mathcal{N}_0$. The product appearing here contains finitely many factors different from 1, i.e. for a given n, the exponent $\alpha_p(n)$ is different from zero for finitely many primes p.

Proof. For $n = \pm 1$ we take $\alpha_p(n) = 0$ for all p. If, on the contrary, n is of the form as in the preceding corollary, then for $p = p_i$ $(i = 1, 2, \ldots, r)$ we take $\alpha_p(n) = \alpha_i,$

and for the remaining primes we put $\alpha_p(n) = 0$. Uniqueness of the representation follows immediately from the preceding corollary. ∎

Corollary 4. *Every rational number w different from zero can be uniquely expressed in the form*

$$w = \operatorname{sgn} w \prod_{p \in \mathscr{P}} p^{\alpha_p(w)},$$

where $\alpha_p(w) \in \mathscr{Z}$. *The product appearing here contains only finitely many factors different from 1.*

Proof. We can write $w = \operatorname{sgn} w \left(\dfrac{a}{b}\right)$, where $a, b \in \mathscr{N}$. Applying the preceding corollary to a and b, we obtain

$$w = \operatorname{sgn} w \prod_{p \in \mathscr{P}} p^{\alpha_p(a) - \alpha_p(b)}.$$

Note that the difference $\alpha_p(a) - \alpha_p(b)$ depends exclusively on w, but not on

the choice of a and b. In fact, if $w = \operatorname{sgn} w \dfrac{a_1}{b_1}$ $(a_1, b_1 \in \mathscr{N})$, then $ab_1 = a_1 b$. Hence we have

$$\prod_{p \in \mathscr{P}} p^{\alpha_p(a)} \prod_{p \in \mathscr{P}} p^{\alpha_p(b_1)} = \prod_{p \in \mathscr{P}} p^{\alpha_p(a_1)} \prod_{p \in \mathscr{P}} p^{\alpha_p(b)},$$

i.e.

$$\prod_{p \in \mathscr{P}} p^{\alpha_p(a) + \alpha_p(b_1)} = \prod_{p \in \mathscr{P}} p^{\alpha_p(a_1) + \alpha_p(b)}.$$

By Corollary 3 we have the equality $\alpha_p(a) + \alpha_p(b_1) = \alpha_p(a_1) + \alpha_p(b)$, for every p, that is $\alpha_p(a) - \alpha_p(b) = \alpha_p(a_1) - \alpha_p(b_1)$, as asserted.

Hence if we define the function $\alpha_p(w)$ for $w \neq 0$ $(w \in \mathscr{Q})$ by the formula

$$\alpha_p(w) = \alpha_p(a) - \alpha_p(b) \qquad \left(|w| = \frac{a}{b}, \ a, b \in \mathscr{N}\right),$$

then we get the representation required.

Suppose that

$$w = \operatorname{sgn} w \prod_{p \in \mathscr{P}} p^{\alpha_p(w)} = \operatorname{sgn} w \prod_{p \in \mathscr{P}} p^{c_p}.$$

Let us show that $c_p = \alpha_p(w)$ must hold for every $p \in \mathscr{P}$.

If $A = \{p \in \mathscr{P} : \alpha_p(w) > c_p\}$ and $B = \{p \in \mathscr{P} : \alpha_p(w) < c_p\}$, then the number

$$m = \prod_{p \in A} p^{\alpha_p(w) - c_p} = \prod_{p \in B} p^{c_p - \alpha_p(w)}$$

is a natural number. If the sum of the sets A and B were non-empty, then m would have two representations in the form of the product of prime powers, which contradicts Corollary 2. Therefore $A = B = \emptyset$, that is for every p we have the equality $\alpha_p(w) = c_p$. ∎

Remark. The decomposition appearing in the above corollaries is called the *canonical decomposition* of the corresponding numbers.

We shall now give fundamental properties of the function $\alpha_p(w)$ appearing in Corollary 4:

Theorem 1.7.

(i) $\alpha_p(ab) = \alpha_p(a) + \alpha_p(b)$, $\alpha_p(a/b) = \alpha_p(a) - \alpha_p(b)$, $\alpha_p(-a) = \alpha_p(a)$.

(ii) Let $w \neq 0$ be a rational number. Then $w \in \mathcal{Z}$ iff we have $\alpha_p(w) \geqslant 0$
      for every p.

(iii) If $m, n \in \mathcal{Z}$, then $m|n$ iff we have $\alpha_p(m) \leqslant \alpha_p(n)$ for every p.

(iv) For $n \in \mathcal{Z}$ the largest power of p dividing n is $p^{\alpha_p(n)}$.

(v) $\alpha_p(a \pm b) \geqslant \min(\alpha_p(a), \alpha_p(b))$, and if $\alpha_p(a) \neq \alpha_p(b)$, then
      $\alpha_p(a \pm b) = \min(\alpha_p(a), \alpha_p(b))$.

Proof.

(i)   The equalities (i) follow at once from the definition of $\alpha_p(n)$.

(ii) If we have $\alpha_p(w) \geqslant 0$, for every p, then $w \in \mathcal{Z}$, as it is the product of
      integers. Conversely, if $w \in \mathcal{Z}$, then from Corollaries 3 and 4 follows
      $\alpha_p(w) \geqslant 0$.

(iii) m divides n iff $n|m \in \mathcal{Z}$. Hence it is enough to apply (i) and (ii).

(iv) It is apparent that $p^{\alpha_p(n)}|n$, and if we have

$$p^{\alpha_p(n)+1}|n = p^{\alpha_p(n)} \prod_{\substack{q \neq p \\ q \in \mathcal{P}}} q^{\alpha_q(n)},$$

then

$$p\Big| \prod_{\substack{q \neq p \\ q \in \mathcal{P}}} q^{\alpha_q(n)},$$

and so p would divide some product of primes different from p, which is impossible because of Corollary 1.

(v)   First consider the case $a, b \in \mathcal{Z}$. Then for $\alpha = \alpha_p(a)$, $\beta = \alpha_p(b)$ we have
      $p^\alpha|a$, $p^\beta|b$, hence $p^{\min(\alpha, \beta)}|a \pm b$, which gives, by (iv), the inequality
      $\min(\alpha, \beta) \leqslant \alpha_p(a \pm b)$.

In the general case we write $a = x/y$, $b = x'/y'$ $(x, y, x', y' \in \mathcal{Z}$ and $\neq 0)$. Then $a \pm b = (xy' \pm x'y)/yy'$, hence, using (i) and that part of (v) already proved, we obtain

$$\alpha_p(a \pm b) = \alpha_p(xy' \pm x'y) - \alpha_p(yy')$$
$$\geq \min\{\alpha_p(x) + \alpha_p(y'),\ \alpha_p(x') + \alpha_p(y)\} - \alpha_p(y) - \alpha_p(y')$$
$$\geq \min\{\alpha_p(x) - \alpha_p(y),\ \alpha_p(x') - \alpha_p(y')\}$$
$$= \min\{\alpha_p(a),\ \alpha_p(b)\}$$

Lastly, if $\alpha_p(a) < \alpha_p(b)$, then we have on the one hand $\alpha_p(a \pm b) \geq \alpha_p(a)$, but, on the other hand, we can write

$$\alpha_p(a) = \alpha_p(a \pm b \mp b) \geq \min(\alpha_p(a \pm b), \alpha_p(b)).$$

Since the inequality $\alpha_p(a) \geq \alpha_p(b)$ is excluded, we should have $\alpha_p(a) = \alpha_p(a \pm b)$. ∎

Corollary 1. *If p is a prime dividing the product $a_1 \ldots a_n$ of integers, then p divides one of the a_i's.*

Proof. The case when $a_1 \ldots a_n = 0$ is trivial, and if $a_1 \ldots a_n \neq 0$, then we have

$$\sum_{i=1}^{n} \alpha_p(a_i) = \alpha_p(a_1 \ldots a_n) > 0,$$

so for some i we have $\alpha_p(a_i) > 0$, that is $p | a_i$. ∎

Corollary 2. *For every prime p and $n \in \mathcal{N}$ we have*

$$\alpha_p(n!) = \sum_{k=1}^{\infty} \left[\frac{n}{p^k}\right].$$

Proof. By (i) we have

$$\alpha_p(n!) = \sum_{k=1}^{n} \alpha_p(k) = \sum_{j=1}^{\infty} j \sum_{\substack{k \leq n \\ \alpha_p(k)=j}} 1,$$

but

$$\sum_{\substack{k \leq n \\ \alpha_p(k)=j}} 1 = \sum_{\substack{k \leq n \\ p^j | k \\ p^{j+1} \nmid k}} 1 = \sum_{\substack{1 \leq n/p^j \\ p \nmid l}} 1 = \left[\frac{n}{p^j}\right] - \left[\frac{n}{p^{j+1}}\right],$$

so

$$\alpha_p(n!) = \sum_{j=1}^{\infty} j \left(\left[\frac{n}{p^j}\right] - \left[\frac{n}{p^{j+1}}\right]\right)$$

$$= \sum_{j=1}^{\infty} j \left[\frac{n}{p^j}\right] - \sum_{j=2}^{\infty} (j-1)\left[\frac{n}{p^j}\right] = \left[\frac{n}{p}\right] + \sum_{i=2}^{\infty} \left[\frac{n}{p^j}\right] = \sum_{j=1}^{\infty} \left[\frac{n}{p^j}\right]. \quad ∎$$

Remark. If $p \in \mathscr{P}$ and $p^t | n$, but $p^{t+1} \nmid n$, then we write $p^t \| n$.

Thus from the above theorem it follows that $p^{\alpha_p(n)} \| n$.

We call the function $\alpha_p(n)$ the *exponent* corresponding to the prime p. A. Ostrowski [1] proved that if a function $f(x)$ is real-valued, defined for rational $x \neq 0$ and satisfies conditions (i) and (v) of Theorem 1.7, then for some constant c and a prime number p we have $f(x) = c\alpha_p(x)$.

It is worthwhile paying attention to the fact that in some subset of the set of natural numbers, closed under multiplication, the analogy of Theorem 1.6 is not true. Indeed, let A_4 be the set of all natural numbers of the form $4n+1$ $(n = 0, 1, 2, \ldots)$. The product of two elements of A_4 belongs to A_4 in view of $(4m+1)(4n+1) = 4(4mn+m+n)+1$. We call a number $n \in A_4$ prime if it cannot be expressed as a product of two elements of A_4 different from unity. Thus for example the numbers 5, 9, 13, 17, 21, 29 are primes in A_4, whereas 25 is not since $25 = 5 \cdot 5$ and $5 \in A_4$. Just as in the proof of Theorem 1.3 we can show that each number in A_4 different from the identity is a product of prime numbers in A_4. This representation is not generally unique as for example the number 693 has two decompositions: $693 = 21 \cdot 33 = 9 \cdot 77$, and moreover the numbers 9, 21, 33 and 77 are primes in A_4.

4. If $a_1, \ldots, a_k$ are integers different from zero, then we call the largest integer d dividing $a_1, \ldots, a_k$ the *greatest common divisor* (GCD) of $a_1, \ldots, a_k$ and denote it by $(a_1, \ldots, a_k)$. Thus we have $d = (a_1, \ldots, a_k)$ iff $d | a_i$ $(i = 1, \ldots, k)$ and from $D | a_i$ $(i = 1, \ldots, k)$, $D \in \mathscr{Z}$, it follows that $D \leqslant d$. It is clear that $(a_1, \ldots, a_k)$ is a natural number.

We call the least natural number divisible by $a_1, \ldots, a_k$ the *least common multiple* (LCM) of these integers and denote it by $[a_1, \ldots, a_k]$. Thus we have $M = [a_1, \ldots, a_k]$ iff $a_i | M$ $(i = 1, \ldots, k)$ and from $a_i | m$ $(i = 1, \ldots, k)$, $m \in \mathscr{N}$ it follows that $m \geqslant M$.

We shall now prove the fundamental properties of GCD and LCM.

Theorem 1.8.

(i) $d = (a_1, \ldots, a_k)$ holds iff for every prime p we have
       $\alpha_p(d) = \min(\alpha_p(a_1), \ldots, \alpha_p(a_k))$.

(ii) $M = [a_1, \ldots, a_k]$ holds iff for every prime p we have
       $\alpha_p(M) = \max(\alpha_p(a_1), \ldots, \alpha_p(a_k))$.

(iii) If $D | a_1, \ldots, a_k$, $D \in \mathscr{Z}$, then $D | (a_1, \ldots, a_k)$.

(iv)   If $a_1 | m, \ldots, a_k | m$, $m \in \mathscr{Z}$, then $[a_1, \ldots, a_k] | m$.

(v)   If $d = (a_1, \ldots, a_k)$ and $b_i = a_i/d$, then $(b_1, \ldots, b_k) = 1$.

(vi) $[a_1, a_2](a_1, a_2) = a_1 a_2$.

Proof. (i) Let $d = (a_1, \ldots, a_k)$ and $D = \prod_{p \in \mathscr{P}} p^{c_p}$, where $c_p = \min(\alpha_p(a_1), \ldots,$

$\alpha_p(a_k)$). By Theorem 1.6 (iii) D divides all the integers $a_1, \ldots, a_k$, so that $0 \leqslant D \leqslant d$, but, on the other hand, $\alpha_p(d) \leqslant \alpha_p(a_i)$ $(i = 1, \ldots, k)$, which gives $\alpha_p(d) \leqslant c_p$, and hence d divides D and $d \leqslant D$. Thus, $d = D$.

Replacing min by max and reversing the inequalities in the above argument, we obtain the proof of (ii).

Part (iii) and part (iv) follow from (i) and (ii) together with Theorem 1.7 (iii). Let $D = (b_1, \ldots, b_k)$. Then $b_i/D = c_i$ are integers, and so $a_i = b_i d = dDc_i$, that is, dD divides a_i for $i = 1, \ldots, k$. Therefore $dD \leqslant d$, which gives $D = 1$.

In order to prove the last part of the theorem, note that min (x,y) + max $(x,y) = x + y$ and apply (i), (ii) together with Theorem 1.7 (i). ∎

Remark. The last part of Theorem 1.8 cannot be generalized for $k \geqslant 3$ integers a_i, since for example

$$(1, 2, 4, \ldots, 2^{k-1})[1, 2, \ldots, 2^{k-1}] = [1, 2, \ldots, 2^{k-1}] = 2^{k-1},$$

whereas $1 \cdot 2 \cdot 4 \cdot \ldots \cdot 2^{k-1} = 2^{k(k-1)/2} > 2^{k-1}$ for $k > 2$.

Corollary. *If* $d = (a_1, \ldots, a_n)$, *then there exist integers* $x_1, \ldots, x_n$ *such that*

$$\sum_{i=1}^{n} a_i x_i = d.$$

Moreover, d is the least natural number expressible in the form $\sum_{i=1}^{n} a_i x_i$ *with* $x_i \in \mathscr{Z}$.

Proof. Let I be the set of all natural numbers expressible in the form

$$\sum_{i=1}^{n} a_i x_i, \quad x_i \in \mathscr{Z}$$

and let D be the least number of this set. In view of Theorem 1.2 we can find integers q_i, r_i satisfying the equalities

$$a_i = q_i D + r_i, \quad 0 \leqslant r_i < D \ (i = 1, \ldots, n),$$

whence follows $r_i \in I$, since I is evidently an ideal in $\mathscr{Z}$. By the choice of D this is possible only in the case $r_1 = \ldots = r_n = 0$. and so $D|a_1, \ldots, a_n$ and from Theorem 1.8 (iii) it follows that $D|d$. On the other hand, we have $d|D$, so that $d = D$. ∎

We say that integers a, b are *relatively prime* if $(a,b) = 1$. Similarly, if $a_1, \ldots, a_k \in \mathscr{Z}$ and for $i \neq j$ we have $(a_i, a_j) = 1$, then we say that the integers $a_1, \ldots, a_k$ are *pairwise (relatively) prime*.

Theorem 1.9.

(i)　*If* $a, b \in \mathscr{Z}$ *are relatively prime, then there exist* $x, y \in \mathscr{Z}$ *such that* $ax + by = 1$.

(ii)　*If* $a, b, c \in \mathscr{Z}$, $(a,b) = 1$ *and* $a|bc$, *then* $a|c$.

(iii) If $a, b_1, \ldots, b_n \in \mathscr{Z}$ and for $i = 1, \ldots, n$ we have $(a, b_i) = 1$, then $(a, b_1 \ldots b_n) = 1$.

(iv) If $a_1, \ldots, a_n \in \mathscr{Z}$ are pairwise relatively prime and $a_1 | a, \ldots, a_n | a$, then $a_1 \ldots a_n | a$.

Proof. (i) is a particular case of the Corollary to Theorem 1.8. To prove (ii) we use (i) and write

$$ax + by = 1$$

with suitable $x, y \in \mathscr{Z}$. Then $a | acx + bcy = c$.

For a proof of (iii) we use Corollary 1 to Theorem 1.7. Let p be a prime dividing a. Then $p \nmid b_1, \ldots, p \nmid b_n$, and so $p \nmid b_1 \ldots b_n$ and we obtain $(a, b_1 \ldots b_n) = 1$.

Lastly, (iv) follows from the remark that $\alpha_p(a_i) > 0$ can occur for at most one index i, so that

$$\alpha_p(a_1 \ldots a_n) = \alpha_p(a_1) + \ldots + \alpha_p(a_n) = \max \{ \alpha_p(a_i) : i = 1, \ldots, n \} \leqslant \alpha_p(a).$$

and we can use Theorem 1.7 (iii). ∎

For finding the GCD of two integers, the following method called the *Euclid algorithm* can be used:

Assume that $a, b \in \mathscr{N}$, $a > b$. Define the sequence $r_{-1}, r_0, r_1, \ldots$ and $q_1, q_2, \ldots$ in the following way: $r_{-1} = a$, $r_0 = b$, and if the numbers $r_{-1}, r_0, \ldots, r_k$, are already defined and $r_k \neq 0$, then determine r_{k+1}, q_{k+1} by

(1.3) $r_{k-1} = q_{k+1} r_k + r_{k+1}, \qquad 0 \leqslant r_{k+1} < r_k$.

In particular, we have $a = q_1 b + r_1$, $b = q_2 r_1 + r_2$. Due to $r_{-1} > r_0 > \ldots \geqslant 0$, we have $r_{n-1} \neq 0$, $r_n = 0$ for some n.

Theorem 1.10. *If the sequence* r_k *is defined by* (1.3), *then* $r_{n-1} = (a, b)$, *where* n *is the first index for which* $r_n = 0$.

Proof. We shall show by induction that r_{n-1} divides r_j for $j = n, n-1, \ldots, 1, 0, -1$. In fact, $r_{n-1} | 0 = r_n$ and $r_{n-1} | r_{n-1}$, and if r_{n-1} divides r_{k+1} and r_k, then by (1.3) it also divides r_{k-1}. In particular, r_{n-1} is a common divisor of a and b, which gives $r_{n-1} | (a,b)$ by Theorem 1.8 (iii).

Now let us show that (a,b) divides $r_{-1}, r_0, \ldots, r_{n-1}$. In fact, (a,b) divides $a = r_{-1}$ and $b = r_0$, and if $(a,b) | r_{k-1}, r_k$, then by (1.3), (a,b) divides r_{k+1}. Hence $(a,b) | r_{n-1}$ and thus, $(a, b) = r_{n-1}$. ∎

Exercises

1.   Show that Theorem 1.1 (i), (ii), (iii) is true in any ring.

2.   Give an example of an integral domain for which part (v) of Theorem 1.1 is false.

3.   Use one of the proofs of Theorem 1.4 to show that $\pi(x) \geqslant c \log x$, where c is some positive constant.

4. We denote by p_n the n-th prime number.
    (i) Prove that
    $$\limsup_{n \to \infty} (p_{n+1} - p_n) = \infty .$$
    (ii) Prove the inequality
    $$p_n < e^{1+n}.$$

5. Prove that if the number $1 + 2^n$ is a prime number, then n is a power of 2.

6. Integers $M_n = 2^n - 1$ are called *Mersenne's numbers*. Prove that
    (i) if M_n is a prime number, then n is also a prime number.
    (ii) M_n is not a k-th power of any natural number > 1 provided that $k > 1$.

7. Show that $\omega(n) \leqslant \dfrac{\log n}{\log 2}$.

8. Prove that if $p \in \mathscr{P}$, $w_1, \ldots, w_n \in \mathscr{Q}$ are non-zero, and for $k = 2, 3, \ldots, n$ we have $\alpha_p(w_k) > \alpha_p(w_1)$, then
    $$\alpha_p(w_1 + \ldots + w_n) = \alpha_p(w_1).$$

9. Prove the identities:
    $$(a, b, c)[ab, bc, ca] = (ab, bc, ca)[a, b, c]$$
    $$= (a, b, c)[a, b, c][(a, b); (b, c), (c, a)] = abc.$$

10. Let u_n denote the n-th term of the Fibonacci sequence that is
    $$u_1 = u_2 = 1, \quad u_{n+2} = u_{n+1} + u_n \quad (n = 1, 2, \ldots).$$
    Prove that $(u_m, u_n) = u_d$, where $d = (m, n)$ and use it to prove the existence of infinitely many primes.

11. Prove that for $n = 1, 2, \ldots$ the formula
    $$u_n = \frac{(1 + \sqrt{5})^n - (1 - \sqrt{5})^n}{2^n \sqrt{5}}$$

holds.

§2. Linear and quadratic Diophantine equations

1. Every equation whose solutions we seek in the set of integers or natural numbers is called a *Diophantine equation*. The theory of such equations makes up an extensive branch of number theory with which the reader can familiarize himself using W. Sierpiński's book "Theory of numbers" and also L. J. Mordell's "Diophantine equations". In this book we confine ourselves therefore to linear and quadratic Diophantine equations.

Let us begin by stating some solutions of the equation

(1.4) $$a_1 x_1 + \ldots + a_n x_n = b ,$$

where $a_1, \ldots, a_n$ are integers different from zero, and b is any positive integer.

A necessary and sufficient condition for the existence of a solution of this equation is given by the theorem below:

Theorem 1.11. *If $a_1, \ldots, a_n \in \mathscr{Z}$, $a_i \neq 0$, then Eq. (1.4) has a solution $x_1, \ldots, x_n \in \mathscr{Z}$ iff $d = (a_1, \ldots, a_n)|b$. If this condition is satisfied, then there exists a solution $x_1, \ldots, x_n$ satisfying*

$$|x_i| \leqslant |b| + (n-1)H \qquad (i = 1, \ldots, n),$$

where $H = \max(|a_1|, \ldots, |a_n|)$.

Proof. If (1.4) has a solution $x_1, \ldots, x_n \in \mathscr{Z}$, then

$$d | a_1 x_1 + \ldots + a_n x_n = b,$$

and so $d|b$ is a necessary condition for the solvability of (1.4). Conversely, if this condition is satisfied, then by the Corollary to Theorem 1.8 there exist $y_1, \ldots, y_n \in \mathscr{Z}$, satisfying

$$a_1 y_1 + \ldots + a_n y_n = d \quad ,$$

and $x_j = y_j b/d$ satisfies (1.4).

Now note that in the case when (1.4) has an integral solution, we can find a solution $x_1, \ldots, x_n$ satisfying the condition

$$0 \leqslant x_i < |a_n| \qquad (i = 1, \ldots, n-1).$$

In fact, if $y_1, \ldots, y_n$ is any integral solution of (1.4) and

$$y_i = q_i a_n + x_i, \qquad 0 \leqslant x_i < |a_n|, \qquad q_i \in \mathscr{Z} \ (i = 1, \ldots, n-1),$$

then, putting

$$x_n = y_n + \sum_{j=1}^{n-1} a_j q_j \quad ,$$

we see that $x_1, \ldots, x_n$ satisfies (1.4), and

$$|a_n x_n| = \left| b - \sum_{j=1}^{n-1} a_j x_j \right| \leqslant |b| + H(n-1)|a_n|.$$

Hence

$$|x_n| \leqslant H(n-1) + \frac{|b|}{|a_n|} \leqslant H(n-1) + |b|. \qquad \blacksquare$$

From the proof of the Corollary to Theorem 1.8 follows a method for the effective determination of solutions of (1.4). Here it suffices to restrict ourselves to the case $b = d$. Choose $z_1, \ldots, z_n \in \mathscr{Z}$ so that the number $d_1 = a_1 z_1 + \ldots + a_n z_n$ should be positive. If $d_1 = d$, then z_i forms a solution of (1.4). Otherwise, from the same corollary follows $d_1 > d$, and hence not all the numbers a_i are divisible by d_1. Without loss of generality we can assume that d_1 does not divide a_1. Then $a_1 = qd_1 + d_2$, $0 < d_2 < d_1$. If we now take $z_i' = -qz_i$ $(i \neq 1)$, $z_1' = 1 - qz_1$, then

$$0 < d_2 = a_1 z_1' + \ldots + a_n z_n' < d_1 \ .$$

Repeating this procedure, we obtain a decreasing sequence $\{d_k\}$ of natural numbers of the form $a_1 X_1 + \ldots + a_n X_n \geq d$, and so after a suitable number of steps we arrive at the representation of d in the required form.

This algorithm was proposed by R. Weinstock [1]. Other algorithms can be found in the papers of W. A. Blankenship [1] and G. H. Bradley [1].

2.   We shall be concerned with the determination of all the solutions of Eq. (1.4) assuming, of course, that the condition $d \,|\, b$ is satisfied. For this purpose it will be convenient to use the vector notation. By $\mathscr{Q}^n$ we denote the linear space comprising all n-tuples $[x_1, \ldots, x_n]$ of rational numbers, and by $\mathscr{Z}^n$ its subspace formed of n-dimensional vectors with integral entries. By $L(x)$ we denote a function with values in $\mathscr{Q}$, defined for $x = [x_1, \ldots, x_n] \in \mathscr{Q}^n$ by the formula

$$L(x) = a_1 x_1 + \ldots + a_n x_n,$$

wherein we suppose that at least one of a_i's does not vanish. Then L is obviously a linear transformation, and Eq. (1.4) takes the form $L(x) = b$.

The following lemma reduces our problem to finding all solutions of the equation $L(x) = 0$ in $x \in \mathscr{Z}^n$.

Lemma 1.1. *Let* $x_0 \in \mathscr{Z}^n$ *satisfy the condition* $L(x_0) = b$. *For the vector* $x \in \mathscr{Z}^n$ *to satisfy the equation* $L(x) = b$, *it is necessary and sufficient that* $x = x_0 + y$, *with* $y \in \mathscr{Z}^n$, $L(y) = 0$.

Proof. If $x = x_0 + y$ and $L(y) = 0$, then $L(x) = L(x_0) + L(y) = b$. Conversely, if $L(x) = b$, then $L(x - x_0) = 0$. ∎

Since L is a linear transformation, its kernel $\mathrm{Ker}\, L = \{[x_1 \ldots, x_n] : x_i \in \mathscr{Q}, L(x) = 0\}$, is a linear subspace of $\mathscr{Q}^n$, and since not all the a_i's are equal to zero, the dimension of $\mathrm{Ker}\, L$ is equal to $n - 1$. We denote the set of all solutions of $L(x) = 0$, $x \in \mathscr{Z}^n$ by $J(L)$. Then

$$J(L) = \mathscr{Z}^n \cap \mathrm{Ker}\, L.$$

Evidently, $J(L)$ is an additive subgroup of $\mathscr{Z}^n$ and so the structure of $J(L)$ follows from the theorem below.

Theorem 1.12. *If A is a non-zero subgroup of $\mathscr{Z}^n$, then there exist* $1 \leq k \leq n$ *elements* $v_1, \ldots, v_k \in A$ *such that every element of A can be expressed uniquely in the form* $m_1 v_1 + \ldots + m_k v_k$ $(m_i \in \mathscr{Z})$, *that is,* $A \simeq \mathscr{Z}^k$.

Proof. We apply induction on n. Let A be any non-zero subgroup of the group $\mathscr{Z}$ and let m be the smallest natural number contained in A. If a is an arbitrary element of A, we write

$$a = qm + r \qquad (0 \leqslant r < m).$$

The number $r = a - qm$ lies in A, and hence by the choice of m, $r = 0$.

Hence $m \mid a$ and $A \subset m \mathcal{Z}$. Since obviously $m \mathcal{Z} \subset A$, we have $A = m \mathcal{Z}$ and we may suppose that $k = 1$, $v_1 = m$.

Now suppose that the theorem is valid for all the groups $\mathcal{Z}^k$ for $k < n$ and let A be any non-zero subgroup of $\mathcal{Z}^n$. We denote by d the smallest natural number for which there exists at least one element of the set A of the form $[x_1, \ldots, x_n]$ such that $x_n = d$. (If there is no such element d, then every element of A is of the form $[x_1, \ldots, x_{n-1}, 0]$ and by means of the isomorphism $[x_1, \ldots, x_{n-1}, 0] \to [x_1, \ldots, x_{n-1}]$ we can regard A as a subgroup of $\mathcal{Z}^{n-1}$ and apply the induction hypothesis). Let z be an arbitrary element of A of the form $[d_1, \ldots, d_{n-1}, d]$, $d_i \in \mathcal{Z}$.

For every $x = [x_1, \ldots, x_n] \in A$ we have $d \mid x_n$. If fact, if we had $x_n = qd + r$ $(0 < r < d)$, then the element $qz - x$ would belong to A, and moreover, would be equal to $[qd_1 - x_1, \ldots, qd_{n-1} - x_{n-1}, r]$, contrary to the choice of d. Therefore every element of A can be expressed in the form

$$x = b + tz \ ,$$

where $t \in \mathcal{Z}$, $b \in B$, and $B = \{b : b \in A, b = [b_1, \ldots, b_{n-1}, 0]\}$, and the expression is unique. Indeed, for $x = [x_1, \ldots, x_n] \in A$ we have $x_n = td$ for a suitable $t \in \mathcal{Z}$, and hence $x - tz \in B$. If $x = b + tz = b_1 + t_1 z$ $(b, b_1 \in B)$, then, comparing the last coefficients, we get $t = t_1$, and hence also $b = b_1$. Applying the induction hypothesis to the group B which is contained in $\mathcal{Z}^{n-1}$, we infer the existence of a system $\{v_1, \ldots, v_{k-1}\} \subset B$ such that any element $b \in B$ can be expressed uniquely in the form $m_1 v_1 + \ldots + m_{k-1} v_{k-1}$ $(m_i \in \mathcal{Z})$, and hence putting $v_k = z$ we see that every element of A is of the form

$$m_1 v_1 + \ldots + m_k v_k \qquad (m_i \in \mathcal{Z})$$

and that this expression is unique. Moreover, $k - 1 \leqslant n - 1$, hence $k \leqslant n$. ∎

Now we can give a description of all the solutions of Eq. (1.4).

Theorem 1.13. *If $a_1, \ldots, a_n$ are integers, not all zero, and x_0 is an arbitrarily chosen solution of Eq. (1.4), then there exist elements $v_1, \ldots, v_{n-1} \in \mathcal{Z}^n$ such that every integral solution of Eq. (1.4) can be expressed in the form*

$$x = m_1 v_1 + \ldots + m_{n-1} v_{n-1} + x_0 \qquad (m_i \in \mathcal{Z})$$

in a unique way. Conversely, every element of this form is a solution of Eq. (1.4).

Proof. Applying the previous theorem and Lemma 1.1, we see that there exist elements $v_1, \ldots, v_k$ $(k \leqslant n)$ such that every solution of Eq. (1.4) is of the form $\sum_{i=1}^{k} m_i v_i + x_0$ and this representation is unique. Furthermore, every such element is a solution of Eq. (1.4). It remains therefore to show that $k = n - 1$.

Let $e_1, \ldots, e_{n-1}$ be any basis of the space Ker L. Since the coordinates of the vectors of this basis are rational numbers, there exists a natural number D such that the vectors $De_1, \ldots, De_{n-1}$ have integral coordinates, and hence lie in Ker $L \cap \mathscr{Z}^n = J(L)$. These vectors are obviously linearly independent. Now let E be a linear space generated by the vectors $v_1, \ldots, v_k$. Its dimension is at most k, but it contains all the elements of $J(L)$, and, in particular $De_1, \ldots, De_{n-1}$. Hence the dimension of E must be at least $n - 1$, and this implies $k \geqslant n - 1$. If $k = n$, then the vectors $v_1, \ldots, v_k$ would be linearly dependent, that is, one of them say v_1, could be expressed in terms of the remaining ones:

$$v_1 = c_2 v_2 + \ldots + c_k v_k \ ,$$

and this contradicts the uniqueness of the expression of elements of $J(L)$ as a linear combination of vectors v_i. ∎

3.   We now pass on to solutions of Diophantine equations of degree 2, restricting ourselves to equations with two unknowns, i.e. equations of the form

(1.5) $$ax^2 + bxy + cy^2 + dx + ey + f = 0 \quad .$$

We call the number $\delta = b^2 - 4ac$ the *discriminant* of Eq. (1.5), and call $H = \max(|a|, |b|, \ldots, |f|)$ its *height*.

Concerning Eq. (1.5), we shall prove the following result due to A. Schinzel [1]:

Theorem 1.14. *If Eq. (1.5) has an integral solution, and the polynomial*

$$\Phi(x, y) = ax^2 + bxy + cy^2 + dx + ey + f$$

is not a product of two linear factors, and moreover if we have $\delta \leqslant 0$ or δ is a square of some natural number, then there exists a solution x, y of this equation satisfying the inequalities

$$\max\{|x|, |y|\} \leqslant \begin{cases} 9H^2 & (\delta < 0), \\ 6H^{\frac{7}{2}} & (\delta = 0), \\ 20H^4 & (\delta = n^2, \ n \in \mathscr{Z}). \end{cases}$$

Remark 1. In the case $\delta \neq 0$ this estimate is true for any solution of Eq. (1.5).

Remark 2. In the case where $\delta > 0$ is not a square of any natural number, one can prove the following estimate

$$\max(|x|, |y|) \leqslant (5H)^{200H^3}$$

(A. Schinzel, loc. cit.).

Proof. First we consider the case $\delta = 0$. In that case we can write the quadratic polynomial $ax^2 + bx + c$ in the form $\gamma(\alpha x + \beta)^2$ with integral α, β, γ.

Now let $t = \alpha x + \beta y$, $\epsilon = \alpha e - \beta d$. Note that $\epsilon \neq 0$. Indeed, if we have $\epsilon = 0$, i.e. $\alpha e = \beta d$, then with a suitable λ we can write

$$\alpha = \lambda d, \qquad \beta = \lambda e ,$$

and hence

$$
\begin{aligned}
\Phi(x, y) &= \gamma(\alpha x + \beta y)^2 + dx + ey + f \\
&= \gamma(\lambda dx + \lambda ey)^2 + dx + ey + f \\
&= \gamma\lambda^2 (dx + ey)^2 + dx + ey + f .
\end{aligned}
$$

Putting $T = dx + ey$, we see that the polynomial $\gamma\lambda^2 T^2 + T + f$ has integral roots, and so it can be expressed as a product of two linear factors. Hence $\Phi(x, y)$ is also a product of linear factors, contrary to the hypothesis.

The numbers $x, y \in \mathscr{Z}$ are the solutions of Eq. (1.5) iff there exists a $t \in \mathscr{Z}$ satisfying the system of equations:

$$
\begin{aligned}
\alpha x + \beta y &= t , \\
dx + ey &= -\gamma t^2 - f .
\end{aligned}
$$

The determinant of this system is equal to $\epsilon \neq 0$ and from Cramer's formula it follows that the condition of its solvability for a given $t \in \mathscr{Z}$ is the divisibility of both determinants

$$
\det \begin{bmatrix} t & \beta \\ -\gamma t^2 - f & e \end{bmatrix} = \beta\gamma t^2 + et + \beta f
$$

and

$$
\det \begin{bmatrix} \alpha & t \\ d & -\gamma t^2 - f \end{bmatrix} = -\alpha\gamma t^2 - dt - \alpha f
$$

by ϵ.

If $t \in \mathscr{Z}$ satisfies this condition, then we can find $t_0 \in \mathscr{Z}$, which also satisfies the condition and lies in the interval $[0, |\epsilon|)$. Indeed, it suffices to take the residue of t in the division by ϵ.

On noting that

$$|\gamma| \epsilon^2 = |e^2 a - bed + cd^2| \leqslant 3H^3 ,$$

we obtain

$$|t_0| < \epsilon \leqslant 3^{\frac{1}{2}} H^{\frac{3}{2}}$$

and

$$|\gamma t_0^2 + f| < |\gamma| \epsilon^2 + |f| \leqslant 4H^3.$$

Now we shall use

$$|\alpha|^2 \leqslant |\alpha^2 \gamma| = |a| \leqslant H$$

and

$$|\beta|^2 \leqslant |\beta^2 \gamma| = |c| \leqslant H,$$

to obtain ultimately

$$\max(|x|, |y|) \leqslant H|t_0| + H^{\frac{1}{2}}|\gamma t_0^2 + f| < 6H^{\frac{7}{2}}.$$

Thus we have proved the assertion in the case $\delta = 0$.

In the remaining cases we shall begin by transforming Eq. (1.5) by the linear transformation

$$X = \delta x + be - 2cd,$$
$$Y = \delta y + bd - 2ae,$$

into the form

$$aX^2 + bXY + cY^2 = 4\delta\Delta,$$

where

$$\Delta = \det \begin{bmatrix} a & \frac{1}{2}b & \frac{1}{2}d \\ \frac{1}{2}b & c & \frac{1}{2}e \\ \frac{1}{2}d & \frac{1}{2}e & f \end{bmatrix}$$

The number 4Δ is evidently an integer, and easy computation gives the estimate: $0 < |\Delta| < 2H^3$.

Now, if $\delta < 0$ and x, y satisfy Eq. (1.5), then

$$a\left[\left(X - \frac{b}{2a}Y\right)^2 + \frac{|\delta|}{4a^2}Y^2\right] = 4\delta\Delta,$$

hence

$$|\delta| Y^2 \leqslant 16|a\delta\Delta|$$

and

$$Y^2 \leqslant 16|a\Delta| \leqslant 32aH^3 \leqslant 32H^4,$$

whence

$$|Y| \leqslant 6H^2.$$

Similarly, using

$$c\left[\left(Y - \frac{b}{2c}X\right)^2 + \frac{|\delta|}{4c^2}X^2\right] = 4\delta\Delta,$$

we obtain

$$|\delta|X^2 \leqslant 16|c\delta\Delta|$$

and

$$|X| \leqslant 6H^2.$$

Hence

$$\max(|X|, |Y|) \leqslant 6H^2$$

and finally

$$|x| \leqslant \frac{1}{|\delta|}(|X| + |be-2cd|) \leqslant 6H^2 + 3H^2 = 9H^2,$$

$$|y| \leqslant \frac{1}{|\delta|}(|Y| + |bd - 2ae|) \leqslant 6H^2 + 3H^2 = 9H^2.$$

It remains to consider the case where $\delta = n^2$ for a suitable $n \in \mathcal{N}$. In this case

$$aX^2 + bXY + cY^2 = (\alpha_1 X + \beta_1 Y)(\alpha_2 X + \beta_2 Y) \quad,$$

where $|\alpha_i|, |\beta_i| < H$, $i = 1, 2$. Moreover,

$$|\alpha_i X + \beta_i Y| \leqslant 4\delta\Delta \quad (i = 1, 2) ,$$

which leads to

$$\max(|X|, |Y|) \leqslant 8\delta|\Delta|H,$$

and hence

$$\max(|x|, |y|) \leqslant 8|\Delta|H + 3H^2 < 20H^4. \quad \blacksquare$$

Exercises

1.   Find all the integral solutions of the equation $2x + 3y + 5z = 11$.

2.   Following the example of Theorem 1.12, formulate and prove a theorem on the integral solutions of a system of equations

$$\sum_{j=1}^{m} a_{ij}x_j = b_i \quad (i = 1, ..., n).$$

3.   Describe all the integral solutions of the system of equations

$$x - 2y + 3z = 6,$$
$$2x + y - 2z = 9.$$

4.   Show that if a, b are relatively prime natural numbers, then every natural number $> ab$ can be expressed in the form $ax + by$ with natural numbers x, y.

§3. Congruences

1.   Let N be a fixed natural number, and $N \mathscr{Z} \subset \mathscr{Z}$ be an ideal composed of all its multiples. We call the quotient ring $\mathscr{Z}/N\mathscr{Z}$ the *ring of residue classes to the modulus N*, or simply the residue class ring (mod N). We call the elements of this ring the residue classes (mod N). In view of Theorem 1.2, the set $\{0, 1, 2, \ldots, N-1\}$ is a set of representatives for $\mathscr{Z}/N\mathscr{Z}$, but we shall use also other such sets.

We denote the canonical homomorphism $\mathscr{Z} \to \mathscr{Z}/N\mathscr{Z}$ by f, i.e. the projection which maps each $n \in \mathscr{Z}$ to the coset $n + N\mathscr{Z}$. We usually write $n(\mathrm{mod}\, N)$ instead of $f(n)$.

If $F(n)$ is a function defined on $\mathscr{Z}$, whose values depend solely on the residue classes $n(\mathrm{mod}\, N)$, then we understand

$$\sum_{n(\mathrm{mod}\, N)} F(n), \qquad \prod_{n(\mathrm{mod}\, N)} F(n)$$

to mean a sum or a product in which n runs through any system of representatives of $\mathscr{Z}/N\mathscr{Z}$.

If $a, b \in \mathscr{Z}$ and $f(a) = f(b)$, then we say that a and b are *congruent to each other with respect to the modulus N* (or simply (mod N)) and write

(1.6) $a \equiv b(\mathrm{mod}\, N)$.

We call the sign $\equiv$ *congruence*, and the formula of type (1.6) a *congruence*. We shall present the fundamental properties of congruences below:

Theorem 1.15.

(i)   If $a_1 \equiv a_2 \,(\mathrm{mod}\, N)$, $b_1 \equiv b_2 \,(\mathrm{mod}\, N)$, then $a_1 \circ b_1 \equiv a_2 \circ b_2 \,(\mathrm{mod}\, N)$ (the sign $\circ$ signifies one of the operations: $+$, $-$, $\cdot$).

(ii) $a \equiv b(\mathrm{mod}\, N)$ *iff* N *divides* $a - b$.

(iii)   If $ab_1 \equiv ab_2 \,(\mathrm{mod}\, N)$ *and* $(a, N) = 1$, *then* $b_1 \equiv b_2 \,(\mathrm{mod}\, N)$.

(iv)   If a is a natural number and $ab_1 \equiv ab_2 \,(\mathrm{mod}\, aN)$, then $b_1 \equiv b_2 \,(\mathrm{mod}\, N)$.

Proof.

(i)   follows immediately from the fact that f is a homomorphism.

(ii)   follows from (i) on noting that $a \equiv b \,(\mathrm{mod}\, N)$ is equivalent to $a - b \equiv 0 \,(\mathrm{mod}\, N)$, and this, in turn, implies the divisibility of $a - b$ by N.

To prove (iii) we observe that from $ab_1 \equiv ab_2 \pmod N$ follows the divisibility of $a(b_1 - b_2)$ by N and we can then use Theorem 1.9 (ii).

Proof of (iv) follows from the fact that if aN divides $a(b_1 - b_2)$, then N divides $b_1 - b_2$. ∎

Corollary 1. *If $W(x_1, \ldots, x_n)$ is a polynomial in n variables with integral coefficients and $a_1, \ldots, a_n, b_1, \ldots, b_n$ are integers satisfying the condition $a_i \equiv b_i \pmod N$ for $i = 1, 2, \ldots, n$, then $W(a_1, \ldots, a_n) \equiv W(b_1, \ldots, b_n) \pmod N$.*

Proof. If

$$W(x_1, \ldots, x_n) = \sum_{i_1, \ldots, i_n} \alpha_{i_1, \ldots, i_n} x_1^{i_1} \ldots x_n^{i_n},$$

then, applying (i), we get successively

$$a_1^{i_1} \ldots a_n^{i_n} \equiv b_1^{i_1} \ldots b_n^{i_n} \pmod N,$$

$$\alpha_{i_1, \ldots, i_n} a_1^{i_1} \ldots a_n^{i_n} \equiv \alpha_{i_1, \ldots, i_n} b_1^{i_1} \ldots b_n^{i_n} \pmod N,$$

$$W(a_1, \ldots, a_n) \equiv W(b_1, \ldots, b_n) \pmod N. \quad ∎$$

Corollary 2. *If $W(X)$ is a polynomial with integral coefficients whose values are prime numbers for all natural X, then $W(X)$ is a constant.*

Proof. Suppose that $W(X)$ is such a polynomial of degree $k \geqslant 1$ and let $W(1) = p$. Now if for $n = 1, 2, 3, \ldots$ we put $m_n = 1 + pn$, then in view of the preceding corollary, $W(m_n) \equiv W(1) \equiv 0 \pmod p$, because $m_n \equiv 1 \pmod p$. Hence $W(m_n)$ is divisible by p for $n = 1, 2, \ldots$. Since for $X \in \mathcal{N}$ the values $W(X)$ are primes, we have $W(m_n) = p$, but the polynomial $W(X)$ cannot take the same value at infinitely many points, because it is not a constant. ∎

The last corollary shows, in particular, that it is impossible to find a formula of the form $p_n = W(n)$, where p_n denotes the n-th prime and $W(x)$ is a polynomial with integral coefficients[a] . As R. C. Buck [1] proved, there does not exist any rational function $R(x)$, different from a constant, which for all natural x takes exclusively prime values. W. Sierpiński [2] showed that there exists a real number a such that for every n, the formula

$$p_n = [10^{2^n} a] - 10^{2^{n-1}} [10^{2^{n-1}} a],$$

holds, but this cannot be used to find successive prime numbers since for finding a it is

---

[a] Added in proof: In the case of polynomials with several variables the situation is different. Recently an example of a polynomial of degree 25 with 26 variables has been given with the property that all its positive values at natural arguments are prime numbers and every prime number can be obtained in this way. (J. P. Jones, D. Sato, H. Wada, D. Wiens, *Amer. Math. Monthly* 83 (1976), pp. 449-464).

already necessary to know these primes, for

$$a = \sum_{n=1}^{\infty} p_n 10^{-2^n}.$$

There is no formula which enables us to find quickly the successive prime numbers; this is due to the great irregularity of their distribution. We shall deal more closely with prime numbers in Chapter III.

Corollary 3. *If $N_1 | N_2$, then the map*

$$\varphi_{N_1, N_2} : x \, (\mathrm{mod} \, N_2) \mapsto x \, (\mathrm{mod} \, N_1)$$

is an epimorphism (i.e. an onto homomorphism) from the ring $\mathscr{Z}/N_2\mathscr{Z}$ onto $\mathscr{Z}/N_1\mathscr{Z}$. Moreover, the map $\varphi_{N, N}$ is the identity map, and, if $N_1 | N_2 | N_3$, then the composition $\varphi_{N_1, N_2} \circ \varphi_{N_2, N_3}$ is equal to φ_{N_1, N_3}.

Proof. To begin with, we note that the map φ_{N_1, N_2} is well-defined, i.e. that $x \, (\mathrm{mod} \, N_1)$ depends only on the value $x \, (\mathrm{mod} \, N_2)$, and does not depend upon the choice of x. Indeed, in view of $N_1 | N_2$ from $x \equiv y \, (\mathrm{mod} \, N_2)$ it follows that $x \equiv y \, (\mathrm{mod} \, N_1)$. From part (i) of the theorem it follows that φ_{N_1, N_2} is a homomorphism, and the remaining parts of the assertion are trivial. ∎

Corollary 4. *If $N_1 | N_2$, then there exist exactly N_2/N_1 distinct residue classes $x \, (\mathrm{mod} \, N_2)$ for which $x \, (\mathrm{mod} \, N_1)$ has a given value.*

Proof. This follows from the previous corollary and the remark that the kernel of the homomorphism φ_{N_1, N_2} has exactly

$$|\mathscr{Z}/N_2\mathscr{Z}| : |\mathscr{Z}/N_1\mathscr{Z}| = N_2/N_1$$

elements. ∎

2.   If $u \in \mathscr{Z}/N\mathscr{Z}$ and there exists an $m \in \mathscr{Z}$ satisfying the conditions $(m, N) = 1$ and $u = m(\mathrm{mod} \, N)$, then we say that u is a *residue class* (mod N) *relatively prime to N* (or a *reduced residue class mod N*). Obviously, in that case every m satisfying $m(\mathrm{mod} \, N) = u$ must be relatively prime to N. We denote the set of all reduced residue classes mod N by $G(N)$.

Theorem 1.16. *The set $G(N)$ is a group with respect to multiplication.*

Proof. Let a, b be elements of $G(N)$ and let $x(\mathrm{mod} \, N) = a$, $y(\mathrm{mod} \, N) = b$. Then $(x, N) = (y, N) = 1$, and hence, according to Theorem 1.9 (iii) we have $(xy, N) = 1$. In view of this, $ab = xy(\mathrm{mod} \, N) \in G(N)$, and so $G(N)$ is closed under multiplication. The element $1(\mathrm{mod} \, N)$ is evidently the identity in $G(N)$ and there remains to show that every element of $G(N)$ has an inverse element.

From Theorem 1.9 (i) it follows that if $a = x(\mathrm{mod} \, N) \in G(N)$ then there exist integers X, Y such that $xX + NY = 1$, and hence $1(\mathrm{mod} \, N) = xX(\mathrm{mod} \, N) = x(\mathrm{mod} \, N) \cdot X(\mathrm{mod} \, N)$ as $NY(\mathrm{mod} \, N) = 0$. ∎

We denote the number of elements of $G(N)$ by $\varphi(N)$. The function defined in this way is called *Euler's function*.

Obviously, for prime numbers p we have $\varphi(p) = p - 1$, because the images of the numbers $1, \ldots, p - 1$ in $\mathscr{Z}/p\mathscr{Z}$ are all distinct and exhaust the elements of $G(p)$. The formula

(1.7) $$\varphi(p^k) = p^{k-1}(p-1) \qquad (p \text{ is a prime}, k \in \mathscr{N})$$

also holds since the set

$$\{1 \leqslant m \leqslant p^k : p \nmid m\},$$

is a set of representatives of $G(p^k)$, and it contains

$$p^k - \sum_{\substack{n \leqslant p^k \\ p \mid n}} 1 = p^k - \sum_{m \leqslant p^{k-1}} 1 = p^k - p^{k-1} = p^{k-1}(p-1)$$

elements.

Theorem 1.17. *If* $(a, N) = 1$, *then* $a^{\varphi(N)} \equiv 1 \pmod{N}$.

Proof. This follows from the fact that if G is any finite group with k elements, e is the identity and $g \in G$, then $g^k = e$. ∎

Corollary. *If* p *is a prime and* $p \nmid a$, *then* $a^{p-1} \equiv 1 \pmod{p}$.

Theorem 1.17 carries the name of *Euler's theorem*, and the Corollary just proven is usually called *Fermat's theorem* or *Fermat's small theorem* (in distinction to the so called Fermat's great theorem appearing in the theory of Diophantine equations).

Theorem 1.18. *If* $N = N_1 \ldots N_t$, *and* N_i's *are pairwise prime, then the map*

$$g : \mathscr{Z}/N\mathscr{Z} \to \bigoplus_{i=1}^{t} \mathscr{Z}/N_i\mathscr{Z}$$

defined by

$$g\big(a\,(\mathrm{mod}\,N)\big) = [a\,(\mathrm{mod}\,N_1),\ a\,(\mathrm{mod}\,N_2),\ \ldots,\ a\,(\mathrm{mod}\,N_t)]$$

is an isomorphism between the rings $\mathscr{Z}/N\mathscr{Z}$ *and* $\bigoplus_{i=1}^{t} \mathscr{Z}/N_i\mathscr{Z}$.

Proof. First note that g is well-defined, that is, the n-tuple $[a(\mathrm{mod}\,N_1), \ldots, a(\mathrm{mod}\,N_t)]$ does not depend upon the choice of the element a in $\mathscr{Z}$, and depends solely upon $a(\mathrm{mod}\,N)$. In fact, if $a(\mathrm{mod}\,N) = b(\mathrm{mod}\,N)$, then the difference $a - b$ is divisible by N, and so, *a fortiori*, by each of N_i, which implies that $a(\mathrm{mod}\,N_i) = b(\mathrm{mod}\,N_i)$ for $i = 1, 2, \ldots, t$. The rings $\mathscr{Z}/N\mathscr{Z}$ and $\bigoplus_{i=1}^{t} \mathscr{Z}/N_i\mathscr{Z}$ have N elements, and hence it suffices to show that g is injective, i.e. that the equality $g\big(a\,(\mathrm{mod}\,N)\big) = g\big(b\,(\mathrm{mod}\,N)\big)$ infers the equality $a\,(\mathrm{mod}\,N) = b\,(\mathrm{mod}\,N)$. To this end note that from $g\big(a\,(\mathrm{mod}\,N)\big) =$

$g\left(b(\operatorname{mod}N)\right)$ results $a(\operatorname{mod}N_i) = b(\operatorname{mod}N_i)$ for $i = 1, \ldots, t$, and so $a \equiv b \ (\operatorname{mod}N_i)$ $(i = 1, \ldots, t)$, which means that each N_i divides $a - b$. From Theorem 1.9 (iv) it follows that N must divide $a - b$, that is $a \equiv b(\operatorname{mod}N)$, which is to be proved. ∎

Corollary 1 (Chinese remainder theorem). *If* $a_1, \ldots, a_t \in \mathscr{Z}$ *and* $N_1, \ldots, N_t$ *are pairwise prime natural numbers, then there exists a solution of the system of congruences*

$$x \equiv a_i(\operatorname{mod}N_i) \qquad (i = 1, \ldots, t).$$

Herein, if x, y are solutions, then

$$x \equiv y(\operatorname{mod}N_1 \ldots N_t).$$

Proof. An integer x satisfies this system of congruences iff

$$g\left(x(\operatorname{mod}N_1 \ldots N_t)\right) = [a_1 (\operatorname{mod}N_1), \ldots, a_t \ (\operatorname{mod}N_t)],$$

but g is bijective, so that $x(\operatorname{mod}N_1 \ldots N_t)$ is uniquely determined. ∎

Corollary 2. *If* $N_1, \ldots, N_t$ *are pairwise prime natural numbers and* N *is their product, then*

$$G(N) \simeq \bigoplus_{i=1}^{t} G(N_i).$$

Proof. Let h be the restriction of g to the set $G(N)$. Then h is bijective and $h(xy) = h(x)h(y)$. Furthermore, for $x \in G(N)$ we have $x = a(\operatorname{mod}N)$, where $(a, N) = 1$, which implies $(a, N_i) = 1$ for $i = 1, \ldots, t$, so that the image of $G(N)$ under h is contained in $\bigoplus_{i=1}^{t} G(N_i)$. It remains to note that if $[a(\operatorname{mod}N_1), \ldots, a(\operatorname{mod}N_t)] \in \bigoplus_{i=1}^{t} G(N_i)$, i.e. $(a, N_i) = 1$, then $(a, N) = 1$ by virtue of Theorem 1.9 (iii), and hence h is an epimorphism. ∎

Corollary 3. *If* $N = p_1^{a_1} \ldots p_t^{a_t}$ *is the canonical decomposition of a natural number N, then* $G(N) \simeq \bigoplus_{i=1}^{t} G(p_i^{a_i})$.

Proof. This is a special case of the preceding corollary. ∎

3.   The structure of the group $G(p^n)$ is described by the following theorem:

Theorem 1.19. *If p is an odd prime, then the group* $G(p^n)$ *is cyclic for* $n = 1, 2, \ldots$. *If, on the contrary, $p = 2$, then* $G(p^n)$ *is cyclic for $n = 1, 2$, and is a direct product of* C_2 *and* $C_{2^{n-2}}$ *for $n = 3, 4, \ldots$.*

Proof. Let us begin with the case $n = 1$. Let $r_1, \ldots, r_k$ be the orders of elements of the group $G(p)$ and denote their LCM by r. Note that r is a divisor of $p - 1$. Actually, each of r_i's divides the order of the group $G(p)$, which is $p - 1$. Let us show that there exists an element in $G(p)$ whose order is equal to r. Let

$$r = p_1^{a_1} \ldots p_t^{a_t}$$

be the canonical decomposition of r. For each j we can choose r_i such that $p_j^{q_j} \| r_i$, i.e. $r_i = p_j^{q_j} \cdot d_i$, $p_j \nmid d_i$.

If g_i is an element of $G(p)$ of order r_i, then $g_i^{d_i}$ is of order $p_j^{q_j}$ and it remains to prove the following fact from the theory of commutative groups:

If $g_1, \ldots, g_n$ are elements of a commutative group and s_j is the order of g_j $(j = 1, \ldots, n)$, where $s_1, \ldots, s_n$ are pairwise prime, then the order of the product $g_1 \ldots g_n$ equals $s_1 \ldots s_n$.

It suffices to prove this in the case $n = 2$ and apply induction. Let e be the identity of the group. If $(g_1 g_2)^k = e$, then $(g_1 g_2)^{k s_i} = e$ for $i = 1, 2$, so that $g_1^{k s_2} = g_2^{k s_1} =$ This shows the divisibility of $k s_2$ by s_1 and of $k s_1$ by s_2, but $(s_1, s_2) = 1$, so that from Theorem 1.9 (ii) follows the divisibility of k by s_1 and s_2, hence by $s_1 s_2$. Therefore the order of $g_1 g_2$ is divisible by $s_1 s_2$. On the other hand, we have evidently $(g_1 g_2)^{s_1 s_2} = e$, and hence the order of $g_1 g_2$ equals $s_1 s_2$. Applying this to our situation, we see that there exists an element g_0 in the group $G(p)$ of order r. There remains to show that $r = p - 1$. We already know that r does not exceed $p - 1$, hence, to prove the reverse inequality we note that every element $g \in G(p)$ satisfies the equality $g^r = e$. Since the ring $\mathscr{Z}/p\mathscr{Z}$ is a field, this equation has at most r roots, so $p - 1 \leqslant r$.

In the case $n = 1$ our theorem has been proven. Now we pass on to the case $n \neq 1$, $p \neq 2$, leaving the case $p = 2$ to the end. We shall require a simple lemma concerning Newton's binomial coefficients:

Lemma 1.2. *If p is a prime, then for $j = 1, 2, \ldots, p - 1$, the binomial coefficients $\binom{p}{j}$ are divisible by p.*

Proof. In the expression

$$\binom{p}{j} = \frac{p!}{j!(p - j)!}$$

the numerator is divisible by p, while the denominator is not. ∎

Corollary 1. *If p is a prime and $a, b \in \mathscr{Z}$, then for $k = 1, 2, \ldots$ we have $(a + b)^{p^k} \equiv a^{p^k} + b^{p^k} \pmod{p}$.*

Corollary 2. *If K is a field containing p^n elements, then for $a, b \in K$ we have $(a + b)^{p^k} = a^{p^k} + b^{p^k}$.*

Let $n \geqslant 2$ and let $g \pmod{p}$ be a generator of the group $G(p)$. We shall seek for a generator of the group $G(p^n)$ among the elements of the form

$$\xi = g + tp \pmod{p^n}$$

for $t = 0, 1, \ldots, p - 1$. We denote by r the order of such an element ξ in the group $G(p^n)$. Then

$$(g + tp)^r \equiv 1 \pmod{p^n},$$

and so

$$g^r \equiv (g+tp)^r \equiv 1 \,(\mathrm{mod}\,p),$$

so that r is divisible by the order of $g(\mathrm{mod}\,p)$, i.e. by $p-1$. On the other hand, r divides $\varphi(p^n) = p^{n-1}(p-1)$, and so we have

$$r = p^j \cdot (p-1)$$

for some j in the interval $[0, n-1]$. Let us show that for a suitably chosen t this order equals to $\varphi(p^n)$. For this purpose we choose t such that

$$(g+pt)^{p-1} \not\equiv 1 \,(\mathrm{mod}\,p^2).$$

To show that such a choice is possible we note that $g^{p-1} \equiv 1 \,(\mathrm{mod}\,p)$, and hence $g^{p-1} = 1+ap$ $(a \in \mathscr{Z})$ and

$$(g+pt)^{p-1} \equiv g^{p-1} + (p-1)ptg^{p-2} \equiv 1 + p(a - tg^{p-2}) \,(\mathrm{mod}\,p^2).$$

If we now take $t \not\equiv ag(\mathrm{mod}\,p)$, then $tg^{p-2} \not\equiv ag^{p-1} \equiv a(\mathrm{mod}\,p)$, so that $(g+pt)^{p-1} \not\equiv 1(\mathrm{mod}\,p^2)$.

Now we may write

$$(g+pt)^{p-1} = 1 + a_1 p \quad (p \nmid a_1),$$

and owing to Lemma 1.2 and the oddness of p, we obtain

$$(g+pt)^{p(p-1)} = (1+a_1 p)^p = 1+a_2 p^2 \quad (p \nmid a_2)$$

and by easy induction we arrive at

$$(g+pt)^{p^j(p-1)} = 1 + a_{j+1} p^{j+1} \quad (p \nmid a_{j+1})$$

for $j = 1, 2, \ldots$.

We find that for $j \leqslant n-2$ we have

$$\xi^{p^j(p-1)} \not\equiv 1 \,(\mathrm{mod}\,p^n),$$

so that the order of ξ is $> p^{n-2}(p-1)$ and must be equal to $p^{n-1}(p-1) = \varphi(p^n)$.

There remains to consider the case $p = 2$. The situation is clear when $n = 1, 2$ because then the corresponding groups $G(2^n)$ have one and two elements respectively and are both cyclic. Therefore let n be equal to 3 at least. In that case the group $G(2^n)$ is not cyclic since for each of the integers 1, 3, 5, 7 its square is congruent to 1(mod 8), and so $G(8) = C_2 \oplus C_2$, and if some group $G(2^n)$ were cyclic for $n \geqslant 3$, then $G(8)$ would also be cyclic as a homomorphic image of $G(2^n)$ under the map $\varphi_{8,2^n}$.

In order to prove that $G(2^n)$ is, in our case, a direct product of groups C_2 and $C_{2^{n-2}}$, we note first that $5(\mathrm{mod}\,2^n)$ is of order 2^{n-2} in the group $G(2^n)$. Indeed, we have

$$5 = 1+2^2, \qquad 5^2 = 1+2^3+2^4, \qquad 5^4 = 1+2^4+a_2 2^5$$

and generally

$$5^{2^k} = 1 + 2^{k+2} + a_k 2^{k+3} \qquad (a_k \in \mathscr{Z}),$$

so that for $k \leqslant n-3$ we have $5^{2^k} \equiv 1 \pmod{2^n}$. Let G_1 denote the cyclic subgroup of $G(2^n)$ generated by $5 \pmod{2^n}$ and let G_2 be the 2-element cyclic subgroup generated by $-1 \pmod{2^n}$. These groups have the intersection consisting of $1 \pmod{2^n}$, for every element of G_1 is of the form $(4k+1) \pmod{2^n}$, and G contains their direct product. Since G_1 contains 2^{n-2} elements and G_2 has two elements, this direct product has 2^{n-1} elements and thus coincides with $G(2^n)$. ∎

Every integer m such that $m \pmod N$ is a generator of the cyclic group $G(N)$ is called a *primitive root* (mod N) or a *primitive root to the modulus N*. The preceding theorem can be expressed as follows:

The integers 2, 4 and all the powers p^k of odd primes have primitive roots.

Corollary. *If p is a prime, then there exist precisely $\varphi(p-1)$ primitive roots (mod p).*

Proof. Let g be a generator of $G(p)$ and let $(d, p-1) = 1$, $1 \leqslant d < p-1$. Then g^d is also a generator of $G(p)$, for, if two powers $(g^d)^k$ and $(g^d)^l$ are equal and $1 \leqslant k \leqslant l \leqslant p-1$, then

$$g^{d(l-k)} = 1 \pmod{p},$$

so that $p-1 \mid d(l-k)$ and $p-1 \mid l-k$, which leads to $k = l$. On the other hand, if $(d, p-1) = \delta > 1$, then the order of the element g^d does not exceed

$$\frac{p-1}{\delta} < p-1 \text{ and } g^d \text{ is not a primitive root (mod } p). ∎$$

In 1927 E. Artin set forth the conjecture that every integer a which is not a square of natural numbers and different from $0, -1$ is a primitive root for infinitely many prime numbers.

This conjecture has not yet been proven. It is only known (C. Hooley [1]) that Artin's conjecture follows from some conjectures concerning the zeros of the Dedekind zeta function (these functions are generalizations of the Riemann zeta-function to which we shall refer in Chapter III). We shall prove a result concerning Artin's conjecture in Chapter IV (see Theorem 4.8).

It was also asked how large the least positive primitive root $r(p)$ can be for a prime number p. The best known estimate for the function $r(p)$ from above has been given by D. A. Burgess [1], viz.

$$r(p) \ll p^{\frac{1}{4}+\varepsilon} \qquad (\varepsilon > 0).$$

On the other hand, it is known (P. Turán [3]) that the estimate

$$r(p) = o(\log p)$$

is false.

4.   We shall now be engaged in solving a congruence of the form

(1.8) $$F(x) \equiv 0 (\text{mod } N) \ ,$$

where $F(X)$ is a polynomial with integral coefficients. Note that this congruence is equivalent to the equation

(1.9) $$G(x) = 0$$

considered in the ring $\mathscr{Z}/N\mathscr{Z}$, where $G(X)$ is a polynomial with coefficients in this ring which arises from $F(X)$ by replacing each coefficient by the corresponding residue class (mod N). Note that the degree of $G(X)$ does not exceed that of $F(X)$, and may be smaller than that, as is illustrated by the example $F(X) = 2X^2 + X - 1$ in the case $N = 2$.

Call the number of solutions of Eq. (1.9) *the number of solutions of the congruence* (1.8), that is, we shall identify those solutions which are congruent to each other. We shall denote this number by $\lambda_F(N)$.

Theorem 1.20. *If $N = p$ is a prime number, and $F(X)$ is a polynomial of degree n with integral coefficients, not all of which are divisible by p, then the congruence (1.8) has at most n solutions.*

Proof. In this case $\mathscr{Z}/N\mathscr{Z}$ is a field, and the polynomial $G(X)$ is non-zero and of degree $\leqslant n$. Hence (1.9) has at most n solutions. ■

Remark. This theorem ceases to be true in the case where N is composite because for example, the congruence

$$x^2 - 1 \equiv 0 \ (\text{mod } 8)$$

has four solutions: $x = 1, 3, 5, 7$.

We shall show that solving the congruence (1.8) can be reduced to the case where N is a prime power. Later we shall be concerned only with that case.

Theorem 1.21. *Let $F(X)$ be a polynomial with integral coefficients and let $N = p_1^{a_1} \ldots p_t^{a_t}$ be the canonical decomposition of a natural number N. Then, if x_i is a solution of the congruence $F(X) \equiv 0 \ (\text{mod } p_i^{a_i}) \ (i = 1, 2, \ldots, t)$, then every integer satisfying the system $x \equiv x_i \ (\text{mod } p_i^{a_i}) \ (i = 1, \ldots, t)$ is a solution of the congruence (1.8) and each solution is of this form. Moreover, the formula*

(1.10) $$\lambda_F(N) = \prod_{i=1}^{t} \lambda_F(p_i^{a_i})$$

holds.

Proof. If $F(x_i) \equiv 0 \pmod{p_i^{a_i}}$ for $i = 1, \ldots, t$ and $x \equiv x_i \pmod{p_i^{a_i}}$ for $i = 1, \ldots, t$, then by Corollary 1 to Theorem 1.15 $F(x) \equiv 0 \pmod{p_i^{a_i}}$, so that $F(x) \equiv 0 \pmod N$ by virtue of Theorem 1.9 (iv). Conversely, if $F(x) \equiv 0 \pmod N$ and $x_i \equiv x \pmod{p_i^{a_i}}$, then $F(x_i) \equiv 0 \pmod{p_i^{a_i}}$. Finally, formula (1.10) follows from Theorem 1.18, since the map g defined there maps the set $\{x \pmod N : F(x) \equiv 0 \pmod N\}$ on the Cartesian product of the sets $\{x_i \pmod{p_i^{a_i}} : F(x_i) \equiv 0 \pmod{p_i^{a_i}}\}$. ∎

The simplest case of the congruence (1.8) is that in which polynomial F is of degree one,

$$F(X) = aX - b \quad (a \neq 0) \quad ,$$

in this case we can write this in an equivalent form

(1.11) $$ax \equiv b \pmod N \quad .$$

Theorem 1.22. *If $a \neq 0$, then the congruence (1.11) has a solution iff b is divisible by (a, N). When this condition is satisfied, the congruence (1.11) has (a, N) solutions and all of them are congruent to one another* $\left(\bmod \dfrac{N}{(a, N)}\right)$.

Proof. The congruence (1.11) is satisfied iff there exists an integer y such that

$$ax = b + Ny \quad ,$$

i.e.

$$ax - Ny = b \; .$$

From Theorem 1.11 it follows that this equation has an integral solution iff (a, N) divides b. This proves the first part of the theorem. To prove the second we write $N = (a, N) \cdot N_1$, $a = (a, N) \cdot a_1$ and $b = (a, N) \cdot b_1$. Then $(a_1, N_1) = 1$ and on the ground of Theorem 1.15 (iv) the congruence (1.11) is equivalent to the congruence

$$a_1 x \equiv b_1 \pmod{N_1} \quad .$$

Let x_0 be its solution. Since $a_1 \pmod{N_1}$ lies in $G(N_1)$, there exists an $a' \in \mathscr{Z}$ such that $a_1 a' \equiv 1 \pmod{N_1}$. Thus $x_0 \equiv a_1 a' x_0 \equiv b_1 a' \pmod{N_1}$, and hence $x_0 \pmod{N_1}$ is uniquely determined. It remains to note that by Corollary 4 to Theorem 1.15 there are exactly $(a, N) = N/N_1$ elements $x_0 \pmod N$ with this property. ∎

Since solving the congruence (1.11) can be reduced to solving the equation $ax - Ny = b$, we may apply here the method given in the preceding section.

Now we shall express $\lambda_F(p^k)$ in terms of $\lambda_F(p)$ and in this way the problem of solving the congruence (1.8) will be reduced to the case in which N is a prime. Let us begin with a simple lemma concerning polynomials with integral coefficients.

Lemma 1.3. *If $F(X)$ is a polynomial with integral coefficients and $F'(X)$ is its derivative, then for any integer x_0 the formula*

$$F(X) = F(x_0) + (X - x_0) F'(x_0) + (X - x_0)^2 R(X)$$

holds where the polynomial $R(X)$ has integral coefficients.

Proof. First consider the case $F(X) = X^k$. We denote $R(X)$ in this case by $R_k(X)$. Then we have

$$R_k(X) = (X - x_0)^{-2} \left(X^k - x_0^k - (X - x_0) k x_0^{k-1} \right)$$

$$= \sum_{j=0}^{k-1} x_0^j \frac{X^{k-1-j} - x_0^{k-1-j}}{X - x_0} \in \mathscr{Z}[X] \quad .$$

In the general case we have

$$F(X) = \sum_{k=0}^{n} a_k X^k \quad ,$$

hence

$$R(X) = \sum_{k=0}^{n} a_k R_k(X) \in \mathscr{Z}[X] \quad . \qquad \blacksquare$$

Corollary. *If $F(X)$ is a polynomial with integral coefficients and p^k is a prime power, then from $x \equiv x_0 \pmod{p^k}$ it follows that*

$$F(x) \equiv F(x_0) + (x - x_0) F'(x_0) \pmod{p^{k+1}} \quad . \qquad \blacksquare$$

We shall now prove the fundamental result of this subsection.

Theorem 1.23. *Let $F(X)$ be a polynomial with integral coefficients, and p^k be a prime power. Let x_0 be any solution of the congruence*

$$F(x) \equiv 0 \pmod{p^k} \quad .$$

Then the congruence

$$F(x) \equiv 0 \pmod{p^{k+1}}$$

has $\epsilon(x_0)$ solutions satisfying $x \equiv x_0 \pmod{p^k}$, where

$$\epsilon(x_0) = \begin{cases} 1 & for \quad F'(x_0) \not\equiv 0 \pmod{p} \ , \\ p & for \quad F'(x_0) \equiv 0 \pmod{p} \ , \ F(x_0) \equiv 0 \pmod{p^{k+1}} \ , \\ 0 & for \quad F'(x_0) \equiv 0 \pmod{p} \ , \ F(x_0) \not\equiv 0 \pmod{p^{k+1}} \ . \end{cases}$$

Moreover,

$$\lambda_F(p^{k+1}) = \sum_{\substack{x_0 (\bmod \, p^k) \\ F(x_0) \equiv 0 (\bmod \, p^k)}} \varepsilon(x_0).$$

Proof. If x_0 satisfies the congruence $F(x_0) \equiv 0 (\bmod \, p^k)$, then we can write $F(x_0) = ap^k$ for a suitable integer a. We can express, in turn, each integer congruent to $x_0 (\bmod \, p^k)$ in the form $x \equiv x_0 + tp^k (\bmod \, p^{k+1})$, where $t (\bmod \, p)$ is uniquely determined by x.

Therefore

$$F(x) \equiv 0 (\bmod \, p^{k+1}) \;\Leftrightarrow\; F(x_0 + tp^k) \equiv 0 (\bmod \, p^{k+1})$$

$$\Leftrightarrow\; F(x_0) + tp^k F'(x_0) \equiv 0 (\bmod \, p^{k+1})$$

$$\Leftrightarrow\; p^k a + tp^k F'(x_0) \equiv 0 (\bmod \, p^{k+1})$$

$$\Leftrightarrow\; tF'(x_0) \equiv -a (\bmod \, p) \;.$$

If $F'(x_0) \not\equiv 0 (\bmod \, p)$, then the last condition is satisfied by precisely one value of $t (\bmod \, p)$, so that $\epsilon(x_0) = 1$. If, on the other hand, $F'(x_0) \equiv 0 (\bmod \, p)$, then in the case $p | a$ (i.e. $F(x_0) \equiv 0 (\bmod \, p^{k+1})$)) we have p possibilities of $t (\bmod \, p)$, and in the opposite case this condition is not satisfied by any value of t. The last part of the assertion follows immediately by adding all the values of $\epsilon(x_0)$. ■

Corollary 1. *If $F(X)$ is a polynomial with integral coefficients and the congruences $F(x) \equiv 0 (\bmod \, p)$, $F'(x) \equiv 0 (\bmod \, p)$ have no common solutions, then for each $n \geq 1$ the equality*

$$\lambda_F(p^n) = \lambda_F(p)$$

holds. ■

Corollary 2. *If k is a given natural number, $a \in \mathscr{Z}$ and p is a prime number not dividing k, then for each natural number n the congruence $x^k \equiv a \, (\bmod \, p^n)$ has the same number of solutions as the congruence $x^k \equiv a \, (\bmod \, p)$.* ■

Without additional conditions on the polynomial F it is hard to say anything concrete about the values of $\lambda_F(p)$ besides the rather banal estimate given in Theorem 1.20. If the polynomial $F(X)$ is irreducible over the field of rational numbers, then it can be shown that for infinitely many primes p the equality $\lambda_F(p) = n$ holds, where n is the degree of the polynomial $F(X)$. This is usually proven using algebraic number theory.

Recently, I. Gerst and J. Brillhart [1] have shown that this can be done using only the elements of field theory. This paper contains relatively simple proofs of many interesting results concerning prime numbers for which $\lambda_F(p) \neq 0$.

Now we prove a result concerning these prime numbers, employing only elementary methods:

Theorem 1.24. *If F(X) is a polynomial of positive degree and has integral coeffi-cients, then there exist infinitely many primes p for which $\lambda_F(p) \neq 0$.*

Proof. (Due to I. Schur [1]). If $F(0) = 0$, then for each prime p we have $\lambda_F(p) \geqslant 1$. Hence we let $F(0) = a \neq 0$. Now suppose that we have $\lambda_F(p) \neq 0$ only for primes $p_1, \ldots, p_r$. We denote their product by A and consider the polynomial

$$g(X) = \frac{F(aAX)}{a}.$$ This polynomial has integral coefficients which are divisible by A,

save the constant term equal to 1. If b is a solution of the congruence $g(x) \equiv 0 \,(\mathrm{mod}\, p)$, then

$$F(aAb) = ag(b) \equiv 0 \,(\mathrm{mod}\, p),$$

so that the prime p must be divisible by one of the p_i's. But, for every integral x we have $g(x) \equiv 1 \,(\mathrm{mod}\, p_i)$ for $i = 1, 2, \ldots, r$ and we see that for any prime p we have $\lambda_g(p) = 0$. Therefore $g(X)$ takes exclusively the values 0, 1, −1 for integral arguments, which contra-dicts our assumption on the polynomial $F(X)$. This contradiction proves the assertion of the theorem. ■

Exercises

1. Prove Wilson's theorem: If p is a prime, then $(p-1)! \equiv -1 (\mathrm{mod}\ p)$.

2. Show that if $(n-1)! \equiv -1 (\mathrm{mod}\ n)$, then n is a prime.

3. Show that for the integers $n > 1$ and $n + 2$ to be both prime it is necessary and sufficient that

$$4\big((n-1)!+1\big) \equiv -n\big(\mathrm{mod}\, n(n+2)\big).$$

(P. A. Clement, Amer. Math. Monthly, 56 (1948), 23-25).

4. Prove Euler's theorem (Theorem 1.17) without the use of group theory.

5. Show that if $2 \nmid N$, then the groups $G(N)$ and $G(2N)$ are isomorphic.

6. Prove that the primitive roots (mod N) exist iff $N = p^k$ or $N = 2p^k$, or lastly $N = 1, 2, 4$ where p is an odd prime, and $k = 1, 2, \ldots$.

7. Prove that if p is a prime, then the congruence

$$x^{p-1} \equiv 1 \,(\mathrm{mod}\, p^n)$$

has exactly $p - 1$ solutions for each n.

§4. Quadratic congruences

1. We shall be concerned here with solving quadratic congruences, i.e. congruences of the form

(1.12) $$ax + bx + c \equiv 0 (\mathrm{mod}\, p^n),$$

where a, b, c are integers, p is a given prime number, $p \nmid a$ and $n \in \mathcal{N}$. First we shall investigate the case $n = 1$.

For a start, suppose that p is an odd prime, since the case $p = 2$ has its own peculiarities and will be considered at the end of this subsection. Under the supposition $p \neq 2$ we can replace (1.12) by an equivalent congruence of a simpler form, with the coefficient zero at the term x. In fact, if d is a solution of the congruence $2ax \equiv 1 \pmod{p}$, then we have

$$ax^2 + bx + c \equiv a(x + bd)^2 - a(b^2 - 4ac)d^2 \pmod{p} ,$$

so that, if $B = a(b^2 - 4ac)d^2$, then x is a solution of the congruence (1.12) iff $y = x + bd$ is a solution of the congruence $ay^2 \equiv B \pmod{p}$ and multiplying both sides of this last congruence by $2d$ we obtain

$$y^2 \equiv 2Bd \pmod{p} .$$

Hence we may restrict ourselves to the congruence of the form

(1.13) $x^2 \equiv A \pmod{p}$.

If A is divisible by p, then the congruence (1.13) has a unique solution $x \equiv 0 \pmod{p}$. If, on the contrary, $p \nmid A$, then it has a solution only in the case where $A \pmod{p}$ is a square in the group $G(p)$. In this case there are even two solutions, for if $a \pmod{p}$ is a solution, then also is $-a \pmod{p}$ wherein these solutions are distinct in view of $p \neq 2$.

Traditionally, we say that $A \not\equiv 0 \pmod{p}$ is a *quadratic residue* $\pmod{p}$ in the case where (1.13) has a solution, and a *quadratic non-residue* $\pmod{p}$ otherwise. Let us notice the set of squares in the group $G(p)$. This forms a group under multiplication, since the product as well as the quotient of squares is a square. We denote this group by $G^2(p)$. Since, owing to Theorem 1.19, the group $G(p)$ is cyclic, $G^2(p)$ as its subgroup, is also cyclic. Moreover, it contains $\frac{1}{2}(p-1)$ elements because it is an image of the group $G(p)$ under the homomorphism $h(x) = x^2$ whose kernel consists of two elements, i.e. of the residue classes $1 \pmod{p}$ and $-1 \pmod{p}$. We have therefore $\frac{1}{2}(p-1)$ quadratic residues giving distinct residue classes $\pmod{p}$ and as many quadratic non-residues.

The set $\{-1, 1\}$ forms a group under multiplication, isomorphic with C_2. Hence there exists a unique homomorphism χ of the group $G(p)$ onto the group $\{-1, 1\}$ with kernel $G^2(p)$. If now a is any integer indivisible by p, then we shall denote the number $\chi(a \pmod{p})$ by $\left(\dfrac{a}{p}\right)$. In order to define thus defined function for all integers, we make a convention in the case $p \mid a$ that $\left(\dfrac{a}{p}\right) = 0$. The function defined in this way

$$\left(\frac{x}{p}\right): \ \mathcal{Z} \to \{1, 0, -1\}$$

is called the *Legendre symbol* to the modulus p (recall that p is an odd prime). The fundamental properties of the Legendre symbol are contained in the following theorem:

Theorem 1.25.

(i) $\left(\dfrac{ab}{p}\right) = \left(\dfrac{a}{p}\right) \cdot \left(\dfrac{b}{p}\right).$

(ii) For $p \nmid a$, $\left(\dfrac{a}{p}\right) = +1 \Leftrightarrow x^2 \equiv a \pmod{p}$ *has a solution,*

$\left(\dfrac{a}{p}\right) = -1 \Leftrightarrow x^2 \equiv a \pmod{p}$ *has no solution.*

(iii) *If* $p \nmid b$, *then* $\left(\dfrac{b^2}{p}\right) = 1$.

(iv) *If* $a \equiv b \pmod{p}$, *then* $\left(\dfrac{a}{p}\right) = \left(\dfrac{b}{p}\right).$

(v) $\left|\left\{x \pmod{p} : \left(\dfrac{x}{p}\right) = 1\right\}\right| = \left|\left\{x \pmod{p} : \left(\dfrac{x}{p}\right) = -1\right\}\right| = \tfrac{1}{2}(p-1).$

Proof.   The results follow from the definition and fundamental properties of homomorphisms. ∎

For the determination of values of the Legendre symbol the following result is useful:

Theorem 1.26. *If* a *is not divisible by* p, *then*

$$\left(\frac{a}{p}\right) \equiv a^{\frac{1}{2}(p-1)} \pmod{p}.$$

Proof. If $b = a^{\frac{1}{2}(p-1)}$, then by virtue of the Corollary to Theorem 1.17 we have $b^2 = a^{p-1} \equiv 1 \pmod{p}$, so that $b \equiv 1$ or $b \equiv -1 \pmod{p}$. If a is a quadratic residue $\pmod{p}$ and c is a solution of the congruence $x^2 \equiv a \pmod{p}$, then we have $b = a^{\frac{1}{2}(p-1)} \equiv c^{p-1} \equiv 1 \pmod{p}$, so that in this case our theorem is true. Since by Theorem 1.20, the congruence $x^{\frac{1}{2}(p-1)} \equiv 1 \pmod{p}$ has at most $\frac{1}{2}(p-1)$ solutions, and we checked just now that every quadratic residue (they are exactly $\frac{1}{2}(p-1)$ in number) is its solution, hence in the case where a is a quadratic non-residue, we must have $b \equiv -1 \pmod{p}$. ∎

Corollary. $\left(\dfrac{-1}{p}\right) = 1$ *iff* $p \equiv 1 \pmod 4$, *that is* $\left(\dfrac{-1}{p}\right) = (-1)^{\frac{1}{2}(p-1)}.$ ∎

Applying Theorem 1.23, we can find the number of solutions of the congruence (1.13) for $n \geqslant 2$.

Theorem 1.27. *If p is an odd prime, then the congruence*

$$X^2 \equiv A \ (\mathrm{mod}\, p^n)$$

has no solutions if $\left(\dfrac{A}{p}\right) = -1$, *and has two solutions if* $\left(\dfrac{A}{p}\right) = 1$. *In the case $p|A$ we*

write $A = p^r a$, *with a indivisible by p. If, then $r \geqslant n$, then we have p^{n-k} solutions, where $k = \frac{1}{2}n$, when n is even, and $k = \frac{1}{2}(n+1)$, when n is odd. Finally, if $r < n$, then*

for $2 \nmid r$ or $2 | r$ and $\left(\dfrac{a}{p}\right) = -1$ *there are no solutions at all, and for an even r and*

$\left(\dfrac{a}{p}\right) = 1$ *we have* $2p^{\frac{1}{2}r}$ *solutions.*

Proof. The case of A indivisible by p is an immediate consequence of Corollary 1 to Theorem 1.23. If $A = p^r a$, $p \nmid a$, and $r > 0$, then each solution x of our congruence must be divisible by p. Write $x = p^j y$, where y is not divisible by p. Then we must have $p^{2j} y^2 \equiv a p^r \ (\mathrm{mod}\, p^n)$. In the case $r \geqslant n$, we obtain $p^{2j} y^2 \equiv 0 \ (\mathrm{mod}\, p^n)$, and further, $p^{2j} \equiv 0 \ (\mathrm{mod}\, p^n)$, which is equivalent to $2j \geqslant n$. Hence every integer x divisible by p^k is a solution, and this gives p^{n-k} elements $x \ (\mathrm{mod}\, p^n)$.

In the case $r < n$ we must have $2j = r$ and $y^2 \equiv a \ (\mathrm{mod}\, p^{n-r})$. Hence for odd r or also for even r with $\left(\dfrac{a}{p}\right) = -1$ we have no solutions, and for even r with $\left(\dfrac{a}{p}\right) = 1$ we have two possibilities of $y(\mathrm{mod}\, p^{n-r})$, and hence of $x(\mathrm{mod}\, p^n) = p^{\frac{r}{2}} y(\mathrm{mod}\, p^n)$ we obtain $2p^{\frac{r}{2}}$ possibilities. ∎

2.   Now we shall prove a theorem, which in the 19th century was considered as a central result in the theory of numbers, the so-called *quadratic reciprocity law*. Let p and q be distinct odd prime numbers. Then the reciprocity law enables us to evaluate $\left(\dfrac{p}{q}\right)$, if $\left(\dfrac{q}{p}\right)$ is known. This result is rather unexpected because it is not clear *a priori* why the information on the fact that p is a quadratic residue (mod q) could allow us to infer whether q is or is not a quadratic residue (mod p). This result was formulated first by L. Euler in 1772, and was proved by C. F. Gauss in 1801. Many different proofs of this theorem are known — Gauss himself gave seven proofs.

Theorem 1.28. *If p, q are distinct primes $\neq 2$, then*

$$\left(\frac{p}{q}\right) \cdot \left(\frac{q}{p}\right) = (-1)^{\frac{1}{2}(p-1) \cdot \frac{1}{2}(q-1)}.$$

We shall give a proof after J. P. Serre [1]. We shall need a lemma easily deducible from elementary facts of algebra:

Lemma 1.4. *Let p, q be prime numbers, p ≠ q. We denote the field with p elements by k. Then there exists a field K containing k and having the following properties: in K one can find an element u such that* $u, u^2, \ldots, u^{q-1}$ *are all distinct, and* $u^q = e$, *where e denotes the identity of the field k.*

Proof. From algebra it is known (see, e.g. B. L. van der Waerden [1], p. 121, Theorem) that there exists a field $K \supset k$ in which the polynomial $X^q - e$ decomposes into linear factors, that is

$$X^q - e = (X - a_1) \ldots (X - a_q) \qquad (a_i \in K).$$

From the second Theorem on p. 85 of B. L. van der Waerden [1], it follows that all the elements a_i are different (here we use the assumption $p \neq q$). Moreover, we see that $\{a_1, \ldots, a_q\}$ is the set of all the elements x of the field K satisfying $x^q = e$, and so it forms a group, since the product as well as the quotient of the elements satisfying this relation also satisfies this relation. This group has q elements, and q is a prime number, hence this is a cyclic group. If we take a generator of this group as u, then the assertion of the lemma will be satisfied. ■

Proof of Theorem 1.28. We denote the finite field of q elements by L and the finite field of p elements by k. Let u be an element of the former field K which satisfies the assertion of the lemma. Lastly let

$$\tau = \sum_{x(\bmod q)} \left(\frac{x}{q}\right) u^x.$$

Obviously, τ is an element of the field K if we identify the natural numbers 0, 1 and -1 with their images in the field k. We now show that

(1.14)
$$\tau^2 = (-1)^{\frac{1}{2}(q-1)} \cdot \bar{q},$$

where $\bar{q} = q \,(\bmod p)$.

Indeed, by virtue of Theorem 1.25

$$\tau^2 = \sum_{x(\bmod q)} \left(\frac{x}{q}\right) u^x \sum_{y(\bmod q)} \left(\frac{y}{q}\right) u^y$$

$$= \sum_{x(\bmod q)} \sum_{y(\bmod q)} \left(\frac{xy}{q}\right) u^{x+y} = \sum_{z(\bmod q)} u^z \sum_{\substack{x(\bmod q) \\ x \not\equiv 0(\bmod q)}} \left(\frac{x(z-x)}{q}\right),$$

since for $x \,(\bmod q) = 0$ we have $\left(\dfrac{x(z-x)}{q}\right) = 0$.

If $x(\bmod q) \neq 0$, then we let $x'(\bmod q)$ be the element satisfying $xx'(\bmod q) = 1(\bmod q)$. Then from Theorem 1.25 and the Corollary to Theorem 1.26 it follows that

$$\left(\frac{x(z-x)}{q}\right) = \left(\frac{-x^2(1-zx')}{q}\right) = \left(\frac{-1}{q}\right)\cdot\left(\frac{1-zx'}{q}\right) = (-1)^{\frac{1}{2}(q-1)}\left(\frac{1-zx'}{q}\right),$$

so that

$$(-1)^{\frac{1}{2}(q-1)}\tau^2 = \sum_{z(\bmod q)} u^z \sum_{\substack{x(\bmod q) \\ x \not\equiv 0(\bmod q)}} \left(\frac{1-zx'}{q}\right)$$

$$= \sum_{\substack{x(\bmod q) \\ x \not\equiv 0(\bmod q)}} \left(\frac{1}{q}\right) + \sum_{\substack{z(\bmod q) \\ z \not\equiv 0(\bmod q)}} u^z \sum_{\substack{x(\bmod q) \\ x \not\equiv 0(\bmod q)}} \left(\frac{1-zx'}{q}\right).$$

Now note that $\left(\dfrac{1}{q}\right) = 1$, and if $x(\bmod q)$ runs through all the non-zero elements of the field L then with $z(\bmod q) \neq 0(\bmod q)$, $1-zx'$ runs through all the elements of this field other than $1(\bmod q)$, so that

$$(-1)^{\frac{1}{2}(q-1)}\tau^2 = \bar{q}-1+ \sum_{\substack{z(\bmod q) \\ z \not\equiv 0(\bmod q)}} u^z \cdot \sum_{\substack{w(\bmod q) \\ w \not\equiv 1(\bmod q)}} \left(\frac{w}{q}\right),$$

but

$$\sum_{\substack{z(\bmod q) \\ z \not\equiv 0(\bmod q)}} u^z = u+u^2+ \ldots +u^{q-1} = -1,$$

and

$$\sum_{\substack{w(\bmod q) \\ w \not\equiv 1(\bmod q)}} \left(\frac{w}{q}\right) = \sum_{\substack{w(\bmod q) \\ w \not\equiv 0(\bmod q)}} \left(\frac{w}{q}\right) - \left(\frac{1}{q}\right)$$

$$= \sum_{\substack{w(\bmod q) \\ \left(\frac{w}{q}\right)=+1}} 1 - \sum_{\substack{w(\bmod q) \\ \left(\frac{w}{q}\right)=-1}} 1 - 1 = -1$$

in view of Theorem 1.25 (v). In the end we obtain

$$(-1)^{\frac{1}{2}(q-1)}\tau^2 = \bar{q}-1+(-1)\cdot(-1) = \bar{q}.$$

which proves formula (1.14).

Using Corollary 2 to Lemma 1.2, we now note that

$$\tau^p = \sum_{x(\mathrm{mod}\,q)} \left(\frac{x}{q}\right)^p u^{px} = \sum_{x(\mathrm{mod}\,q)} \left(\frac{x}{q}\right) u^{px} = \sum_{y(\mathrm{mod}\,q)} \left(\frac{yp'}{q}\right) u^y$$

$$= \left(\frac{p'}{q}\right) \sum_{y(\mathrm{mod}\,q)} \left(\frac{y}{q}\right) u^y = \left(\frac{p'}{q}\right)\tau = \left(\frac{p}{q}\right)\tau$$

as p and p' are either both quadratic residues or both non-residues (mod q). From (1.14) it follows that τ is non-zero, and so we obtain

(1.15) $$\tau^{p-1} = \left(\frac{p}{q}\right).$$

This implies that $\tau^{p-1} \in k$. From formula (1.14) we also obtain

$$\left(\frac{p}{q}\right) = \tau^{2 \cdot \frac{1}{2}(p-1)} = (-1)^{\frac{1}{2}(p-1) \cdot \frac{1}{2}(q-1)} q^{\frac{1}{2}(p-1)},$$

so that

$$\left(\frac{p}{q}\right) \equiv (-1)^{\frac{1}{2}(p-1) \cdot \frac{1}{2}(q-1)} q^{\frac{1}{2}(p-1)} (\mathrm{mod}\,p),$$

which, in view of Theorem 1.26, leads to

$$(-1)^{\frac{1}{2}(p-1) \cdot \frac{1}{2}(q-1)} \left(\frac{q}{p}\right) \cdot \left(\frac{p}{q}\right) \equiv (-1)^{\frac{1}{2}(p-1) \cdot \frac{1}{2}(q-1)} \left(\frac{q}{p}\right) \cdot (-1)^{\frac{1}{2}(p-1) \cdot \frac{1}{2}(q-1)} q^{\frac{1}{2}(p-1)}$$

$$\equiv \left(\frac{q}{p}\right) q^{\frac{1}{2}(p-1)} \equiv q^{\frac{1}{2}(p-1)} \cdot q^{\frac{1}{2}(p-1)} \equiv q^{p-1} \equiv 1 \,(\mathrm{mod}\,p),$$

which, however, is equivalent to our assertion. ∎

Let us now evaluate the value of the Legendre symbol $\left(\dfrac{a}{p}\right)$ in the case $a = 2$.

Theorem 1.29. *If p is an odd prime, then*

$$\left(\frac{2}{p}\right) = (-1)^{\frac{1}{8}(p^2-1)}.$$

Proof. Let K be an extension of the field $\mathscr{Z}/p\mathscr{Z}$, in which the polynomial $X^8 - 1$ decomposes into linear factors. On the basis of §6.6 of B. L. van der Waerden [1], there exists in K a primitive 8-th root of 1, that is, there exists an element $u \in K$ which is a root of this polynomial and the remaining roots are powers of u. Let $z = u + u^{-1}$. Then $z^2 = u^2 + 2 + u^{-2}$, but $u^4 = -1$ (we must have $u^4 = \pm 1$, for $u^8 = 1$, and the equality $u^4 = 1$ would mean that u is not a primitive root), hence, dividing by u^2 we obtain $u^2 = -u^{-2}$, which gives $z^2 = 2$. Furthermore, we have $z^p = u^p + u^{-p}$ in virtue of Lemma 1.3. If $p = 8k + r$, with integral k, r and $r = \pm 1, \pm 3$, then clearly $z^p = u^r + u^{-r}$. Now if $r = \pm 1$, then we can see at once that $z^p = z$, and this implies

$$2^{\frac{1}{2}(p-1)} = z^{p-1} = 1, \text{ and thus } \left(\frac{2}{p}\right) = 1. \text{ If, on the contrary, } r = \pm 3, \text{ then in view of}$$

$$u + u^3 + u^5 + u^7 = u(u^8 - 1)(u^2 - 1)^{-1} = 0$$

we obtain $z^p = u^5 + u^{-5} = -(u + u^{-1}) = -z$, and so we obtain in the same way as in the former case, $\left(\dfrac{2}{p}\right) = -1.$ ■

Theorems on the Legendre symbol obtained above enable us to compute its value readily. Indeed, if we have to compute, for instance, the value $\left(\frac{131}{257}\right)$, then using the reciprocity law, we obtain

$$\left(\tfrac{131}{257}\right) = \left(\tfrac{257}{131}\right) = \left(\tfrac{-5}{131}\right) = -\left(\tfrac{5}{131}\right) = -\left(\tfrac{131}{5}\right) = -\left(\tfrac{1}{5}\right) = -1.$$

Similarly, using, in addition, Theorem 1.29, we obtain

$$\left(\tfrac{102}{167}\right) = \left(\tfrac{2}{167}\right) \cdot \left(\tfrac{51}{167}\right) = \left(\tfrac{51}{167}\right) = \left(\tfrac{3}{167}\right)\left(\tfrac{17}{167}\right) = -\left(\tfrac{167}{3}\right)\left(\tfrac{167}{17}\right) = -\left(\tfrac{2}{3}\right)\left(\tfrac{14}{17}\right)$$

$$= \left(\tfrac{14}{17}\right) = \left(\tfrac{2}{17}\right)\left(\tfrac{7}{17}\right) = \left(\tfrac{7}{17}\right) = \left(\tfrac{17}{7}\right) = \left(\tfrac{3}{7}\right) = -\left(\tfrac{7}{3}\right) = -\left(\tfrac{1}{3}\right) = -1.$$

3.   Returning to the congruence (1.12), we consider the case $p = 2$. Here we shall confine ourselves to the solutions of the congruence

(1.16) $X^2 \equiv a(\mathrm{mod}\ 2^n)$

in the case of odd a, leaving the general case of (1.12) for the reader's own discussion.

For $n = 1$ the congruence (1.16) takes the form $X^2 \equiv 1(\mathrm{mod}\ 2)$, and hence has a unique solution $1(\mathrm{mod}\ 2)$. Since $(4k \pm 1)^2 = 16k^2 \pm 8k + 1 \equiv 1(\mathrm{mod}\ 8)$, for $n = 2, 3$ we see that the congruence (1.16) has a solution only when $a \equiv 1(\mathrm{mod}\ 2^n)$, wherein the number of solutions are 2 and 4 respectively.

We define the *Legendre symbol* to the modulus 2 by the formula

$$\left(\frac{x}{2}\right) = 1, \qquad \text{when} \qquad x \equiv 1 \pmod 8,$$

$$\left(\frac{x}{2}\right) = -1, \qquad \text{when} \qquad x \equiv 3, 5, 7 \pmod 8,$$

$$\left(\frac{x}{2}\right) = 0, \qquad \text{when} \qquad x \equiv 0, 2, 4, 6 \pmod 8.$$

It is easy to check that the formulas given in Theorem 1.25 (i), (iii), (iv) are

also true in the case $p = 2$. The relation of the symbol $\left(\dfrac{x}{2}\right)$ with the congruence (1.16) is given by the following theorem:

Theorem 1.30. *If a is an odd integer, then the congruence (1.16) with $n \geqslant 3$ has a*

solution iff $\left(\dfrac{a}{2}\right) = 1$. If this condition is satisfied, then it has precisely four solutions.

Proof. For $n = 3$ we already know this fact, so let $n \geqslant 4$. The condition $\left(\dfrac{a}{2}\right) = 1$

is a necessary condition for the solvability of (1.16) for $n \geqslant 3$, because every one of its solution satisfies the congruence $x^2 \equiv 1 \pmod 8$, and we have seen above that this

implies $a \equiv 1 \pmod 8$, that is $\left(\dfrac{a}{2}\right) = 1$. Now suppose that $\left(\dfrac{a}{2}\right) = 1$. From Theorem

1.19 it follows that $G(2^n) = C_2 \oplus C_{2^{n-2}}$. Therefore we infer the existence of elements $g_1, g_2 \in G(2^n)$ such that every element u of this group is written uniquely in the form $u = g_1^a g_2^b$ ($a = 0, 1$, $b = 0, 1, \ldots, 2^{n-2}-1$), and $g_1^2 = g_2^{2^{n-2}} = 1 \pmod{2^n}$. From this it follows that u is a square in $G(2^n)$ iff $a = 0$ and b is even. Hence we have 2^{n-3} squares in $G(2^n)$. Define the map $f: G(2^n) \to G(2^{n-1})$ by the formula

$$f(x \pmod{2^n}) = x \pmod{2^{n-1}}.$$

We see immediately that f is a homomorphism, and hence it maps squares onto squares. Thus every square in $G(2^n)$ must be of the form $f^{-1}(a^2)$ for some a in $G(2^{n-1})$. For the element a^2 in $G(2^{n-1})$ let $e(a^2)$ denote the number of squares in $f^{-1}(a^2)$, i.e.

$$e(a^2) = \sum_{\substack{b^2 \in G(2^n) \\ f(b^2) = a^2}} 1.$$

Since Ker f has two elements, viz. $1 \pmod{2^n}$ and $2^{n-1} + 1 \pmod{2^n}$, we have $e(a^2) \leqslant 2$. Furthermore we have

$$\sum_{a^2 \in G(2^{n-1})} \left(2 - e(a^2)\right) = 2|G(2^{n-1})^2| - \sum_{a^2 \in G(2^{n-1})} e(a^2)$$

$$= 2|G(2^{n-1})^2| - |G(2^n)^2| = 2 \cdot 2^{n-4} - 2^{n-3} = 0.$$

Since all the summands in the sum on the left-hand side are non-negative, all of them must vanish, that is for every $a^2 \in G(2^{n-1})$, $e(a^2) = 2$ and hence $f^{-1}(a^2)$ contains exclusively the squares for any $a \in G(2^{n-1})$. This can be formulated as follows: $a \in G(2^n)$ is a square iff $f(a)$ is a square in $G(2^{n-1})$.

Hence the congruence $x^2 \equiv a \pmod{2^n}$ has a solution iff the congruence $x^2 \equiv a \pmod{2^{n-1}}$ has a solution.

Now we evaluate the number of solutions. For $n = 3$ we already know that there are four solutions. Suppose that for some n there are four solutions, say $x_1, \ldots, x_4 \pmod{2^n}$. From Theorem 1.23 it follows that from every solution $x_i \pmod{2^n}$ we obtain two solutions of $X^2 \equiv a \pmod{2^{n+1}}$ if only $x_i^2 \equiv a \pmod{2^{n+1}}$. Here we know that there exists at least one such solution x_i, but $2^n - x_i$ is a second solution with this same property. Hence our congruence has at least four solutions arising, say from $x_i \pmod{2^n}$, $i = 1, 2$. Applying the same reasoning to the congruence $X^2 \equiv a + 2^n \pmod{2^{n+1}}$, we find that it has also at least four solutions, but $a + 2^n \equiv a \pmod{2^n}$, so they must arise from the solutions x_3, x_4 of the congruence $X^2 \equiv a \pmod{2^n}$. Since every solution of $X^2 \equiv a \pmod{2^{n+1}}$ is obtained from one of the solutions x_i ($i = 1, 2, 3, 4$), we see ultimately that $X^2 \equiv a \pmod{2^{n+1}}$ has exactly four solutions. ∎

Exercises

1. Let p, q be odd primes satisfying $p \equiv 1 \pmod{q}$ and let $G(p)^q = \{a^q : a \in G(p)\}$.
    (i) Show that $G(p)^q$ is a group, and that $G(p)/G(p)^q$ is isomorphic with the group A_q of the q-th roots of unity.

    (ii) Let $\varphi : G(p) \to A_q$ be a homomorphism arising from the composition of

    $$\psi_1 : G(p) \to G(p)/G(p)^q \quad \text{and} \quad \psi_2 : G(p)/G(p)^q \to A_q . \quad \text{Let} \quad \varphi(x) = \left(\frac{x}{p}\right)_q .$$

    Prove that for $\left(\dfrac{x}{p}\right)_q$ the analogues of Theorems 1.25 and 1.26 are true.

2. Let p be an odd prime and k be any natural number. Show that the congruence

    $$x^k \equiv -1 \pmod{p}$$

has a solution iff

    $$(k, p-1) \left| \frac{p-1}{2} \right. .$$

3. Give a condition of the solvability of the congruence (1.12) for $p = 2$.

4. Prove that the integer 3 is a primitive root for every Fermat's prime number greater than 3.

5. Show that if p and $4p + 1$ are both primes, then 2 is a primitive root for $4p + 1$.

§5. An application of trigonometrical sums in the theory of numbers

1. In this section we shall give a formula for the number of solutions of the congruence

$$(1.17) \qquad f(X_1, \ldots, X_n) \equiv 0 (\mathrm{mod}\, N) \ ,$$

where N is any natural number. We suppose that the function f is periodic (mod N) with respect to each variable, that is if $X_i \equiv X_i' \pmod N$ holds for $i = 1, 2, \ldots, n$, then

$$f(X_1, \ldots, X_n) \equiv f(X_1', \ldots, X_n') \pmod N \ .$$

By Corollary 1 to Theorem 1.15 every polynomial in n variables with integral coefficients has this property.

Let us begin with the elementary identity:

Lemma 1.5. *If N is a natural number, then*

$$S_a(N) = \sum_{t(\mathrm{mod}\, N)} \exp\left(\frac{2\pi i a t}{N}\right) = \begin{cases} N, & \text{if} & N | a, \\ 0, & \text{if} & N \nmid a. \end{cases}$$

Proof. If N divides a, then all the summands of the sum are equal to one, and so the sum is equal to N. On the contrary, if $N \nmid a$, then we see that the summands form a geometric progression with the initial term equal to one and the common ratio $z = \exp(2\pi i a/N)$, so that

$$S_a(N) = \frac{z^N - 1}{z - 1} = 0. \qquad \blacksquare$$

Theorem 1.31. *The number of solutions of the congruence (1.17) equals*

$$\frac{1}{N} \sum_{t(\mathrm{mod}\, N)} \sum_{\substack{X_1(\mathrm{mod}\, N) \\ \cdots\cdots \\ X_n(\mathrm{mod}\, N)}} \exp\left\{\frac{2\pi i t}{N} f(X_1, \ldots, X_n)\right\}$$

Proof. In the above sum we change the order of summation and note that by Lemma 1.5 we have

$$\frac{1}{N} \sum_{t(\mathrm{mod}\, N)} \exp\left(\frac{2\pi i t}{N} f(X_1, \ldots, X_n)\right) = \begin{cases} 1, & \text{if} & f(X_1, \ldots, X_n) \equiv 0 \,(\mathrm{mod}\, N), \\ 0, & \text{if} & f(X_1, \ldots, X_n) \not\equiv 0 \,(\mathrm{mod}\, N). \end{cases} \qquad \blacksquare$$

We apply the formula obtained to the calculation of the number of solutions of the congruence

(1.18) $F_1(x_1) + \ldots + F_s(x_s) \equiv a \pmod{N}$,

where $F_1(t), \ldots, F_s(t)$ are polynomials with integral coefficients of degree $k_1, \ldots, k_s$, respectively and a is an integer. From Theorem 1.31 it follows that the number λ_N of solutions of (1.18) is equal to

$$\frac{1}{N} \sum_{t(\mathrm{mod}\,N)} \sum_{\substack{x_1(\mathrm{mod}\,N) \\ \cdots\cdots\cdots \\ x_s(\mathrm{mod}\,N)}} \exp\left\{\frac{2\pi it}{N}\left(\sum_{j=1}^{s} F_j(x_j) - a\right)\right\}$$

$$= \frac{1}{N} \sum_{t(\mathrm{mod}\,N)} \left\{\prod_{j=1}^{s} \sum_{x(\mathrm{mod}\,N)} \exp\left\{\frac{2\pi it F_j(x)}{N}\right\}\right\} \exp\left\{-\frac{2\pi iat}{N}\right\}.$$

Extracting the term corresponding to $t = 0$ in this sum, we obtain

(1.19) $\lambda_N = N^{s-1} + \dfrac{1}{N} \displaystyle\sum_{\substack{t(\mathrm{mod}\,N) \\ t \not\equiv 0\,(\mathrm{mod}\,N)}} \prod_{j=1}^{s} \sum_{x(\mathrm{mod}\,N)} \exp\left\{\frac{2\pi it F_j(x)}{N}\right\} \exp\left\{-\frac{2\pi iat}{N}\right\}.$

If we succeed in showing that the second term of this formula is less than the first in absolute value, then we obtain $\lambda_N > 0$.

Obviously, this does not hold in every case, since for example the congruence $x_1^2 + x_2^2 \equiv 3 \pmod 4$ has no solutions, but we shall try to find sufficient conditions for $\lambda_N > 0$.

2.   From formula (1.19) one can see that one must estimate in a non-trivial way the sum

(1.20) $\displaystyle\sum_{x(\mathrm{mod}\,N)} \exp\left\{\frac{2\pi it F_j(x)}{N}\right\}$

for its trivial estimate by N leads to $\lambda_N = N^{s-1} + O(N^s)$, which does not give anything. L. J. Mordell [1] proved such an estimate for prime N, and L. K. Hua [1] proved it in the general case. Now we shall prove this estimate, restricting ourselves to the result of Mordell.

Theorem 1.32. *If* $W(X) = a_k X^k + \ldots + a_0$ *is a polynomial of degree* k *with integral coefficients and* p *is a prime not dividing* a_k, *then for the sum*

(1.21) $S_p(W) = \displaystyle\sum_{X(\mathrm{mod}\,p)} \exp\left\{\frac{2\pi i W(X)}{p}\right\}$

the estimate

$$|S_p(W)| \leqslant kp^{1-\frac{1}{k}}$$

holds.

Proof. To begin with, note that it suffices to consider a polynomial $W(X)$ of degree $k \geqslant 2$ satisfying the condition $W(0) = 0$, that is $a_0 = 0$. Indeed,

$$S_N\big(W - W(0)\big) = S_N(W) \exp\left\{\frac{-2\pi i W(0)}{N}\right\},$$

and so

$$|S_N\big(W - W(0)\big)| = |S_N(W)|,$$

and for $k = 1$ the assertion follows from Lemma 1.5. We may also suppose that $p \nmid k$, for in the case $p \mid k$ we have $p \leqslant k$, and so

$$|S_p(W)| \leqslant p \leqslant k < kp^{1-\frac{1}{k}}.$$

We denote the polynomial $W(aX + b) - W(b)$ by $W_{a,b}(X)$, where $a = 1, 2, \ldots, p-1$ and $b = 0, 1, \ldots, p-1$. If X runs through a complete set of representatives $(\bmod \, p)$, then $aX + b$ does the same, and so $S_p(W) = S_p(W_{a,b})$, which gives

$$|S_p(W)|^{2k} = |S_p(W_{a,b})|^{2k}$$

and we obtain

$$|S_p(W)|^{2k} = \frac{1}{p(p-1)} \sum_{a=1}^{p-1} \sum_{b=0}^{p} |S_p(W_{a,b})|^{2k}.$$

In each of the polynomials $W_{a,b}$ we replace the coefficients by the integers in $[0, p-1]$ congruent to them $(\bmod \, p)$. We shall then obtain a new polynomial which we denote by $T(W_{a,b})$. Note that $S_p(W_{a,b}) = S_p(T(W_{a,b}))$. If we denote by R the set of all polynomials with integral coefficients lying in $[0, p-1]$, and for a given $f \in R$ we denote the number of pairs $\langle a, b \rangle$ for which $T(W_{a,b}) = f$, by $r(f)$, then we obtain

$$|S_p(W)|^{2k} = \frac{1}{p(p-1)} \sum_{f \in R} r(f) |S_p(f)|^{2k}.$$

Now we estimate $r(f)$. If $f(X) = \sum_{i=0}^{k} b_i X^i$ and $T(W_{a,b}) = f$, then by

$$W_{a,b}(X) = a_k a^k X^k + (ka_k b + a_{k-1})a^{k-1}X^{k-1} + \cdots$$

we obtain $a_k u^k \equiv b_k \pmod{p}$, so that in accordance with Theorem 1.20 we have at most k possibilities of a. Analogously, we have $(ka_k b + a_{k-1})a^{k-1} \equiv b_{k-1} \pmod{p}$ and by $p \nmid ka_k$ we conclude that there are at most k possibilities for b. Thus, we see that $r(f)$ does not exceed k^2, and that from $f(0) \neq 0$ follows $r(f) = 0$, hence formula (1.21) leads to

$$|S_p(W)|^{2k} \leqslant \frac{k^2}{p(p-1)} \sum_{\substack{f \in R \\ f(0)=0}} |S_p(f)|^{2k}.$$

Let us show that the sum appearing on the right-hand side gives the number of solutions of some system of congruences, and then estimate this quantity directly. If $f(X) = b_k X^k + \ldots + b_0$, then this sum will be equal to

$$\sum_{\substack{b_1 (\bmod p) \\ \cdots \\ b_k (\bmod p)}} |S_p(b_k X^k + \ldots + b_1 X)|^{2k}$$

$$= \sum_{\substack{b_1(\bmod p)\, X_1(\bmod p)\, Y_1(\bmod p) \\ \cdots \cdots \cdots \\ b_k(\bmod p)\, X_k(\bmod p)\, Y_k(\bmod p)}} \exp\left\{\frac{2\pi i}{p} \sum_{j=1}^{k} b_j \left(\sum_{s=1}^{k} X_s^j - \sum_{s=1}^{k} Y_s^j\right)\right\}$$

$$= \sum_{\substack{X_1(\bmod p)\, Y_1(\bmod p) \\ \cdots \cdots \cdots \\ X_k(\bmod p)\, Y_k(\bmod p)}} \prod_{j=1}^{k} \sum_{b_j(\bmod p)} \exp\left\{\frac{2\pi i}{p} b_j \left(\sum_{s=1}^{k} X_s^j - \sum_{s=1}^{k} Y_s^j\right)\right\}.$$

On the basis of Lemma 1.5 the inner sum is equal to p in the case when $X_1^j + \ldots + X_k^j \equiv Y_1^j + \ldots + Y_k^j \pmod{p}$ and is equal to zero otherwise. Thus we obtain

$$(1.22) \qquad |S_p(W)|^{2k} \leqslant \frac{k^2}{p(p-1)} p^k \cdot S \quad,$$

where we denote by S the number of solutions of the system of congruences

$$(1.23) \qquad X_1^j + \ldots + X_k^j \equiv Y_1^j + \ldots + Y_k^j \pmod{p} \qquad (j = 1, 2, \ldots, k) \quad.$$

Therefore all amounts to the estimation of S. To this end we regard (1.23) as a system of equations in the field of p elements (replacing X_i, Y_j by their images in $\mathscr{Z}/p\mathscr{Z}$, which we shall denote by X_i', Y_j' respectively) and note that if $X_1', \ldots, X_k'$, $Y_1', \ldots, Y_k'$ satisfy this system, then by Newton's formula (see, e.g. B. L. van der Waerden [1], Exercise 5.18) the equality

$$\prod_{j=1}^{k} (X - X_i') = \prod_{j=1}^{k} (X - Y_i') \quad.$$

holds. This means that the sequence $Y'_1, \ldots, Y'_k$ is a permutation of the sequence $X'_1, \ldots, X'_k$. Returning to $X_1, \ldots, Y_k$ we see that for fixed $X_1, \ldots, X_k$ we have $k!$ possibilities of $Y_1, \ldots, Y_k$ and herein we have p^k possibilities of $X_1, \ldots, X_k$ so that $S = p^k k!$. Substituting this value of S in (1.22), we finally obtain

$$|S_p(W)|^{2k} \leqslant \frac{k^2 k! p^{2k-1}}{p-1} \leqslant 2k^2 k! p^{2k-2} \leqslant k^{2k} p^{2k-2}$$

and

$$|S_p(W)| \leqslant kp^{1-1/k} \quad . \qquad \blacksquare$$

Remark. From the results of A. Weil [1] one can get, under the same suppositions, a remarkably better estimate of the sum $S_p(W)$, viz.

$$|S_p(W)| \leqslant (k-1)p^{\frac{1}{2}}.$$

3. Let us return to the congruence (1.18). Using Theorem 1.32 we can prove the following theorem on it:

Theorem 1.33. *Let $F_1(t), \ldots, F_s(t)$ be non-constant polynomials of degree $k_1, \ldots, k_s$ respectively and with integral coefficients. If*

$$\eta = \sum_{j=1}^{s} \frac{1}{k_j} - 1 > 0$$

and p is a prime not dividing any leading coefficients of the polynomials $F_1(t), \ldots, F_s(t)$, and greater than $\prod_{i=1}^{s} k_i^{1/\eta}$, then the congruence (1.18) has a solution for $N = p$. If λ_p denotes the number of these solutions, then

$$\lambda_p = p^{s-1} + R.$$

where

$$|R| \leqslant \left(1 - \frac{1}{p}\right) p^{s-1-\eta} \prod_{i=1}^{s} k_i.$$

Proof. Applying Theorem 1.32 to the second term of formula (1.19), we get the second assertion of the theorem. Herein we have for $p > \prod_{i=1}^{s} k_i^{1/\eta}$

$$\lambda_p \geqslant p^{s-1} - |R| \geqslant p^{s-1}\left(1 - \left(1 - \frac{1}{p}\right)p^{-\eta}\prod_{i=1}^{s}k_i\right)$$

$$\geqslant p^{s-1}\left(1 - \left(1 - \frac{1}{p}\right)\right) = p^{s-2} > 0. \quad \blacksquare$$

Of particular importance is the case $F_i(t) = t^k$ $(i = 1, 2, \ldots, s)$. In this case we shall proceed to prove some improvements of Theorem 1.33 by algebraic methods:

Theorem 1.34. *If p is a prime and a is an arbitrary integer, then the congruence*

(1.24) $$X_1^k + \ldots + X_s^k \equiv a \,(\mathrm{mod}\, p)$$

has a solution if only $s \geqslant k$.

Proof. We denote by A_i the set of non-zero elements of the field $\mathscr{Z}/p\mathscr{Z}$, which are the sums of at most i k-th powers of elements of this field. We have evidently

$$A_1 \subset A_2 \subset \ldots ,$$

and from the finiteness of $\mathscr{Z}/p\mathscr{Z}$ it follows that there exists an index t such that $A_t = A_{t+1} = \ldots$. Denote by r the smallest such index. We prove that r does not exceed k. For this purpose let us consider the group $G(p)^k$ made up of the k-th powers of elements of $G(p)$. The factor group $G(p)/G(p)^k$ is also cyclic as it is an image of the cyclic group $G(p)$, and since each of its elements satisfies $X^k = 1$, it has at most k elements. Note that every one of the sets A_i is a union of some number of cosets of $G(p) \,\mathrm{mod}\, G(p)^k$. Actually, if $x(\mathrm{mod}\, p)$ lies in A_i and $y(\mathrm{mod}\, p)$ lies in the same coset $\mathrm{mod}\, G(p)^k$ as $x(\mathrm{mod}\, p)$, then we have on the one hand $x \equiv x_1^k + \ldots + x_i^k(\mathrm{mod}\, p)$, and on the other hand, $y \equiv xz^k \,(\mathrm{mod}\, p)$, hence

$$y \equiv (x_1 z)^k + \ldots + (x_i z)^k \,(\mathrm{mod}\, p) \ .$$

From this it follows that r cannot exceed the number of cosets, that is $r \leqslant k$.

It remains to show that A_r contains all the elements $x(\mathrm{mod}\, p)$, $x \not\equiv 0(\mathrm{mod}\, p)$, but this is very easy. In fact, suppose $a(\mathrm{mod}\, p) \neq 0$ does not line in A_r. Then in view of

$$a(\mathrm{mod}\, p) = 1^k\,(\mathrm{mod}\, p) + \ldots + 1^k\,(\mathrm{mod}\, p) \qquad (a \text{ times})$$

we have $a(\mathrm{mod}\, p) \in A_a$, so that $a > r$, but $A_r = A_{r+1} = \ldots = A_a \ni a(\mathrm{mod}\, p)$, a contradiction. $\quad \blacksquare$

In the case $a = 0$, the congruence (1.24) always has a solution, viz. $x_1 = \ldots = x_s = 0$. If we wish to have a solution in which, for at least one x_i, we have $x_i \not\equiv 0(\mathrm{mod}\, p)$, then it suffices to take $s \geqslant 1 + k$ due to the following reason: We have $p - 1 \equiv x_1^k + \ldots + x_s^k(\mathrm{mod}\, p)$. Here not all the x_i's are divisible by p, hence $0 \equiv 1^k + x_1^k + \ldots + x_s^k(\mathrm{mod}\, p)$. Thus we obtain

Corollary. *If $s = 1 + k$, then for every a the congruence (1.24) has a solution in which not all the x_i's are divisible by p.* $\quad \blacksquare$

We shall now be concerned with a congruence somewhat more general than (1.24).

Theorem 1.35. *If*

$$s \geqslant \begin{cases} 4k & for & 2 \mid k \quad, \\ 2k & for & 2 \nmid k \quad, \end{cases}$$

then for any integral a *and any* p^m $(p \in \mathscr{P})$ *the congruence*

$$(1.25) \qquad x_1^k + \ldots + x_s^k \equiv a \,(\mathrm{mod}\, p^m)$$

has a solution in which not all the x_i*'s are divisible by* p.

Proof. Let us begin with a lemma which allows us to reduce the problem to the case in which m is specially chosen.

Lemma 1.6. *We denote by* t *the number* $\alpha_p(k) + 1$ *in the case* $p \neq 2$ *and* $\alpha_p(k) + 2$ *in the case* $p = 2$. *If* $p \nmid a$ *and the congruence* $X^k \equiv a(\mathrm{mod}\, p^t)$ *has a solution, then for all* r *the congruence* $X^k \equiv a(\mathrm{mod}\, p^r)$ *has a solution, too.*

Proof. Obviously we may suppose that $r \geqslant t$. Denote by $H(2^r)$ the subgroup of $G(2^r)$ consisting of those elements $x(\mathrm{mod}\, 2^r)$ for which $x \equiv 1(\mathrm{mod}\, 4)$. This group is cyclic generated by $5(\mathrm{mod}\, 2^r)$. Indeed, we have seen in the proof of Theorem 1.19 that the order of the element $5(\mathrm{mod}\, 2^r)$ in $G(2^r)$ is equal to 2^{r-2}, so that this element generates a group of 2^{r-2} elements, and $|H(2^r)| = 2^{r-2}$, and moreover $H(2^r)$ contains all powers of $5(\mathrm{mod}\, 2^r)$. Let $g(\mathrm{mod}\, p^r)$ denote a generator of the group $G(p^r)$ for odd p and a generator of the group $H(2^r)$ for $p = 2$. It is clear that $g(\mathrm{mod}\, p^k)$ generates the group $G(p^k)$ respectively $H(2^k)$ for $k = 1, 2, \ldots$. Now note that if $p \nmid k$, then the assertion of the lemma follows immediately from Theorem 1.23. Therefore we can suppose further that $p \mid k$. In particular, in the case $p = 2$, k is even, so that from the assumptions of the lemma it follows that $a(\mathrm{mod}\, 2)$ lies in $H(2^t)$, since a, being a square in $G(4)$, satisfies $a \equiv 1(\mathrm{mod}\, 4)$.

From the assumption it follows that $a \equiv g^{kb}(\mathrm{mod}\, p^t)$ for some integer b, since g is a generator of $G(p^t)$ respectively of $H(2^t)$. Hence, if we determine the integer c by the condition $a \equiv g^c(\mathrm{mod}\, p^r)$, then we shall also have $a \equiv g^c(\mathrm{mod}\, p^t)$, so that

$$c \equiv kb(\mathrm{mod}\, \varphi(p')) \qquad \mathrm{or} \qquad c \equiv kb(\mathrm{mod}\, p^{r-1}(p-1))$$

In particular,

$$(1.26) \qquad c \equiv kb(\mathrm{mod}\, (p-1)) \quad,$$

and since $p^{t-1} \mid k$, we have $p^{t-1} \mid c$. Hence we may write $k = p^{t-1}k_1$ $(p \nmid k_1)$ and $c = p^{t-1}c_1$.

The lemma will be established if we can show that the congruence

$$(1.27) \qquad \cdot \qquad c \equiv Xk(\mathrm{mod}\, p^{r-1}(p-1))$$

has a solution. Indeed, we then get

$$a \equiv g^c \equiv g^{Xk} \equiv (g^X)^k \pmod{p^r} \ .$$

Dividing both sides and the modulus of the congruence (1.27) by p^{t-1}, we obtain

$$c_1 \equiv Xk_1 \pmod{p^{r-t}(p-1)} \ .$$

By virtue of Theorem 1.21 its solution amounts to that of two separate congruences:

$$c_1 \equiv X_1 k_1 \pmod{p^{r-t}}$$

and

$$c_1 \equiv X_2 k_1 \pmod{(p-1)} \ .$$

The first of them is solvable by $p \nmid k_1$, and so is the second by virtue of (1.26), since it suffices to take $X_2 = bp^{t-1}$. ∎

Corollary. *If the congruence*

(1.28) $$X_1^k + \ldots + X_s^k \equiv a \pmod{p^t}$$

has a solution in which not all the X_i's are divisible by p (where t is defined as in the lemma), then for each m the congruence (1.25) *has a solution in which not all the x_i's are divisible by p.*

Proof. Without loss of generality we may suppose that X_1 is not divisible by p. Thus the congruence

$$X^k \equiv a - X_2^k - \ldots - X_s^k \pmod{p^t}$$

has a solution not divisible by p, viz. $X = X_1$. By the lemma, the congruence

$$X^k \equiv a - X_2^k - \ldots - X_s^k \pmod{p^m}$$

has a solution, where $p \nmid X$. ∎

To prove the theorem it is now necessary to show the solvability of (1.25) for $m = t$, where at least one of the numbers x_i must be indivisible by p.

First consider the case $p = 2$. If k is odd, then $t = 2$ and $2^t = 4$, and the integers $0 = 0^k$, $1 = 1^k$, $3 \equiv (-1)^k \pmod 4$ are k-th powers (mod 4) and already for $s = 2 \leqslant 2k$ the congruence (1.28) has a solution, wherein not all the x_i's are even. On the contrary, if k is even, then in the representation of the integers $1 \leqslant a \leqslant 2^t$ in the form

$$a \equiv 1^k + \ldots + 1^k \pmod{2^t}$$

we have at most $a \leqslant 2^t = 2^{\alpha_p(k)+2} \leqslant 4k$ terms.

The case where p is odd is somewhat more complicated.

For a given $a \in [1, p^t - 1]$ we denote by $N(a)$ the least s for which the congruence (1.28) has a solution. Note that $N(a)$ depends solely upon the coset of $G(p^r) \bmod G(p^r)^k$. Indeed, it suffices to repeat the argument made in the special case $r = 1$ in the proof of Theorem 1.34. From this it follows that if l is the number of distinct possible values of $N(a)$, then l does not exceed k. Let $N_1 < N_2 < \ldots < N_l$ be those values and let a_i be the least number for which $N(a_i) = N_i$. Clearly, $N_1 = 1$, $a_1 = 1$. We shall prove by induction the inequality $N_i = N(a_i) \leqslant 2i - 1$, $i = 1, 2, \ldots, l$. For $i = 1$ this is clear, and if we assume its validity for $i = 1, 2, \ldots, j$, then in view of the fact that at least one of the integers $a_{j+1} - 1$, $a_{j+1} - 2$ is not divisible by p, we obtain

$$N(a_{j+1} - 1) \leqslant 2j - 1 \qquad \text{or} \qquad N(a_{j+1} - 2) \leqslant 2j - 1 \quad .$$

In the first case a_{j+1} will be a sum of $2j$ k-th powers $(\bmod p^t)$, and in the second, of $2j + 1$ k-th powers $(\bmod p^t)$, and this gives $N(a_{j+1}) \leqslant 2(j + 1) - 1$.

Applying the inequality obtained for $i = l$ we see that

$$\max N_i = N_l \leqslant 2(l + 1) - 1 \leqslant 2k - 1 \quad .$$

So far we have considered only those integers a indivisible by p. If a is divisible by p, then $p \nmid (a - 1)$ and we find that a is congruent to a sum of at most $2k$ k-th powers $(\bmod p^t)$. ∎

The least integer s such that for any $a_1, \ldots, a_s$, $N = p^n$, and $a = 0$ the congruence (1.18) has a solution for $F_i(x) = a_i x^k$ where at least one of the x_i's is not divisible by p is denoted by $\Gamma^*(k)$. The problem of its estimation was first considered by H. Davenport and D. J. Lewis [1], who showed that

$$\Gamma^*(k) \leqslant 1 + k^2 \quad ,$$

where in the case when $k + 1$ is a prime, the equality holds. When k is odd, A. Tietäväinen [1] proved that

$$\limsup \Gamma^*(k)/k \log k = 1/\log 2 \quad ,$$

and S. Chowla and G. Shimura [1] showed that for infinitely many odd k the following holds:

$$\Gamma^*(k) \geqslant 1 + k(\log[2k + 1]/\log 2) \quad .$$

K. K. Norton set forth a conjecture that for all odd k the reverse inequality holds. This conjecture has not been settled yet.

Recently, M. Dodson [1] proved that if $k + 1$ is a composite number other than 8 and 32, then

$$\Gamma^*(k) \leqslant \frac{k^2}{2} \left(1 + \frac{2}{1 + \sqrt{4k+1}}\right) + 1 \quad,$$

and, if furthermore $k = p(p-1)$, where p is a prime, then the equality holds here.

Regarding the estimation of $\Gamma^*(k)$ from below we only know (M. Dodson, loc. cit.) that $\Gamma^*(k) \geqslant 1 + k$ for all k. Dodson set forth a conjecture that $\lim \Gamma^*(k)/k = \infty$.

We list below the exact values of $\Gamma^*(k)$ which are not special cases of the theorems cited above:

$$\Gamma^*(3) = 7, \qquad \Gamma^*(5) = 16, \qquad \Gamma^*(7) = 22,$$

$$\Gamma^*(8) = 39, \qquad \Gamma^*(9) = 37, \qquad \Gamma^*(11) = 45.$$

(J. Bovey [1], H. Davenport, D. J. Lewis [1], M. Dodson [1]).

Exercises

1.   Let $F(x_1, \ldots, x_n)$ be a polynomial in n variables with integral coefficients. Moreover suppose that the degree of this polynomial is less than n. Prove that if the congruence

$$F(x_1, \ldots, x_n) \equiv 0 \,(\mathrm{mod}\, p)$$

has at least one solution, then it also has another solution.

2.   Prove Chevalley's theorem: If $F(x_1, \ldots, x_n)$ is a form (i.e. a homogeneous polynomial) of degree $d < n$, then the congruence

$$F(x_1, \ldots, x_n) \equiv 0 \,(\mathrm{mod}\, p)$$

always has a solution in which at least one of x_i's is not divisible by p.

3.   Show that in Theorem 1.34 the condition $s \geqslant k$ cannot be replaced by $s \geqslant k - 1$ for any odd prime p, that is to say, for every such p there exist a number k and a natural number n which is not congruent to any sum of $\leqslant k - 1$ k-th powers.

4.   Denote by $\gamma^*(k, p)$ the smallest number s with the property that if $a_1, \ldots, a_g$ are not divisible by p, then the congruence

$$a_1 x_1^k + \ldots + a_s x_s^k \equiv 0 \,(\mathrm{mod}\, p)$$

has a non-trivial solution (that is, a solution in which not all x_i's are divisible by p). Show that for $k = \frac{1}{2}(p-1)$ we have

$$\gamma^*(k, p) = 1 + \left[\frac{\log p}{\log 2}\right].$$

CHAPTER II

ARITHMETICAL FUNCTIONS

§1. Fundamental properties

1. We shall call a complex-valued function defined on the set $\mathcal{N}$ of natural numbers or on the set $\mathcal{N}_0$, an *arithmetical function*. In number theory there appear many functions relating to such problems as the divisibility of integers, decomposition into factors, etc. and we shall mainly be concerned with such functions. In the preceding chapter we have already encountered Euler's function $\phi(n)$ giving the number of elements in the group $G(n)$. Other such functions are: the function $d(n)$, the number of natural divisors of a natural number n, the function $\sigma(n)$, the sum of these divisors, the function $\sigma_j(n)$, the sum of j-th powers of these divisors, the function $\omega(n)$, the number of distinct prime divisors of n, the function $\Omega(n)$, the number of factors in the decomposition of n into prime factors, and lastly, the function $\mu(n)$ (the so-called *Möbius function*) defined by

$$\mu(n) = \begin{cases} 0, & \text{if } n \text{ is divisible by a square of a prime,} \\ (-1)^k, & \text{if } n = p_1 \ldots p_k, \quad p_i \text{ being distinct primes} \\ 1, & \text{if } n = 1. \end{cases}$$

In this chapter we shall study the properties of some arithmetical functions.

We shall begin by defining some operations in the set of all arithmetical functions.

It is quite evident that the set of all arithmetical functions defined on $\mathcal{N}$ forms a ring with the ordinary addition and multiplication of functions as its operations. It is also possible, while preserving the ordinary addition, to introduce in this set another multiplication and again obtain a ring. An example of such an operation is the *Dirichlet convolution* of arithmetical functions defined by the formula

$$f*g(n) = \sum_{d|n} f(d)g\left(\frac{n}{d}\right) = \sum_{d_1 d_2 = n} f(d_1)g(d_2).$$

It follows at once from the definition of the Dirichlet convolution that if $I(n)$ is the constant function with the value 1 and $N(n) = n$, then

$$I*I = d, \quad I*N = \sigma, \quad I*N^k = \sigma_k.$$

Algebraic properties of the convolution are described by the following theorem:

Theorem 2.1. *The set of arithmetical functions defined on $\mathcal{N}$ with the ordinary addition and the Dirichlet convolution as multiplication forms a commutative ring with the unity*

$$e(n) = \begin{cases} 1, & \text{if} \quad n = 1, \\ 0, & \text{if} \quad n \neq 1. \end{cases}$$

This ring does not have any divisors of zero and its element f has an inverse element iff $f(1) \neq 0$.

Proof. Commutativity and distributivity follow at once from the definition and associativity of multiplication is a consequence of the identity

$$(f_1 * f_2) * f_3(n) = \sum_{d_1 d_2 d_3 = n} f_1(d_1) f_2(d_2) f_3(d_3)$$
$$= f_1 * (f_2 * f_3)(n),$$

and the function $e(n)$ is the unity in view of

$$e * f(n) = \sum_{d_1 d_2 = n} e(d_1) f(d_2) = f(n).$$

We shall now show that from $f*g = 0$ it follows that one of the functions f, g equals zero. Otherwise, if k is the least natural number satisfying $f(k) \neq 0$ and similarly l is the least natural number satisfying $g(l) \neq 0$, then

$$0 = f*g(kl) = \sum_{d_1 d_2 = kl} f(d_1)g(d_2)$$

$$= \sum_{\substack{d_1 d_2 = kl \\ d_1 < k}} f(d_1)g(d_2) + f(k)g(l) + \sum_{\substack{d_1 d_2 = kl \\ d_2 < l}} f(d_1)g(d_2) = f(k)g(l),$$

a contradiction. To prove the last part of the assertion we observe first that if $f*g = e$, then $f(1)g(1) = e(1) = 1$, and hence $f(1) \neq 0$. Conversely, if $f(1) \neq 0$, then defining $g(n)$ inductively by

$$g(1) = f(1)^{-1},$$

$$g(n) = -f(1)^{-1} \sum_{\substack{d_1 d_2 = n \\ d_1 \neq 1}} f(d_1) g(d_2),$$

we obtain a function g satisfying $f*g = e$. ∎

Another operation with similar properties is the so-called *Abel convolution* defined in the set of arithmetical functions defined on the set $\mathcal{N}_0$. This convolution is defined by the formula

$$f \times g(n) = \sum_{k=0}^{n} f(k)\, g(n-k).$$

and its properties are contained in the theorem whose proof is entirely analogous to that of Theorem 2.1 and we leave it to the reader.

Theorem 2.2. *The set of functions defined on $\mathcal{N}_0$ with the ordinary addition and Abel convolution as multiplication forms a commutative ring with unity*

$$e'(n) = \begin{cases} 1, & \text{if} \quad n = 0, \\ 0, & \text{if} \quad n \neq 0. \end{cases}$$

This ring does not have any zero-divisors and its element f has an inverse element iff $f(0) \neq 0$. ∎

Other convolutions with similar properties have also been studied. The reader will encounter the unitary convolution introduced by E. Cohen [2] at the end of this section. The reader may find other kinds of convolutions in the papers of T. M. K. Davison [1], L. M. Fredman [1], W. Narkiewicz [1] and M. V. Subbarao [2]. The reader who is interested in algebraic properties of the Dirichlet convolution may find many interesting results on this subject in the work of H. N. Shapiro [3].

2.   In number theory there often arises the necessity of estimating the sum of values of arithmetical functions in some set, mostly an interval. We shall give, in this subsection, two elementary methods which enable us to solve this problem in many cases. The first of them is the method of *partial summation*, also called *Abel's (summation) theorem*:

Theorem 2.3. *Let $a_1, \ldots, a_n$ and $b_1, \ldots, b_n$ be complex numbers. Then, with the notation*

$$A(m) = a_1 + \ldots + a_m$$

and

$$c_m = b_{m+1} - b_m,$$

the identity

$$\sum_{i=1}^{n} a_i b_i = -\sum_{m=1}^{n-1} A(m) c_m + A(n) b_n$$

holds.

Proof. We have

$$\sum_{i=1}^{n} a_i b_i = \sum_{i=2}^{n} \left(A(i) - A(i-1) \right) b_i + a_1 b_1$$

$$= A(n)b_n + \sum_{i=2}^{n-1} A(i)b_i - \sum_{i=2}^{n} A(i-1)b_i + a_1 b_1$$

$$= A(n)b_n + \sum_{i=2}^{n-1} A(i)b_i - \sum_{i=1}^{n-1} A(i)b_{i+1} + a_1 b_1$$

$$= A(n)b_n + \sum_{i=2}^{n-1} A(i)(b_i - b_{i+1}) + A(1)b_1 - A(1)b_2$$

$$= A(n)b_n - \sum_{i=1}^{n-1} A(i)c_i. \quad \blacksquare$$

Corollary 1. *If* $a_1, \ldots, a_n$ *are complex numbers and* $b_1 \geq b_2 \geq \ldots \geq b_n \geq 0$, *then with the notation*

$$M = \max_{1 \leq m \leq n} |a_1 + \ldots + a_m|,$$

the inequality

$$\left| \sum_{i=1}^{n} a_i b_i \right| \leq M b_1$$

holds.

Proof. If $c_m = b_{m+1} - b_m$, then

$$\left| \sum_{i=1}^{n} a_i b_i \right| \leq M \left(b_n - \sum_{m=1}^{n-1} c_m \right) = M b_1. \quad \blacksquare$$

Corollary 2 (Dirichlet's test). *If* $z_1, z_2, \ldots$ *is a sequence of complex numbers and* $b_1 \geq b_2 \geq \ldots$, $\lim_{n \to \infty} b_n = 0$, *and furthermore for every* n, $\left| \sum_{j=1}^{n} z_j \right| < c$, *where* c *is a constant, then the series* $\sum_{n=1}^{\infty} z_n b_n$ *is convergent.*

Proof. By the preceding corollary we have

$$\lim_{\substack{m \to \infty \\ N \to \infty}} \left| \sum_{n=m}^{N} z_n b_n \right| = 0. \quad \blacksquare$$

Quite similarly, we obtain

Corollary 3. *If* $\{f_n(z)\}$ *is a sequence of complex-valued functions defined in a set* X *and there exists a constant* C *such that for* $z \in X$ *and* $N = 1, 2, \ldots$ *we have*

$$\left| \sum_{n=1}^{N} f_n(z) \right| \leqslant C,$$

and moreover $b_1 \geqslant b_2 \geqslant \ldots$, $\lim_{n \to \infty} b_n = 0$, *then the series* $\sum_{n=1}^{\infty} b_n f_n(z)$ *is uniformly convergent on* X. $\blacksquare$

The second method relies on replacing a sum by the corresponding integral.

Theorem 2.4.

(i)   *If* A *is a natural number and a function* $f(t) > 0$ *is monotone in the closed interval* $[A, x]$, *then*

$$\sum_{A \leqslant n \leqslant x} f(n) = \int_A^x f(t) \, dt + O(f(A)) + O(f(x)).$$

(ii)   *If* A *is a natural number and a function* $f(x)$ *has a continuous derivative in the interval* $[A, x]$, *then*

$$\sum_{A \leqslant n \leqslant x} f(n) = \int_A^x f(t) \, dt + \int_A^x \{t\} f'(t) \, dt + f(A) - \{x\} f(x).$$

(iii)   *If* A *is a natural number and a function* $f(t)$ *has a continuous second derivative in the interval* $[A, x]$, *then*

$$\sum_{A \leqslant n \leqslant x} f(n) = \int_A^x f(t) \, dt + (\tfrac{1}{2} - \{x\}) f(x) + \tfrac{1}{2} f(A)$$

$$- f'(x) \int_0^x (\tfrac{1}{2} - \{t\}) \, dt + \int_A^x f''(t) \int_0^t (\tfrac{1}{2} - \{u\}) \, du \, dt.$$

Proof.

(i)   It suffices to consider the case of a non-decreasing function $f(t)$. If $N = [x]$, then

$$\int_A^x f(t) \, dt = \sum_{k=A}^{N-1} \int_k^{k+1} f(t) \, dt + \int_N^x f(t) \, dt.$$

In view of

$$f(k) \leqslant \int_k^{k+1} f(t) \, dt \leqslant f(k+1)$$

we obtain

$$\sum_{k=A}^{N-1} f(k) \leqslant \sum_{k=A}^{N-1} \int_k^{k+1} f(t)\,dt \leqslant \sum_{k=1+A}^{N} f(k),$$

and so

$$\sum_{k=A}^{N-1} \int_k^{k+1} f(t)\,dt = \sum_{k=A}^{N} f(k) + O(f(A)) + O(f(N))$$

and

$$\int_A^x f(t)\,dt = \sum_{k=A}^{N} f(k) + O\left(f(A)\right) + O\left(f(N)\right) + O\left(f(x)\right),$$

and this gives the assertion, since $0 < f(N) \leqslant f(x)$.

(ii) We have

$$([x] - A + 1)f(x) - \sum_{A \leqslant k \leqslant x} f(k) = \sum_{A \leqslant k \leqslant x} (f(x) - f(k)) = \sum_{A \leqslant k \leqslant x} \int_k^x f'(t)\,dt$$

$$= \int_A^x \left(\sum_{A \leqslant k \leqslant t} 1\right) f'(t)\,dt = \int_A^x ([t] - A + 1)f'(t)\,dt,$$

hence

$$\sum_{A \leqslant k \leqslant x} f(k) = ([x] - A + 1)f(x) - \int_A^x ([t] - A + 1)f'(t)\,dt$$

$$= [x]f(x) - (A - 1)f(A) - \int_A^x [t]f'(t)\,dt,$$

but

$$\int_A^x f(t)\,dt = xf(x) - Af(A) - \int_A^x tf'(t)\,dt,$$

therefore

$$\int_A^x [t]f'(t)\,dt = xf(x) - Af(A) + \int_A^x ([t] - t)f'(t)\,dt - \int_A^x f(t)\,dt$$

and in the end

$$\sum_{A \leqslant k \leqslant x} f(k) = \int_A^x f(t)\,dt - \int_A^x ([t] - t)f'(t)\,dt + f(A) - (x - [x])f(x).$$

(iii) Let

$$a(t) = \tfrac{1}{2} - \{t\}$$

and

$$\beta(t) = \int_0^t a(u)\,du\,.$$

Then the function $\alpha(t)$ has its derivative equal to -1 at every point $t \notin \mathscr{X}$, so that if $n \in \mathscr{X}$ and $n < x_1 < x_2 < n+1$, then

$$(2.1) \qquad \int_{x_1}^{x_2} f(t)\,dt = -\int_{x_1}^{x_2} a'(t) f(t)\,dt$$

$$= -\int_{x_1}^{x_2} \beta(t) f''(t)\,dt + a(x_1) f(x_1) - a(x_2) f(x_2)$$

$$+ \beta(x_2) f'(x_2) - \beta(x_1) f'(x_1),$$

so that as $x_1 \to n$ and $x_2 \to n+1$ we obtain

$$\int_n^{n+1} f(t)\,dt = -\int_n^{n+1} \beta(t) f''(t)\,dt + \frac{f(n) + f(n+1)}{2},$$

because $\lim_{x \to n} \beta(x) = 0$. Hence, with $N = [x]$,

$$\sum_{A \leqslant n \leqslant x} f(n) = \sum_{n=A}^{N} f(n) = f(A) + f(N) + \sum_{n=A+1}^{N-1} f(n)$$

$$= \tfrac{1}{2} \sum_{n=A}^{N-1} (f(n) + f(n+1)) + \frac{f(A) + f(N)}{2}$$

$$= \sum_{n=A}^{N-1} \int_n^{n+1} f(t)\,dt + \sum_{n=A}^{N-1} \int_n^{n+1} \beta(t) f''(t)\,dt + \frac{f(A) + f(N)}{2}$$

$$= \int_A^N f(t)\,dt + \int_A^N \beta(t) f''(t)\,dt + \frac{f(A) + f(N)}{2}.$$

The right-hand side of this formula differs from that which we wish to have by the term

$$\int_N^x f(t)\,dt + \int_N^x f''(t) \beta(t)\,dt - f'(x)\beta(x) + f(x)a(x) - \tfrac{1}{2} f(N),$$

which, by (2.1) (with $x_2 = x$ and $x_1 \to N$), vanishes. ∎

Corollary 1. *If a function $f(x)$ decreases to zero and has the continuous derivative in the interval $[A, \infty)$ $(A \in \mathcal{N})$, then for every $x \geqslant A$ we have*

$$\sum_{A \leqslant k \leqslant x} f(k) = \int_A^x f(t)\,dt + C + O(f(x)),$$

where C does not depend on x.

Proof. Making use of part (ii) of the theorem and putting

$$C = \int_A^\infty ([t] - t)f'(t)\,dt$$

(where the integral is convergent, being majorated by $\int_A^\infty -f'(t)\,dt = f(A)$), we see that

$$\sum_{A \leqslant k \leqslant x} f(k) = \int_A^x f(t)\,dt + C + O\left(\int_x^\infty (-f'(t))\,dt \right) + O(f(x))$$

$$= \int_A^x f(t)\,dt + C + O(f(x)). \quad \blacksquare$$

Corollary 2. *If a function $f(t)$ has the continuous second derivative for $t \geqslant A$, where $A \in \mathcal{N}$, and $\int_A^\infty |f''(t)|\,dt < \infty$, then for $x \geqslant A$ we have*

$$\sum_{A \leqslant n \leqslant x} f(n) = \int_A^x f(t)\,dt + (\tfrac{1}{2} - \{x\})f(x) - f'(x) \int_0^x (\tfrac{1}{2} - \{t\})\,dt$$

$$- \int_A^\infty f''(t) \int_0^t (\tfrac{1}{2} - \{u\})\,du\,dt + C,$$

where C is a constant independent of x.

Proof. Follows immediately from part (iii) of the theorem. $\blacksquare$

Corollary 3. *For $x \geqslant 2$ the formulas*

(a)
$$\sum_{n \leqslant x} \frac{1}{n} = \log x + C + O\left(\frac{1}{x}\right),$$

(b)
$$\sum_{n \leqslant x} \frac{1}{n \log n} = \log\log x + C_1 + O\left(\frac{1}{x \log x}\right),$$

(c)
$$\sum_{n \leqslant x} \frac{\log n}{n} = \tfrac{1}{2}\log^2 x + C_2 + O\left(\frac{\log x}{x}\right),$$

hold where C, C_1, C_2 are constants. $\blacksquare$

Remark. The constant C appearing here is called *Euler's constant*. It is not known at present whether $C = 0.57721\ldots$ is a rational number or not.

Exercises

1.   a) Check whether Theorem 2.1 remains valid for the unitary convolution defined by

$$f \circ g(n) = \sum_{\substack{d|n \\ (d, n/d) = 1}} f(d) g(n/d).$$

b) Find the inverse function of $I(n)$ in the ring of functions with unitary convolution.

2.   Let $K(m, n)$ be a function in two natural variables assuming values $0, 1$. Define the convolution $f \times g$ by

$$(f \times g)(n) = \sum_{d_1 d_2 = n} f(d_1) g(d_2) \cdot K(d_1, d_2).$$

Characterize the functions K for which this convolution is associative and commutative.

3 (Césaro).   Prove that if $g = 1 * f$ and both the series

$$\sum_{n=1}^{\infty} g(n) x^n \qquad \text{and} \qquad \sum_{n=1}^{\infty} \frac{f(n) x^n}{1 - x^n}$$

are absolutely convergent at a point x, then their sums are equal.

4.   Give an example of an ideal in the ring of functions with Dirichlet convolution that does not have a finite set of generators.

5.   Prove that if

$$\sum_{n=1}^{\infty} |f(n)| < \infty \qquad \text{and} \qquad \sum_{n=1}^{\infty} |g(n)| < \infty$$

and $h = f * g$, then

$$\sum_{n=1}^{\infty} |h(n)| < \infty.$$

6.   Prove that the ring of arithmetical functions with Dirichlet convolution with values in a field K is isomorphic with the ring of formal power series in countably many variables over K.

7 (v. Sterneck's theorem).   Let

$$h(n) = \sum_{[k, l] = n} f(k) g(l)$$

and

$$F = 1 * f, \qquad G = 1 * g, \qquad H = 1 * h.$$

Show that $H(n) = F(n) G(n)$.

§ 2. Additive and multiplicative functions

1.   If a function $f(n)$ defined on the set of natural numbers satisfies for all integers m, n relatively prime to each other, the condition

(2.2) $f(mn) = f(m) + f(n)$,

then we say that f is an *additive function*. If this condition is satisfied for all integers m, n without exceptions, then f is called a *totally additive or completely additive function*. Similarly, if a function $f(n)$, not identically zero, satisfies for all integers m, n relatively prime to each other, the condition

$$(2.3) \qquad f(mn) = f(m)f(n) ,$$

then we say that f is a *multiplicative function*. If the condition (2.3) is satisfied for all m, n, then f is called a *totally* (or *completely*) *multiplicative function*.

These conceptions are closely connected with each other. Indeed, if f is an additive function and c is any positive number, then the function $g(n) = c^{f(n)}$ is multiplicative, and if $f(n)$ is multiplicative and positive, then the function $g(n) = \log f(n)$ is an additive function.

Additive and multiplicative functions are determined by their values at the points $n = p^k$ (where p is a prime number, $k = 1, 2, \ldots$). Indeed, the following theorem holds:

Theorem 2.5. *If $f(n)$ is a multiplicative function, $g(n)$ is an additive function*

and $n = \prod p^{\alpha_p(n)}$ is the canonical decomposition of n, then

$$f(n) = \prod_p f(p^{\alpha_p(n)}) ,$$

$$g(n) = \sum_p g(p^{\alpha_p(n)}) ,$$

and $f(1) = 1$, $g(1) = 0$.

Proof. From the definition it follows that if $n_1, \ldots, n_k$ are pairwise relatively prime integers, then $f(n_1 \ldots n_k) = f(n_1) \ldots f(n_k)$. Applying this remark to the set of numbers $\{p^{\alpha_p(n)}\}$, we obtain the first assertion. If $f(n_0) \neq 0$, then $f(n_0) = f(1)f(n_0)$, so that $f(1) = 1$.

From the additivity of the function g follows the multiplicativity of the function $f(n) = 2^{g(n)}$ and applying to it that part of the theorem already proven, we immediately obtain the validity of the assertion concerning $g(n)$. ∎

From this theorem it follows that if an additive function f takes the value zero at all points p^k (where p is a prime, $k = 1, 2, \ldots$), then $f \equiv 0$, and if f is totally additive, then $f(p^k) = kf(p)$, and hence it suffices to assume the vanishing of $f(p)$ for every prime p. It is worth noting here the following result due to P. D. T. A. Elliott [1]:

If f is an additive function and for every prime p we have $f(1+p) = 0$, then $f \equiv 0$.

This result answers a question raised by I. Kátai [4].

Among the arithmetical functions that we met already, the functions $\omega(n)$ and $\Omega(n)$ are additive. This follows from the unique factorization theorem. The functions

$I(n)$, $N(n)$ and the Möbius function $\mu(n)$ are clearly multiplicative. It happens that Euler's function $\varphi(n)$ is also multiplicative.

Theorem 2.6. *The function $\varphi(n)$ is a multiplicative function.*

Proof. If $(m, n) = 1$, then by Corollary 2 to Theorem 1.18 we have $G(mn) = G(m) \oplus G(n)$, and since $\varphi(N) = |G(N)|$, we have $\varphi(mn) = \varphi(m)\varphi(n)$. ∎

2.   Multiplicativity of many arithmetical functions can be proved on the basis of the following theorem:

Theorem 2.7. *Multiplicative functions form a group under Dirichlet convolution. This means that if f, g are multiplicative, then so is $f*g$, and that if f is multiplicative, then there exists a multiplicative function g such that $f*g = e$.*

Proof. We shall need a simple lemma.

Lemma 2.1. *If m, n are relatively prime integers, then every divisor d of mn may be expressed uniquely in the form $d = d_1 d_2$, where $d_1 | m$, $d_2 | n$, $(d_1, d_2) = 1$.*

Proof of the lemma. Since d can be divisible only by those primes that divide m or n, we may write

$$d = \prod_{p|m} p^{a_p} \cdot \prod_{p|n} p^{b_p}.$$

Since we have the inequalities $a_p \leqslant \alpha_p(m)$, $b_p \leqslant \alpha_p(n)$ in view of $(m, n) = 1$, it follows that the first factor d_1 of this product divides m, and the second factor d_2 divides n. The uniqueness of the representation follows from the observation that d_1 respectively d_2 must be the largest divisor of m respectively, of n dividing d. ∎

Now let f, g be multiplicative and let $(m, n) = 1$. Then we have

$$(f_*g)(mn) = \sum_{dd' = mn} f(d)g(d')$$

$$= \sum_{\substack{d_1|m, d_2|n \\ d_1 d_2 = d \\ d_1'|m, d_2'|n \\ d_1' d_2' = d'}} f(d)g(d') = \sum_{\substack{d_1|m, d_2|n \\ d_1'|m, d_2'|n}} f(d_1)g(d_1')f(d_2)g(d_2')$$

$$= (f*g)(m)(f*g)(n),$$

because $d_1 d_1' = m$, $d_2 d_2' = n$.

This proves the first part of the theorem. Now let f be a multiplicative function. Then $f(1) = 1 \neq 0$, so that by Theorem 2.1 there exists a function g satisfying $f*g = e$. Let us show that it must be a multiplicative function. Let $k = mn$, $(m, n) = 1$ and suppose that for $r < k$, from $r = ab$, $(a, b) = 1$ follows $g(r) = g(a)g(b)$. Then

$$0 = \sum_{d \mid mn} f(d) g(mn/d) = \sum_{\delta_1 \mid m} \sum_{\delta_2 \mid n} f(\delta_1) f(\delta_2) g(mn/\delta_1 \delta_2)$$

$$= \sum_{\substack{\delta_1 \mid m \ \delta_2 \mid n \\ \delta_1 \delta_2 \neq 1}} \{ f(\delta_1) f(\delta_2) g(m/\delta_1) g(n/\delta_2) \} + g(mn).$$

But

$$0 = \sum_{\delta_1 \mid m} f(\delta_1) g(m/\delta_1) \sum_{\delta_2 \mid n} f(\delta_2) g(n/\delta_2)$$

$$= \sum_{\substack{\delta_1 \mid m \ \delta_2 \mid n \\ \delta_1 \delta_2 \neq 1}} \{ f(\delta_1) f(\delta_2) g(m/\delta_1) g(n/\delta_2) \} + g(m) g(n).$$

Comparing these equalities, we obtain the assertion. ∎

Corollary 1. *If the function $h = f*g$ is multiplicative, then either both the functions f, g are multiplicative or none of them.*

Proof. If f is multiplicative, then denoting by f^{-1} the inverse function of f, we obtain $g = h*f^{-1}$, and so g must also be multiplicative. ∎

(This fact was first noted by E. T. Bell [1]).

Corollary 2. *The functions $d(n)$, $\sigma(n)$, $\sigma_j(n)$ are multiplicative.*

Proof. Indeed, we have $d = I*I$, $\sigma = I*N$, $\sigma_j = I*N^j$, and all the functions I, N, N^j are multiplicative. ∎

Corollary 3. *The Möbius function $\mu(n)$ is the inverse function of $I(n)$. In particular,*

$$\sum_{d \mid n} \mu(d) = \begin{cases} 1 & \text{for} \quad n = 1, \\ 0 & \text{for} \quad n > 1. \end{cases}$$

Proof. Since μ and I are multiplicative, $\mu*I$ is also multiplicative and it remains to check that for prime powers p^k the equality

$$\mu*I(p^k) = 0$$

holds. But

$$\mu*I(p^k) = \mu(1) + \mu(p) + \ldots + \mu(p^k) = 1 - 1 + 0 + \ldots + 0 = 0. ∎$$

Corollary 4 (Möbius inversion formula). *If f, g are arbitrary arithmetical functions, then the following three formulas are equivalent:*

(2.4) $$\sum_{d \mid n} f(d) = g(n), \qquad \text{for all } n \in \mathcal{N}$$

$$(2.5) \qquad f(n) = \sum_{d|n} \mu(d)g(n/d), \qquad \text{for all } n \in \mathcal{N}$$

$$(2.6) \qquad \sum_{n \leqslant N} g(n) = \sum_{d \leqslant N} f(d)\left[\frac{N}{d}\right], \qquad \text{for all real } N \geqslant 1.$$

Proof. Formula (2.4) gives $I*f = g$, and formula (2.5) means $f = \mu*g$, therefore their equivalence follows from Corollary 3. Moreover, if formula (2.4) holds, then we have

$$\sum_{n \leqslant N} g(n) = \sum_{n \leqslant N} \sum_{d|n} f(d) = \sum_{d \leqslant N} f(d) \sum_{\substack{n \leqslant N \\ d|n}} 1,$$

and the number of integers $n \leqslant N$ divisible by d is $[N/d]$, hence we obtain formula (2.6).

Lastly, if formula (2.6) holds, then tracing back the above equalities in the reverse direction, we obtain

$$\sum_{n \leqslant N} g(n) = \sum_{n \leqslant N} \sum_{d|n} f(d).$$

and so we have also

$$\sum_{n \leqslant N-1} g(n) = \sum_{n \leqslant N-1} \sum_{d|n} f(d).$$

Subtracting the last equality from the previous equality, we obtain formula (2.4). ∎

Corollary 5. *If a function $f(x)$ is defined for $x \geqslant 1$ and*

$$g(x) = \sum_{n \leqslant x} f\left(\frac{x}{n}\right)\log x,$$

then

$$f(x)\log x + \sum_{n \leqslant x} f\left(\frac{x}{n}\right) \varLambda(n) = \sum_{d \leqslant x} \mu(d)g\left(\frac{x}{d}\right),$$

where

$$\Lambda(n) = \begin{cases} \log p, & \text{if } n = p^k, \ p \text{ prime}, \\ 0 & \text{otherwise}. \end{cases}$$

Proof. We observe that $\sum_{d|n} \Lambda(d) = \log n$, for the left-hand side is equal to

$$\sum_{p^k|n} \log p = \sum_{p^r||n} r\log p = \sum_{p^r||n} \log p^r = \log \prod_{p^r||n} p^r = \log n.$$

Therefore we have from the preceding corollary

$$\Lambda(n) = \sum_{d|n} \mu(d) \log \frac{n}{d},$$

and so

$$f(x)\log x + \sum_{n \leq x} f\left(\frac{x}{n}\right) \Lambda(n) = \sum_{n \leq x} f\left(\frac{x}{n}\right) \log \frac{x}{n} \sum_{d|n} \mu(d) + \sum_{n \leq x} f\left(\frac{x}{n}\right) \sum_{d|n} \mu(d) \log \frac{n}{d}$$

$$= \sum_{n \leq x} f\left(\frac{x}{n}\right) \sum_{d|n} \mu(d) \log \frac{x}{d}$$

$$= \sum_{d \leq x} \mu(d) \log \frac{x}{d} \sum_{m \leq x/d} f\left(\frac{x}{dm}\right) = \sum_{d \leq x} \mu(d) g\left(\frac{x}{d}\right). \quad \blacksquare$$

Corollary 6. *If f is a multiplicative function, then*

$$\sum_{d|n} \mu(d)f(d) = \prod_{p|n} (1-f(p)) \quad and \quad \sum_{d|n} \mu^2(d)f(d) = \prod_{p|n} (1+f(p)).$$

Proof. In each of these formulas both sides are multiplicative, hence it suffices to consider the case $n = p^\alpha$, but then these formulas are valid, for both sides are equal to $1 \pm f(p)$. $\quad \blacksquare$

We may now give explicit formulas for the functions $\varphi(n)$, $\sigma_j(n)$ and $d(n)$, making use of their multiplicativity.

Theorem 2.8. *For natural n the formulas*

(i) $\quad \varphi(n) = n \prod_{p|n} (1-1/p) = \sum_{d|n} \mu(d) n d^{-1}.$

　　　that is, $\varphi = \mu * N$,

(ii) $\quad d(n) = \prod_{p} (1 + a_p(n)),$

(iii) $\quad \sigma_j(n) = \prod_{p|n} (1 + p^j + p^{2j} + \cdots + p^{\alpha_p(n)j}).$

hold.

Proof. In all three cases we have multiplicative functions on both sides, hence it suffices to check their validity in the case $n = p^k$, where p is a prime.

(i)　　Formula (1.7) gives

$$\varphi(p^k) = p^{k-1}(p-1) = p^k(1-1/p).$$

To prove the second equality concerning Euler's function it is sufficient to use Corollary 6 to the preceding theorem.

(ii) In this case it suffices to observe that $d(p^k) = 1+k$, the numbers $1, p, p^2, \ldots, p^k$ are the only divisors of p^k.

(iii) Here we make use of the preceding remark to get

$$\sigma_j(p^k) = 1 + p^j + \ldots + p^{kj} . \quad \blacksquare$$

Corollary. *For every n we have* $\sum_{d|n} \varphi(d) = n$, *i.e.* $\varphi*I = N$.

Proof. We have just seen that $\varphi = \mu*N$ and it suffices to use Corollary 4 to Theorem 2.7. $\blacksquare$

3. Let us now consider more closely the asymptotic behaviour of the functions $\varphi(n)$, $\sigma(n)$ and $d(n)$. We begin with Euler's function.

Theorem 2.9. *The formulas*

$$\limsup_{n \to \infty} \frac{\varphi(n)}{n} = 1, \qquad \liminf_{n \to \infty} \frac{\varphi(n)}{n} = 0$$

hold.

Proof. For any n we have the inequality $\varphi(n) \leqslant n-1$ and for primes p the equality holds here, which proves the first part of the assertion. To prove the second, we put $n_k = p_1 p_2 \ldots p_k$, where p_i denotes the i-th successive prime. By Theorem 2.8 (i) we have

$$\frac{\varphi(n_k)}{n_k} = \prod_{p \leqslant p_k} \left(1 - \frac{1}{p}\right),$$

and this tends to zero by the divergence of the series $\sum_p \frac{1}{p}$. $\blacksquare$

Remark. It can be proved (see, e.g. G. H. Hardy, E. M. Wright [1], Theorem 328) that

$$\liminf_{n \to \infty} \frac{\varphi(n) \log \log n}{n} = e^{-C},$$

where C is Euler's constant.

The following theorem gives information on the average value of Euler's function in the interval $[1, x]$.

Theorem 2.10. *For $x \geqslant 2$ we have*

$$\sum_{n \leqslant x} \varphi(n) = \frac{3}{\pi^2} x^2 + O(x \log x).$$

Proof. Making use of Theorem 2.8 (i), we obtain

$$\sum_{n \leqslant x} \varphi(n) = \sum_{n \leqslant x} \sum_{d \mid n} \mu(d) \frac{n}{d} = \sum_{d \leqslant x} \frac{\mu(d)}{d} \sum_{\substack{n \leqslant x \\ d \mid n}} n$$

$$= \sum_{d \leqslant x} \frac{\mu(d)}{d} \sum_{m \leqslant x/d} dm = \sum_{d \leqslant x} \mu(d) \sum_{m \leqslant x/d} m,$$

so that, using

$$\sum_{m \leqslant t} m = \tfrac{1}{2} t^2 + O(t),$$

we obtain

$$\sum_{n \leqslant x} \varphi(n) = \sum_{d \leqslant x} \mu(d) \left(\frac{1}{2} \cdot \frac{x^2}{d^2} + O\left(\frac{x}{d}\right) \right) = \frac{x^2}{2} \sum_{d \leqslant x} \frac{\mu(d)}{d^2} + O\left(x \sum_{d \leqslant x} \frac{1}{d}\right).$$

In view of

$$\sum_{d \leqslant x} \frac{\mu(d)}{d^2} = \sum_{d=1}^{\infty} \frac{\mu(d)}{d^2} + O\left(\sum_{d > x} \frac{1}{d^2}\right) = \sum_{d=1}^{\infty} \frac{\mu(d)}{d^2} + O\left(\frac{1}{x}\right)$$

and

$$\frac{\pi^2}{6} \sum_{d=1}^{\infty} \frac{\mu(d)}{d^2} = \sum_{n=1}^{\infty} \frac{1}{n^2} \sum_{d=1}^{\infty} \frac{\mu(d)}{d^2} = \sum_{d, n=1}^{\infty} \frac{\mu(d)}{(dn)^2} = \sum_{m=1}^{\infty} \frac{1}{m^2} \sum_{d \mid m} \mu(d) = 1$$

(which follows from Corollary 3 to Theorem 2.7) we obtain now, using Corollary 3 to Theorem 2.4,

$$\sum_{n \leqslant x} \varphi(n) = \frac{3}{\pi^2} x^2 + O(x) + O\left(x \sum_{d \leqslant x} \frac{1}{d}\right) = \frac{3}{\pi^2} x^2 + O(x \log x). \qquad \blacksquare$$

Using analytic methods one can improve the estimate for the error term obtained here. The best known result is due to A. I. Saltykov [1]:

$$\sum_{n \leqslant x} \varphi(n) = \frac{3}{\pi^2} x^2 + O\left(x (\log x)^{2/3} (\log\log x)^{1+\epsilon}\right).$$

Let us now consider the function $d(n)$. Since for every prime p we have $d(p) = 2$, hence $\liminf_{n \to \infty} d(n) = 2$.

The following theorem bounds $d(n)$ from above:

Theorem 2.11. *For every positive* δ *we have* $d(n) = o(n^{\delta})$.

Proof. Here we shall apply the following lemma which can be applied in other analogous situations:

Lemma 2.2. *If f is a multiplicative function whose values tend to zero as the argument runs through the sequence formed by all prime powers, then*

$$\lim_{n \to \infty} f(n) = 0 \ .$$

Proof. Fix a positive number $\epsilon < 1$ arbitrarily and choose N so that for $p^k \geqslant N$ (p prime) we have

$$|f(p^k)| < \epsilon.$$

By M we denote

$$\prod_{\substack{p^z \\ |f(p^z)| \geqslant 1}} |f(p^z)| \ .$$

Then for $n = \prod_p p^{\alpha_p}$ we obtain

$$f(n) = \prod_p f(p^{\alpha_p}) = \prod_{p^{\alpha_p} < N} f(p^{\alpha_p}) \cdot \prod_{p^{\alpha_p} \geqslant N} f(p^{\alpha_p}).$$

If n is sufficiently large, then in the second product there appears at least one factor, hence it will not be greater than ϵ. Moreover, the first product is absolutely bounded by M, and so $|f(n)|$ does not exceed ϵM, whence the assertion follows. ∎

Now we apply this lemma to the function $f(n) = \dfrac{d(n)}{n^\delta}$. Then

$$f(p^k) = \frac{d(p^k)}{p^{k\delta}} = \frac{1+k}{p^{k\delta}} \to 0,$$

so that

$$\lim_{n \to \infty} f(n) = 0 . \quad ∎$$

Corollary. *For every positive j and any positive δ we have*

$$\sigma_j(n) = o(n^{j+\delta}).$$

Proof. Indeed,

$$\sigma_j(n) = \sum_{d|n} d^j \leqslant n^j d(n). \quad ∎$$

An asymptotic formula for the sum of values of the function $d(n)$ was first given by P. G. Lejeune-Dirichlet in 1838.

Theorem 2.12. *For $x \geqslant 2$ we have*

$$\sum_{n \leqslant x} d(n) = x \log x + (2C-1)x + \Delta(x),$$

where $\Delta(x) = O(x^{\frac{1}{2}})$ and C is Euler's constant.

Proof.

$$\sum_{n \leqslant x} d(n) = \sum_{n \leqslant x} \sum_{ab=n} 1 = \sum_{ab \leqslant x} 1$$

$$= \sum_{a \leqslant \sqrt{x}} \sum_{b \leqslant x/a} 1 + \sum_{b \leqslant \sqrt{x}} \sum_{a \leqslant x/b} 1 - \sum_{\substack{ab \leqslant x \\ a \leqslant \sqrt{x}, b \leqslant \sqrt{x}}} 1$$

$$= 2 \sum_{a \leqslant \sqrt{x}} \left[\frac{x}{a}\right] - [\sqrt{x}]^2 = 2x \sum_{a \leqslant \sqrt{x}} \frac{1}{a} - x + O(\sqrt{x}).$$

Using Corollary 3 to Theorem 2.4, we obtain the assertion. ∎

It is worth noting here that the method applied in the proof of Theorem 2.10, that is, the interchange of order of summation in the sum

$$\sum_{n \leqslant x} d(n) = \sum_{n \leqslant x} \sum_{d \mid n} 1,$$

would provide a weaker result. Indeed, applying it in this case we would obtain

$$\sum_{n \leqslant x} d(n) = \sum_{d \leqslant x} \sum_{\substack{n \leqslant x \\ d \mid n}} 1 = \sum_{d \leqslant x} \left[\frac{x}{d}\right] = x \sum_{d \leqslant x} \frac{1}{d} + O(x) = x \log x + O(x).$$

The point is that although in both methods we replace $[x/d]$ by x/d, committing an error $O(1)$ in every term, we have $[x]$ terms in the second method, and only $[\sqrt{x}]$ terms in the first.

By slightly more complicated methods one can improve the estimate of the error here. We shall do this in Chapter V. By analytic methods one can obtain the estimate $\Delta(x) = O(x^a \log^b x)$ with $a = 346/1067$, $b = 211/100$ (G. A. Kolesnik [1]).* On the other hand it is known (G. H. Hardy [1], A. E. Ingham [1]) that $\Delta(x)$ is not $o(x^{\frac{1}{4}})$. The problem of determining the lower bound of those numbers t for which $\Delta(x) = O(x^t)$ is called the *Dirichlet divisor problem*.

Using the formula ($j > 0$)

$$\sum_{n \leqslant x} n^j = \frac{1}{1+j} x^{j+1} + O(x^j) \quad ,$$

---

*Added in proof by the translator: G. A. Kolesnik has improved his own estimate recently by proving $\Delta(x) = O(x^{\frac{35}{108}} \log^\varepsilon x)$ *(Pacific J. Math.* **98** (1982), 107-122).

we may now prove

Theorem 2.13. *If $j > 0$ and $x > 1$, then*

$$\sum_{n \leqslant x} \sigma_j(n) = C_j x^{1+j} + R(x),$$

where

$$C_j = \frac{1}{1+j} \sum_{n=1}^{\infty} \frac{1}{n^{1+j}},$$

and

$$R(x) = \begin{cases} O(x^j) & (j > 1) \\ O(x \log x) & (j = 1) \\ O(x) & (0 < j < 1) \end{cases}$$

Proof. Using the above equality, we obtain

$$\sum_{n \leqslant x} \sigma_j(n) = \sum_{ab \leqslant x} a^j = \sum_{b \leqslant x} \left(\sum_{a \leqslant x/b} 1 \right) a^j$$

$$= \sum_{b \leqslant x} \left\{ \left(\frac{x}{b} \right)^{1+j} \frac{1}{1+j} + O\left(\left(\frac{x}{b} \right)^j \right) \right\}$$

$$= \frac{\zeta(1+j)}{1+j} x^{1+j} + O\left(x^{1+j} \sum_{b > x} \frac{1}{b^{1+j}} \right) + O\left(x^j \sum_{b \leqslant x} \frac{1}{b^j} \right)$$

$$= \frac{\zeta(1+j)}{1+j} x^{1+j} + \begin{cases} O(x^j) & \text{if } j > 1, \\ O(x \log x) & \text{if } j = 1, \\ O(x) & \text{if } 0 < j < 1, \end{cases}$$

since

$$x^{j+1} \sum_{b > x} \frac{1}{b^{1+j}} = O(x) \qquad \text{if} \qquad j > 0,$$

and

$$\sum_{b \leqslant x} \frac{1}{b^j} = \begin{cases} O(1) & \text{if } j > 1 \\ O(\log x) & \text{if } j = 1 \\ O(x^{1-j}) & \text{if } 0 < j < 1, \end{cases}$$

where $\zeta(1 + j) = \displaystyle\sum_{n=1}^{\infty} \frac{1}{n^{1+j}}$ (for this, see (3.6)).

Corollary. *For $x > 1$ we have*

$$\sum_{n \leqslant x} \sigma(n) = \frac{\pi^2}{12} x^2 + O(x \log x) . \quad \blacksquare$$

Remark. As proved by A. Walfisz [2] the error term $O(x \log x)$ in this corollary can be replaced by $O(x \log^{\frac{2}{3}} x)$.

4.   Euler noticed that the sum of a series whose terms are consecutive values of a multiplicative function can often be expressed as an infinite product. We shall now prove Euler's result, which we shall often use in the sequel.

Theorem 2.14. *If $f(n)$ is a multiplicative function and the series*

$$S = \sum_{n=1}^{\infty} f(n)$$

is absolutely convergent, then the infinite product

$$P = \prod_{p} \left(\sum_{k=0}^{\infty} f(p^k) \right)$$

(in which p runs through all the primes) is also absolutely convergent and the equality $S = P$ holds.

Proof. Let N be a given natural number and let $p_1 < p_2 < \ldots < p_r$ be all the primes $\leqslant N$. Denote by A_N the set of all natural numbers of the form

$$p_1^{\alpha_1} \ldots p_r^{\alpha_r} \qquad (\alpha_i \geqslant 0).$$

Making use of the unique factorization theorem and the absolute convergence of S we may write

$$\prod_{p \leqslant N} \sum_{k=0}^{\infty} f(p^k) = \sum_{\alpha_1, \ldots, \alpha_r \geqslant 0} f(p_1^{\alpha_1}) \ldots f(p_r^{\alpha_r}) = \sum_{\alpha_1, \ldots, \alpha_r \geqslant 0} f(p_1^{\alpha_1} \ldots p_r^{\alpha_r}) = \sum_{n \in A_N} f(n),$$

so that

$$\left| \sum_{n=1}^{\infty} f(n) - \prod_{p \leqslant N} \sum_{k=0}^{\infty} f(p^k) \right| = \left| \sum_{n \notin A_N} f(n) \right| \leqslant \sum_{n > N} |f(n)| \to 0.$$

since every number not lying in A_N must be greater than N. This already proves the equality $S = P$. Applying it to the function $|f(n)|$ we obtain the absolute convergence of P. $\blacksquare$

Corollary 1. *If* $f(n)$ *is a totally multiplicative function and the series* S *appearing in the above theorem is absolutely convergent, then*

$$S = \prod_p \frac{1}{1-f(p)}.$$

Proof. In this case we must have $|f(p)| < 1$ for every prime p, for if for some p we have $|f(p)| \geqslant 1$, then

$$\sum_{n=1}^{\infty} |f(n)| \geqslant \sum_{k=0}^{\infty} |f(p^k)| = \sum_{k=0}^{\infty} |f(p)|^k = \infty.$$

Hence

$$\sum_{k=0}^{\infty} f(p^k) = \sum_{k=0}^{\infty} f(p)^k = \frac{1}{1-f(p)}. \quad \blacksquare$$

Corollary 2. *If* $f(n)$ *is a multiplicative function and the series*

$$\prod_p \sum_{k=0}^{\infty} |f(p^k)|$$

is convergent, then S *is absolutely convergent and* $S = P$.

Proof. Since

$$\sum_{n \leqslant N} |f(n)| \leqslant \sum_{n \in A_N} |f(n)| = \prod_{p \leqslant N} \sum_{k=0}^{\infty} |f(p^k)| \leqslant \prod_p \sum_{k=0}^{\infty} |f(p^k)|,$$

the sequence of partial sums of the series $\sum_{n=1}^{\infty} |f(n)|$ is bounded and this series is convergent. The equality $S = P$ now follows from the theorem. $\blacksquare$

Corollary 3. *If* $f(n)$ *is a multiplicative function and* s *is a complex number such that the series*

$$\sum_{n=1}^{\infty} f(n) n^{-s}$$

is absolutely convergent, then

$$\sum_{n=1}^{\infty} f(n) n^{-s} = \prod_p \sum_{k=0}^{\infty} f(p^k) p^{-ks}.$$

Proof. It suffices to note that $f(n)n^{-s}$ is a multiplicative function and to apply the theorem. $\blacksquare$

The series appearing in Corollary 3 carry the name of *Dirichlet series* and are very useful in studying the asymptotic behaviour of multiplicative functions. We shall be concerned with their fundamental properties and applications in Chapter III.

5.   Composing a table of values of functions such as $\omega(n)$, $d(n)$, $\varphi(n)$ or $\sigma(n)$, we notice that they vary in a rather irregular way. For instance, the function $\omega(n)$ assumes the value 1 infinitely many times and moreover it is not bounded, for e.g. $\omega(p_1 \ldots p_r) = r$ if p_i's are distinct primes. Similarly, the equality $d(n) = 2$ holds for every prime, but again for $n = p_1 \ldots p_r$ we have $d(n) = 2^r$. Finally, Euler's function $\varphi(n)$ takes the value $p - 1$ for primes p and its value for $n = 1 + p$ is surely not greater than $\frac{1}{2}(1 + p)$, since $p + 1$ is an even number. From this it follows that none of these functions is monotone. It is, of course, clear that one can find additive or multiplicative functions that are monotone, for example $f(n) = c \log n$, or also $f(n) = n^c$, but, as we shall soon prove, they are rare.

Theorem 2.15 (P. Erdös [4]). We give a proof according to L. Moser and J. Lambek [1]):

   (i)   *If $g(n) > 0$ is a multiplicative and monotone function, then there exists a constant c such that $g(n) = n^c$.*

   (ii) *If $f(n)$ is an additive and monotone function, then there exists a constant c such that $f(n) = c \log n$.*

Proof.   It suffices to prove (i), and then apply it to the function $f(n) = \log g(n)$. Hence let $g(n)$ be a multiplicative and monotone function. Replacing, if necessary the function $g(n)$ by $1/g(n)$, we may in addition suppose that g is non-decreasing. In view of $g(1) = 1$ we have therefore $g(n) \geqslant 1$ for all n.

For each natural $a > 2$ and $t \in \mathscr{N}$ we consider

and
$$R_t = a^t + a^{t-1} + \ldots + a + 1 = (a^{t+1} - 1)/(a - 1)$$

$$S_t = a^t - a^{t-1} - a^{t-2} - \ldots - a - 1 .$$

We have clearly $(a, R_t) = 1$ and $R_t - 1 = aR_{t-1}$, hence, using multiplicativity and monotonicity we have

$$g(R_t) \geqslant g(R_t - 1) = g(a)g(R_{t-1})$$

and by simple induction we obtain

$$g(R_t) \geqslant g(a)^{t-1}g(R_1) = g(a)^{t-1}g(1 + a) \geqslant g(a)^t.$$

Since we have also $(a, S_t) = 1$ and $S_t + 1 = aS_{t-1}$, we have in the same way

$$g(S_t) \leqslant g(S_t + 1) = g(a)g(S_{t-1}) \quad ,$$

whence

$$g(S_t) \leqslant g(a)^{t-1}g(S_1) = g(a)^{t-1}g(a-1) \leqslant g(a)^t \quad .$$

Now fix $n \in \mathcal{N}$ and let $r \in \mathcal{Z}$ be determined by the inequality

$$a^r < n \leqslant a^{r+1},$$

that is

$$r < \frac{(\log n)}{(\log a)} \leqslant r+1.$$

Since $R_{r-1} \leqslant a^r < n$ and $S_{r+2} > a^{r+2} - 2a^{r+1} \geqslant a^{r+1} \geqslant n$ (we recall that $a > 2!$), we have

$$g(n) \geqslant g(R_{r-1}) \geqslant g(a)^{r-1} \geqslant g(a)^{(\log n)/\log a - 2}$$

and

$$g(n) \leqslant g(S_{r+2}) \leqslant g(a)^{r+2} \leqslant g(a)^{(\log n)/\log a + 2},$$

so that if $a, b > 2$,

$$g(a)^{1/\log a + 2/\log n} \geqslant g(n)^{1/\log n} \geqslant g(b)^{1/\log b - 2/\log n}.$$

Since n is any natural number, the last inequality implies

$$g(a)^{1/\log a} \geqslant g(b)^{1/\log b},$$

and by changing the role of a and b, we have also

$$g(b)^{1/\log b} \geqslant g(a)^{1/\log a}.$$

Hence $g(a)^{1/\log a} = g(b)^{1/\log b} = C$, if only $a, b > 2$. The number C being positive, we may write it in the form $C = \exp c$, whence we conclude that $g(n) = n^c$ for $n = 3, 4, \ldots$. But $g(2) = g(6)/g(3) = 6^c/3^c = 2^c$, and clearly $g(1) = 1 = 1^c$. Therefore, for every natural n we have $g(n) = n^c$. ∎

Of many theorems characterizing the functions $c \log n$ among additive functions, we shall prove one:

Theorem 2.16 (P. Erdös [4]). *If f is an additive function and* $\lim_{n \to \infty} (f(n+1) - f(n)) = 0$, *then there exists a constant c such that $f(n) = c \log n$.*

Proof (due to A. Rényi [1]). Let p^k be any prime power and the function $g(n)$ be defined by the formula

$$g(n) = f(n) - \frac{f(p^k) \log n}{\log(p^k)}.$$

This function is additive, satisfies the condition $\epsilon_n = g(n+1) - g(n) \to 0$, and takes the value 0 at the point p^k. Now fix n and define the sequence $n_0, n_1, n_2, \ldots$ as follows: $n_0 = n$, and if $n_0, n_1, \ldots, n_{j-1}$ are already defined, then we put

$$n_j = \begin{cases} [n_{j-1} \, p^{-k}], & \text{if} \quad p \nmid [n_{j-1} \, p^{-k}] \quad \text{or} \quad [n_{j-1} \, p^{-k}] = 0, \\ [n_{j-1} \, p^{-k}] - 1, & \text{if} \quad p \mid [n_{j-1} \, p^{-k}] \neq 0. \end{cases}$$

Observe that the sequence thus defined satisfies the following conditions:

(a)   If $j \geqslant 1$ and $n_{j-1} \geqslant p^k$,   then $(p, n_j) = 1$.

(b)   If $j \geqslant 1$,   then $n_j \leqslant n_{j-1}\, p^{-k}$.

(c) $n_j = p^k n_{j+1} + r_j$,   where $0 \leqslant r_j \leqslant 2p^k$.

From the condition (a) we see that as long as $n_j \geqslant p^k$, we have

$$g(p^k n_{j+1}) = g(p^k) + g(n_{j+1}) = g(n_{j+1}),$$

and so

$$g(n_j) = g(n_{j+1}) + g(n_j) - g(p^k n_{j+1}) = g(n_{j+1}) + \sum_{i=p^k n_{j+1}}^{n_j-1} \varepsilon_i,$$

whence, denoting by r the largest natural number for which $n_{r-1} \geqslant p^k$ (and from condition (b), which gives the inequality $n_j \leqslant np^{-jk}$, it follows that $r \leqslant \dfrac{\log n}{\log p^k} + 1$), we have

$$g(n) = \big(g(n_0) - g(n_1)\big) + \; \cdots \; + \big(g(n_{r-1}) - g(n_r)\big) + g(n_r)$$

(2.7)
$$= g(n_r) + \sum_{j=1}^{r} \sum_{i=n_{j+1}p^k}^{n_j-1} \varepsilon_i .$$

By the choice of r we have $n_r < p^k$, and hence the quantity $|g(n_r)|$ is bounded by

$$B = \max_{1 \leqslant i \leqslant p^k} |g(i)| \; ,$$

which does not depend on the choice of n. Since we may write the equality (2.7) in the form

$$g(n) = g(n_r) + \sum_{j=1}^{N_r} \varepsilon_{m_j},$$

where $m_1 < m_2 < \ldots$, $\lim \varepsilon_{m_j} = 0$, and also

$$N_r \leqslant \left(1 + \frac{\log n}{\log(p^k)}\right) \max_{j \leqslant r} (n_j - p^k n_{j+1})$$

$$\leqslant \left(1 + \frac{\log n}{\log(p^k)}\right) 2p^k \leqslant C \log n$$

with a suitable C, and

$$\lim_{n \to \infty} \frac{1}{N_r} \sum_{j=1}^{N_r} |\varepsilon_{m_j}| = 0,$$

we have

$$0 \leqslant \limsup_{n\to\infty} \frac{|g(n)|}{\log n} \leqslant \lim_{n\to\infty} \frac{B}{\log n} + \lim_{n\to\infty} \frac{1}{\log n} \sum_{j=1}^{N_r} |\varepsilon_{m_j}| = 0.$$

Hence we see that $g(n)/\log n$ tends to zero, which means that $f(p^k)/\log p^k = \lim f(n)/\log n = c$ does not depend on the choice of p^k, that is $f(p^k) = c \log p^k$ and from Theorem 2.5 it follows that $f(n) = c \log n$. $\blacksquare$

This theorem has applications in the information theory cf. A. Rényi [3], Chapter IX.

Many theorems are known which characterize the logarithm among additive functions. We now give some of them without proof:

(a) *If f is additive and* $x^{-1} \sum_{n=0}^{x} |f(n+1) - f(n)| \to 0$, *then* $f(n) = c \log n$. (I. Kátai [5]).

This was conjectured by P. Erdös and was also proved by E. Wirsing [1].

(b) *If f is additive and there exists a non-decreasing function $\epsilon(n)$ tending to zero as $n \to \infty$ such that*

$$f(n+1) - f(n) \geqslant -\varepsilon(n),$$

then $f(n) = c \log n$. (I. Kátai [1]). This theorem generalizes both Theorems 2.15 and 2.16.

(c) *If $f(n)$ is an additive function and there exists a set A consisting of natural numbers such that if we denote by $A(x)$ the number of elements of A not exceeding x, then $A(x)/x$ tends to 1, and moreover we have*

$$f(n + 1) \geqslant f(n) , \qquad n \in A ,$$

then $f(n) = c \log n$. (I. Kátai [2]).

(d) *If $f(n)$ is additive and $f(2n + 1) - f(n)$ tends to zero, then* $f(n) = c \log n$. (J. L. Mauclaire [1]). (See also I. Kátai [3], C. Ryavec [1]).

Exercises

1. Prove that a function f not identically zero is completely additive iff $f*f = fd$.

2. Is it possible to choose for each sequence a_n tending to infinity an additive function f such that $f(n) \geqslant a_n$ holds for $n = 2, 3, \ldots$?

3. (J. Lambek [1]). Prove that a function f not identically zero is completely multiplicative iff for any functions g, h we have

$$f(g * h) = fg * fh.$$

4. (a) Show that the unitary convolution of multiplicative functions is a multiplicative function.

(b) (M. V. Subbarao [1]). Show that if $\circ$ denotes the unitary convolution and the functions f_1, f_2, f_3, f_4 are multiplicative, then

$$(f_1 \circ f_2)(f_3 \circ f_4) = (f_1 f_3) \circ (f_1 f_4) \circ (f_2 f_3) \circ (f_2 \circ f_4) \quad .$$

5.   (a) We denote by P_K the group of formal power series over a field K of the form $1 + a_1 X + a_2 X^2 + \ldots$ $(a_i \in K)$ under multiplication and by $M(K)$, the group of multiplicative functions with values in K under multiplication. Prove that the group $M(K)$ is isomorphic with the product of countably many copies of P_K.

(b) Show that in the group $M(\mathfrak{Z})$ the equation $f^n = g$ has a unique solution for every n and g. Deduce from this that the group $M(\mathfrak{Z})$ is isomorphic with the additive group of real numbers.

6.   Prove the identity

$$\sum_{d \mid n} \omega(d) = \sum_{p \mid n} d\left(\frac{n}{p}\right)$$

(where p runs through prime divisors of n).

7.   (S. S. Pillai). Show that if $b(n) = \sum_{r=1}^{n} (r, n)$, then $(b * 1)(n) = nd(n)$.

8.   Show that for positive integral x the equality

$$\sum_{n \leqslant x} \mu(n)\left[\frac{x}{n}\right] = 1$$

holds and deduce that

$$\sum_{n \leqslant x} \frac{\mu(n)}{n} = O(1).$$

9.   Prove that

$$\sum_{n \leqslant x} \frac{\varphi(n)}{n} = \frac{6}{\pi^2} x + O(\log x).$$

10.   Find an asymptotic formula for

$$\sum_{n \leqslant x} \frac{\varphi(n)}{n^2}.$$

11.   Show that

$$\sum_{\substack{n \leqslant x \\ (m, n) = 1}} 1 = \frac{\varphi(m)}{m} x + O(d(m)).$$

12.   (F. Mertens, 1874, cf. also E. Cohen [1]). Prove that if $\circ$ denotes the unitary convolution and $d^* = 1 \circ 1$, then

$$\sum_{n \leqslant x} d^*(n) = \frac{6}{\pi^2} x \log x + O(x).$$

13.   (E. Landau [1]). Prove the identity

$$\frac{1}{\varphi(n)} = \frac{1}{n} \sum_{d \mid n} \frac{\mu^2(d)}{d}$$

and deduce from it the formula

$$\sum_{n \leqslant x} \frac{1}{\varphi(n)} = C_1 \log x + C_2 + O\left(\frac{\log x}{x}\right).$$

14. Prove the estimate $\sigma(n) = o(n \log n)$.

15. (H. Davenport [1]). For real a, let

$$\varphi_a(n) = \sum_{\substack{m \leqslant n \\ (m,n)=1}} m^a \quad \text{and} \quad S_a(n) = \sum_{m \leqslant n} m^a$$

Prove that

$$\varphi_a(n) = n^a \sum_{d \mid n} \frac{S_a(d)}{d^a} \mu\left(\frac{n}{d}\right).$$

and that for $a \geqslant 0$ we have

$$\varphi_a(n) = \frac{n^a}{1+a} \varphi(n) + O(n^a).$$

16. (P. J. McCarthy [1]). Let $T_r(n)$ denote the number of integers $k \leqslant n$ for which (k, n) is not divisible by the r-th power of any natural number greater than 1. Prove that

$$T_r(n) = n \sum_{d^r \mid n} \mu(d) d^{-r}$$

and

$$\sum_{n \leqslant x} T_r(n) = \tfrac{1}{2} x^2 \sum_{k=1}^{\infty} \frac{1}{k^{2r}} + O(x).$$

§3. The natural density. Average values of arithmetical functions

1. For any set A of natural numbers we shall denote by $A(x)$ the number of its elements not exceeding x, i.e.

$$A(x) = |A \cap [1, x]| = \sum_{\substack{n \leqslant x \\ n \in A}} 1 .$$

If we shall deal with sets denoted by B, C, X, $\ldots$, then $B(x)$, $C(x)$, $X(x), \ldots$ will have the same meaning, i.e.

$$B(x) = \sum_{\substack{n \leqslant x \\ n \in B}} 1, \quad \ldots$$

We call the number defined by

$$d^*(A) = \limsup_{x \to \infty} A(x) x^{-1},$$

the *upper density of the set* A and similarly we call

$$d_*(A) = \liminf_{x \to \infty} A(x) x^{-1}.$$

the *lower density of the set* A. Clearly, every set of natural numbers has both upper and lower density and moreover, the inequality

$$0 \leqslant d_*(A) \leqslant d^*(A) \leqslant 1 .$$

holds. In the case when $d_*(A) = d^*(A)$ we call the common value of upper and lower density *the natural density of the set* A or in short, *the density of the set A* and denote it by $d(A)$. Simple properties of this notion are contained in the following theorem:

Theorem 2.17.

(i)   *If A, B are disjoint sets of natural numbers and there exist densities $d(A)$ and $d(B)$, then their union $C = A \cup B$ has a density and the equality $d(C) = d(A) + d(B)$ holds.*

(ii)   *If $A \subset B$, then $d_*(A) \leqslant d_*(B)$ and $d^*(A) \leqslant d^*(B)$.*

(iii) $d^*(A \cup B) \leqslant d^*(A) + d^*(B)$.

Proof. For a proof of (i) it suffices to note that $C(x) = A(x) + B(x)$, and (ii) and (iii) follow immediately from the definition. ∎

Corollary. *If A, B have density zero, then so does $A \cup B$.* ∎

Thus we see that the density is an additive set function. However, it is not countably additive, and hence is not a measure, since the set of all natural numbers has density 1 and it is a disjoint union of one-point sets of density zero.

When a set A has density 1, we say that it contains almost all (a.a.) natural numbers. Similarly, we shall say that a.a. natural numbers have a property W if the set of those natural numbers having this property has density 1, that is, the set $\{n: n$ does not have the property $W\}$ has density 0. We shall say also that W holds almost every-where or for almost every integer (a.e.).

The simplest example of a set having density is an arbitrary arithmetic progression $A = \{an + b: n = 0, 1, \ldots\} \cap \mathscr{N}$ with $a > 0$. Indeed, in this case we have

$$A(x) = \sum_{0 < an+b \leqslant x} 1 = \left[\frac{x-b}{a}\right] = \frac{x}{a} + O(1),$$

and therefore $d(A) = 1/a$. Another class of examples may be obtained using the following theorem:

Theorem 2.18. *Let $a_1, a_2, \ldots$ be a sequence of natural numbers satisfying the condition $(a_i, a_j) = 1$ for $i \neq j$. Denote by A the set of all natural numbers that are not divisible by any of the a_i's. Then the set A has a density, and*

$$d(A) = \begin{cases} \prod_{i=1}^{\infty} (1 - 1/a_i), & \text{when the series } \sum_{i=1}^{\infty} 1/a_i \text{ converges,} \\ \\ 0, & \text{when the series } \sum_{i=1}^{\infty} 1/a_i \text{ diverges.} \end{cases}$$

Proof. Denote by $A_i(x)$ the number of natural numbers that are not divisible by any of $a_1, a_2, \ldots, a_i$. Then we have clearly

(2.8) $A_i(x) \geqslant A(x), \qquad \lim_{i \to \infty} A_i(x) = A(x)$.

The following lemma enables us to find $d(A_i)$.

Lemma 2.3. *For each x we have the identity:*

$$(2.9) \qquad A_i(x) = [x] - \sum_{j=1}^{i} \left[\frac{x}{a_j} \right] + \sum_{1 \leqslant j_1 < j_2 \leqslant i} \left[\frac{x}{a_{j_1} a_{j_2}} \right] + \cdots$$

$$+ (-1)^k \sum_{1 \leqslant j_1 < \ldots < j_k \leqslant i} \left[\frac{x}{a_{j_1} \ldots a_{j_k}} \right] + \cdots + (-1)^i \left[\frac{x}{a_1 \ldots a_i} \right].$$

Proof. To obtain the identity (2.9) we first count all the natural numbers $n \leqslant x$. They are $[x]$ in number. Now it is necessary to delete all the numbers that are divisible by some a_j. There are $[x/a_j]$ of them, but when we count the number of cancellations which is equal to $\sum_{j=1}^{i} \left[\frac{x}{a_j} \right]$, we find that some natural numbers have been cancelled several times. In fact, the numbers divisible by a_{j_1} and a_{j_2} have been cancelled at least twice. Since $(a_{j_1}, a_{j_2}) = 1$, they are exactly those integers that are divisible by $a_{j_1} a_{j_2}$, and there are $[x/a_{j_1} a_{j_2}]$ of them. Therefore these integers must be added, and we do this for every $j_1 < j_2$, but this time we get too many integers, since if n is divisible by $a_{j_1}, a_{j_2}, a_{j_3}$, then it has been first cancelled thrice, and later added three times, but it should be cancelled. Hence we cancel the natural numbers divisible by $a_{j_1} a_{j_2} a_{j_3}$, and this gives us the successive term in the sum (2.9). Continuing this process which must end at the i-th step, we obtain finally the formula (2.9). This process is a typical example of a sieve which we shall consider in Chapter IV.

And now a formal proof of the equality (2.9): Its right-hand side may be written in the form

$$\sum_{n \leqslant x} 1 + \sum_{k=1}^{i} (-1)^k \sum_{\substack{1 \leqslant j_1 < \ldots < j_k \leqslant i}} \sum_{\substack{n \leqslant x \\ a_{j_1} \ldots a_{j_k} | n}} 1 = \sum_{n \leqslant x} 1 + \sum_{n \leqslant x} \sum_{k=1}^{i} (-1)^k \sum_{\substack{1 \leqslant j_1 < \ldots < j_k \leqslant i \\ a_{j_1} \ldots a_{j_k} | n}} 1 .$$

If n is divisible by $a_{j_1}, \ldots, a_{j_s}$ and not divisible by any other a_i, then we write $s(n) = s$. Clearly, $s(n) = 0$ means that $n \in A$. Note that the sum

$$\sum_{\substack{1 \leqslant j_1 < \ldots < j_k \leqslant i \\ a_{j_1} \ldots a_{j_k} \mid n}} 1$$

is equal to 0 in the case $k > s(n)$, and to $\binom{s(n)}{k}$ in the case $k \leqslant s(n)$, since all the subsequences with k elements of the sequence $a_{j_1}, \ldots, a_{j_k}$ are counted in it. In the case $s(n) = 0$ this sum is clearly empty, and hence is equal to zero. Hence we see that the right-hand side of (2.9) is equal to

$$\sum_{n \leqslant x} 1 + \sum_{\substack{n \leqslant x \\ s(n) \geqslant 1}} \sum_{k=1}^{s(n)} (-1)^k \binom{s(n)}{k} = \sum_{n \leqslant x} 1 - \sum_{\substack{n \leqslant x \\ s(n) \geqslant 1}} 1 = \sum_{\substack{n \leqslant x \\ s(n) = 0}} 1 = A(x)$$

in view of the identity

$$\sum_{k=1}^{N} (-1)^k \binom{N}{k} = -1 \qquad (N \geqslant 1). \qquad \blacksquare$$

From the Lemma it follows now that

$$A_i(x) = x \prod_{j=1}^{i} \left(1 - \frac{1}{a_j}\right) + O(2^i),$$

since replacing $[z]$ by z we commit an error of order $O(1)$, and the sum in the equality (2.9) has 2^i terms. From this follows the equality

(2.10) $$\lim_{x \to \infty} \frac{A_i(x)}{x} = \prod_{j=1}^{i} \left(1 - \frac{1}{a_j}\right)$$

and using (2.8), we deduce that

$$d^*(A) \leqslant \prod_{j=1}^{i} \left(1 - \frac{1}{a_j}\right) \qquad (i = 1, 2, \ldots),$$

and hence also

$$d^*(A) \leqslant \inf_{i} \prod_{j=1}^{i} \left(1 - \frac{1}{a_j}\right).$$

If the series $\sum_{j=1}^{\infty} 1/a_j$ is divergent, then the right-hand side of the last equality is equal to zero, and so in this case $d(A) = 0$. If, on the contrary, this series is convergent, then so is the product $C = \prod_{i=1}^{\infty} (1 - 1/a_i)$ and we have $d^*(A) \leqslant C$. It remains to prove that $d_*(A) \geqslant C$. For this purpose we note that for any i

$$A(x) \geqslant A_i(x) - \sum_{n \leqslant x}' 1$$

(where we sum over those $n \leqslant x$ for which there exists a $j > i$ such that $a_j | n$), and since the second term on the right-hand side does not exceed

$$\sum_{\substack{j=1+i}}^{\infty} \sum_{\substack{n \leqslant x \\ a_j | n}} 1 \leqslant x \sum_{j=1+i}^{\infty} \frac{1}{a_j},$$

we have by (2.10)

$$d_*(A) \geqslant \lim_{x \to \infty} \frac{A_i(x)}{x} - \sum_{j=1+i}^{\infty} \frac{1}{a_j} = \prod_{j=1}^{\infty} \left(1 - \frac{1}{a_j}\right) - \sum_{j=1+i}^{\infty} \frac{1}{a_j}.$$

Now letting i tend to infinity, we obtain $d_*(A) \geqslant C$. ∎

Corollary. *If K is a set of all natural numbers that are not divisible by any square > 1, then*

$$d(K) = \left(\sum_{n=1}^{\infty} \frac{1}{n^2}\right)^{-1} = \frac{6}{\pi^2}.$$

Proof. We apply the Theorem in the case where $a_i = p_i^2$ is a square of the i-th prime. Since the series $\sum_p 1/p^2$ is convergent,

$$d(K) = \prod_p \left(1 - \frac{1}{p^2}\right).$$

Applying now Corollary 1 to Theorem 2.14 to the function $f(n) = \dfrac{1}{n^2}$, we obtain

$$\frac{\pi^2}{6} = \sum_{n=1}^{\infty} \frac{1}{n^2} = \prod_p \frac{1}{1 - \dfrac{1}{p^2}} = \frac{1}{d(K)}. \quad \blacksquare$$

Remark. We call the elements of the set K *square-free integers*.

2.   The notion of the natural density may be treated as an application, in the special case of a characteristic function, of that of the mean value of a function which we shall now introduce.

If f is a real arithmetic function, then by its *upper* and *lower mean values* we shall understand the limits

$$M^*(f) = \limsup_{x \to \infty} x^{-1} \sum_{n \leqslant x} f(n)$$

and

$$M_*(f) = \liminf_{x \to \infty} x^{-1} \sum_{n \leqslant x} f(n),$$

respectively. (It is irrelevant here, whether the summation starts from zero or from other fixed natural number.)

If we have $M_*(f) = M^*(f)$, then we call the common value of these numbers the *mean value of the function* f and denote it by $M(f)$. Thus

$$M(f) = \lim_{x \to \infty} \frac{\sum_{n \le x} f(n)}{x}.$$

Observe that if A is a set of natural numbers and f is its characteristic function, i.e. $f(n) = 1$ when $n \in A$ and $f(n) = 0$ when $n \notin A$, then $d(A)$ exists iff $M(f)$ exists and then $d(A) = M(f)$.

It is easily seen that if there exists the limit $a = \lim_{n \to \infty} f(n)$, then the average value $M(f) = a$ also exists.

Other conditions sufficient or necessary for the existence of $M(f)$ are in general very complicated, even in the case when the functions considered are additive or multiplicative. We shall now present a simple theorem of this type.

Theorem 2.19. *If the series* $\sum_{n=1}^{\infty} |g(n)|/n$ *is convergent,* $f = g * h$, *and the mean value* $M(h)$ *exists, then* $M(f)$ *also exists and is equal to*

$$M(h) \sum_{n=1}^{\infty} \frac{g(n)}{n}.$$

Proof. We make use of a lemma which is a discrete variant of Lebesgue's theorem on termwise integration:

Lemma 2.4. *If* $[a_{k,n}]_{k,n=1,2,\ldots}$ *is an infinite matrix with the following properties:*

(a) *there exists a constant* C *such that* $|a_{k,n}| \le C$ *for all* k *and* n,

(b) *for every* k *there exists the limit* $\lim_{n \to \infty} a_{k,n} = a_k$,

then for every absolutely convergent series $\sum_{k=1}^{\infty} x_k$ *we have*

$$\lim_{n \to \infty} \sum_{k=1}^{\infty} a_{k,n} x_k = \sum_{k=1}^{\infty} a_k x_k.$$

Proof of the Lemma. Select a natural number M so that

$$\sum_{k > M} |x_k| < \frac{\varepsilon}{4C}$$

and N so that for all $k = 1, 2, \ldots, M$ and $n \ge N$ we have

$$|a_{k,n} - a_k| < \frac{\varepsilon}{2 \sum_{k \le M} |x_k|}.$$

Then for $n \geqslant N$ we have

$$\left| \sum_{k=1}^{\infty} a_{k,n} x_k - \sum_{k=1}^{\infty} a_k x_k \right| \leqslant \sum_{k \leqslant M} |a_{k,n} - a_k| \cdot |x_k| + \sum_{k > M} |a_{k,n} - a_k| \cdot |x_k|$$

$$\leqslant \frac{\varepsilon}{2 \sum_{k \leqslant M} |x_k|} \cdot \sum_{k \leqslant M} |x_k| + 2C \cdot \frac{\varepsilon}{4C} = \varepsilon. \quad \blacksquare$$

Proof of the Theorem. From the definition of f we have

$$\frac{1}{x} \sum_{n \leqslant x} f(n) = \frac{1}{x} \sum_{n \leqslant x} \sum_{k \mid n} g(k) h\left(\frac{n}{k}\right) = \frac{1}{x} \sum_{k \leqslant x} g(k) \sum_{m \leqslant x/k} h(m)$$

$$= \sum_{k \leqslant x} \frac{g(k)}{k} \cdot \frac{k}{x} \sum_{m \leqslant x/k} h(m) = \sum_{k=1}^{\infty} \frac{g(k)}{k} \cdot \frac{k}{x} \sum_{m \leqslant x/k} h(m).$$

Putting now $x_k = g(k)/k$, $a_{k,n} = \dfrac{k}{n} \sum_{m \leqslant n/k} h(m)$ we see that the assumptions

of Lemma 2.4 are satisfied, and thus $\lim\limits_{n \to \infty} a_{k,n} = M(h)$. This proves the theorem. $\quad \blacksquare$

Corollary 1. *The function* $f(n) = \sigma(n)/n$ *has the average value equal to* $\frac{1}{6}\pi^2$.

Proof. We put $g(n) = 1/n$. Then $g * I = f$ and it is easy to observe that an application of the theorem leads to the required value $M(f)$. $\quad \blacksquare$

Corollary 2. *The function* $f(n) = \varphi(n)/n$ *has the average value equal to* $6/\pi^2$.

Proof. Here it suffices to take $g(n) = \mu(n)/n$, $h = I$. $\quad \blacksquare$

3. We say that a sequence $f_1, f_2, \ldots$ of arithmetical functions is *convergent to a function* f *relative to the density* if for every $\epsilon > 0$ the upper density $d^*(\{n: |f_k(n) - f(n)| > \epsilon\})$ tends to zero as k increases. We write then $d - \lim f_k = f$ or $f_k \overset{d}{\to} f$. The reader has certainly perceived already the analogy between the density and the measure, the mean value and the integral. Here appears an analogous notion to the convergence relative to the measure. However, one should remember that the density is not a countably additive set function, and this may cause a given sequence f_k to be convergent relative to the density to functions that differ from one another even at a set of positive density. Indeed, let $f_k = I$ for all k and let f be an arbitrary function satisfying the condition $\lim\limits_{n \to \infty} f(n) = 1$. Then for every $\epsilon > 0$ the set $\{n: |f_k(n) - f(n)| > \epsilon\}$ is finite,

and so has measure zero and the sequence f_k is convergent relative to density to the function f. Now we note that we may easily find two functions f satisfying the condition $f(n) \to 1$ which are not equal at any point.

More generally, if the condition

(2.11) $d(\{n: |f(n) - g(n)| > \epsilon\}) = 0$,

is satisfied for every $\epsilon > 0$, then every sequence convergent relative to density to f is again convergent to g and vice versa. This remark motivates the following definition:

We say that the functions f and g are *equivalent in the sense of density* (or more shortly, *d-equivalent*) if the condition (2.11) is satisfied. We then write $f \sim g$. We adopt this definition also in the case when the functions f, g are defined only on some subset of the set of natural numbers of density 1.

Fundamental properties of this concept are contained in the following theorem:

Theorem 2.20.

(i)   *The relation $\sim$ is an equivalence relation.*

(ii)   *If $f_1 \sim g_1$ and $f_2 \sim g_2$, then $f_1 \pm f_2 \sim g_1 \pm g_2$, and moreover, if the functions f_1 and f_2 (or g_1 and g_2) are bounded on some set A of density 1, then $f_1 g_1 \sim f_2 g_2$.*

(iii) *If $f \sim g$ and there exists a positive constant B such that for n in some set A of density 1 we have $|f(n)| \geqslant B$, $|g(n)| \geqslant B$, then $1/f \sim 1/g$.*

Proof.

(i)   Follows immediately from definition.

(ii)   Let $f = f_1 + f_2$, $g = g_1 + g_2$. Since $|f(n) - g(n)| \leqslant |f_1(n) - g_1(n)| + |f_2(n) - g_2(n)|$, from $|f(n) - g(n)| > \epsilon$ it follows that $|f_i(n) - g_i(n)| > \frac{1}{2}\epsilon$ for $i = 1$ or $i = 2$. Each of these inequalities may hold on a set of density zero, and hence $|f(n) - g(n)| > \epsilon$ also holds at most on a set of density zero in view of Theorem 2.17.

In the case of $f = f_1 - f_2$, $g = g_1 - g_2$, the proof is practically identical.

Note that if a function f is bounded a.e. and $f \sim g$, then so is g (by the same constant). Hence from assumption it follows that the functions f_1, f_2, g_1, g_2 are bounded a.e. by some constant B. From the inequality

$$|f_1(n)f_2(n) - g_1(n)g_2(n)| = |f_2(n)\left(f_1(n) - g_1(n)\right) + g_1(n)\left(f_2(n) - g_2(n)\right)|$$

$$\leqslant B\left(|f_1(n) - g_1(n)| + |f_2(n) - g_2(n)|\right)$$

holding for $n \in A$, we derive that for every n satisfying $|f_1 f_2(n) - g_1 g_2(n)| > \epsilon$ we have $n \in A$ or also $|f_i(n) - g_i(n)| > \epsilon/2B$ for $i = 1$ or $i = 2$. Therefore such n's lie in a union of three sets of density zero, i.e., lie in a set of density zero.

(iii) For $n \in A$ we have

$$\left|\frac{1}{f(n)} - \frac{1}{g(n)}\right| = \frac{|f(n) - g(n)|}{|f(n)g(n)|} \leqslant \frac{|f(n) - g(n)|}{B^2},$$

and so if $|1/f(n) - 1/g(n)| > \epsilon$, then $|f(n) - g(n)| > \epsilon B^2$ and again we see that such numbers form a set of density zero. ∎

Closely related to the concepts introduced above is the notion of the normal order of an arithmetical function. We say that g is the *normal order* of the function f if the following condition is satisfied:

For every positive ϵ

(2.12) $$d(\{n\colon |f(n) - g(n)| > \epsilon |g(n)|\}) = 0.$$

We see that if for some positive B we have $|f(n)| \geqslant B$ and $|g(n)| \geqslant B$ a.e. and $f/g \sim 1$, then g is the normal order for f, and at the same time f is the normal order for g.

This concept was first used by G. H. Hardy and S. Ramanujan in their work [1] in which they showed that $\log \log n$ is a normal order of the function $\omega(n)$. We shall prove their result in Chapter III. Various authors have been concerned with the question of which functions have a non-decreasing normal order. We shall give an answer to this question for some class of additive functions in Chapter VII. Presently, we shall prove Birch's theorem showing that multiplicative functions do not have in general a non-decreasing normal order.

Theorem 2.21 (B. J. Birch [1]). *If f is a multiplicative, positive, unbounded function and having a non-decreasing normal order, then there exists a positive constant a such that $f(n) = n^a$ $(n = 1, 2, \ldots)$.*

Proof. Let g be a non-decreasing normal order of the function f. Note first that we may suppose without loss of generality that $g(n)$ is positive. Indeed, it is non-decreasing, and if for all n we have $g(n) \leqslant 0$, then for a.e. n the inequality

$$f(n) - g(n) \leqslant \tfrac{1}{2}|g(n)| = -\tfrac{1}{2}g(n) \ ,$$

holds and hence

$$0 < f(n) \leqslant \tfrac{1}{2}g(n) \leqslant 0 \ .$$

Therefore, for infinitely many n, $g(n)$ is positive, and from the monotony it follows that for sufficiently large n we have $g(n) > 0$. If $g(n) \geqslant c > 0$ for $n > n_0$, then replacing $g(n)$ by $g_1(n) = \max(c, g(n))$, we see that g_1 is the non-decreasing normal order of the function f, and $g_1 > 0$ for all n.

Let $b(n) = \log f(n)$, $c(n) = \log g(n)$. Then for a.e. n the inequality

(2.13) $$|b(n) - c(n)| < \epsilon \ .$$

holds. We need to prove that for some constant c we have $b(n) = c \log n$. We divide its proof into several steps.

Lemma 2.5. *We have*

$$\lim_{n \to \infty} c(n) = \infty \ .$$

Proof. If the assertion of the lemma were false, then the function $c(n)$ would be bounded and there would exist the limit $a = \lim g(n) > 0$. Select arbitrarily positive numbers ϵ and η. Then for a.e. n the inequalities

$$a - \eta < g(n) < a$$

and

$$|f(n) - g(n)| \leqslant \epsilon g(n)$$

hold and hence for these n we have

$$|f(n) - a| \leqslant |f(n) - g(n)| + |g(n) - a| \leqslant \epsilon g(n) + \eta \leqslant \epsilon a + \eta = \epsilon_1 .$$

Hence there exist positive constants A, B such that for a.e. n we have

(2.14) $$A \leqslant f(n) \leqslant B .$$

Since $f(n)$ is not bounded, we may find for any M a number n_M such that $f(n_M) \geqslant M/A$. Now observe that the set K of numbers m satisfying the conditions:

$$m = nn_M , \qquad (n, n_M) = 1, \qquad f(n) \geqslant A$$

has positive lower density since

$$\sum_{\substack{m \leqslant x \\ m \in K}} 1 \geqslant \sum_{\substack{n \leqslant x/n_M \\ n \equiv 1(\mathrm{mod}\, n_M) \\ f(n) \geqslant A}} 1 = \frac{x}{n_M^2} + o(x) .$$

For $m \in K$ we have clearly $f(m) = f(n) f(n_M) \geqslant M$, but this contradicts the inequality (2.14) if we take $M = 2B$. ∎

Changing if necessary the value of $g(n)$ on some finite set we may suppose that for all n we have $c(n) \geqslant 0$.

Lemma 2.6. *Let m, n satisfy (2.13) and let η be a given positive number. Then there exists a number S such that for every number $R > S$ we may find numbers s, t so that*

(i) $(1 - \eta)R < s < R < t < (1 + \eta)R$,

(ii) $s \equiv t \equiv 1 \pmod{mn}$,

(iii) $|b(s) - c(s)| < \epsilon$, $|b(t) - c(t)| < \epsilon$,

   $|b(ms) - c(ms)| < \epsilon$, $|b(nt) - c(nt)| < \epsilon$.

Proof. From (2.13) it follows that for every positive number η and every sufficiently large R, the number of $s \leqslant mnR$ for which $|b(s) - c(s)| \geqslant \epsilon$ does not exceed $\eta R/4mn$. We denote by A the set

$$\{s: (1 - \eta) R < s < R; \; s \equiv 1 \pmod{mn}\}$$

It has $\eta R/mn + o(R)$ elements, and hence

$$\sum_{\substack{s \in A \\ |b(s) - c(s)| \geqslant \epsilon}} 1 \leqslant \frac{\eta R}{4mn} \quad \text{and} \quad \sum_{\substack{s \in A \\ |b(ms) - c(ms)| \geqslant \epsilon}} 1 \leqslant \frac{\eta R}{4mn} .$$

From this we deduce that

$$\sum_{\substack{s \in A \\ |b(s)-c(s)| < \varepsilon \\ |b(ms)-c(ms)| < \varepsilon}} 1 \geqslant \frac{\eta R}{mn} - \frac{\eta R}{2mn} + o(R),$$

and this is positive for sufficiently large R. Therefore there exists a number s satisfying the requirement of the lemma. To obtain the existence of the number t, we apply the result obtained replacing η by $2\eta/(1-\eta)$ and R by $(1+\eta)R$.   ∎

Corollary. *There exist sequences $s_0 < s_1 < \ldots$ and $t_0 < t_1 < \ldots$ satisfying the conditions:*

(i) *for every j, k we have $s_j \equiv t_k \equiv 1 \pmod{mn}$,*

(ii) $(1-\eta)S < s_0 < S < t_0 < (1+\eta)S$,

$(1-\eta)ms_i < s_{i+1} < ms_i$,

$nt_j < t_{j+1} < n(1+\eta)t_j$,

(iii) $|b(s_i) - c(s_i)| < \epsilon$, $|b(t_j) - c(t_j)| < \epsilon$,

$|b(ms_i) - c(ms_i)| < \epsilon$, $|b(nt_j) - c(nt_j)| < \epsilon$.

Lemma 2.7. *If $|c(m) - b(m)| < \epsilon$, $|c(n) - b(n)| < \epsilon$,   then*

$$\left| \frac{c(m)}{\log m} - \frac{c(n)}{\log n} \right| < 3\varepsilon \left(\frac{1}{\log m} + \frac{1}{\log n} \right).$$

Proof. Since $b(ms_i) = b(m) + b(s_i)$, hence

$$|c(ms_i) - c(m) - c(s_i)| \leqslant |c(ms_i) - b(ms_i)| + |b(m) - c(m)| + |b(s_i) - c(s_i)| \leqslant 3\epsilon.$$

Using the monotonicity of $c(x)$ we obtain

$$c(s_i) > c(ms_i) - c(m) - 3\epsilon \geqslant c(s_{i+1}) - c(m) - 3\epsilon,$$

and hence

$$c(s_{i+1}) - c(s_i) < c(m) + 3\epsilon,$$

which establishes the inequality

$$c(s_h) < c(S) + hc(m) + 3h\epsilon,$$

valid for $h = 1, 2, 3, \ldots$.

Furthermore, the inequality

(2.15) $s_h > (1-\eta)^{1+h} m^h S$.

clearly holds.

The same reasoning leads us to the inequality

$$c(t_k) > c(S) + kc(n) - 3k\epsilon$$

and

(2.16) $t_k < (1+\eta)^{1+k}\, n^k\, S$.

Now choose h, k and η so that

$$(1-\eta)^{h+1} m^h > (1+\eta)^{k+1} n^k.$$

Note that we may choose η arbitrarily small. Then the inequalities (2.15) and (2.16) leads to $s_h > t_k$, and so $c(s_h) \geqslant c(t_k)$ and $hc(m) + 3h\epsilon > kc(n) - 3k\epsilon$. Since η is a positive number as small as we please, we see that the inequality $h \log m > k \log n$ implies

$$hc(m) > kc(n) - 3(k+h)\,\epsilon,$$

therefore

$$\frac{\log n}{\log m} \geqslant \frac{c(n) - 3\epsilon}{c(m) + 3\epsilon}$$

and finally

$$c(m)\log n - c(n)\log m \geqslant 3\epsilon\,(\log m + \log n).$$

Reversing the roles of m and n, we obtain

$$c(m)\log n - c(n)\log m \geqslant -3\epsilon\,(\log n + \log m),$$

whence the assertion.   ∎

Corollary. *For every pair of integers m, n we have the inequality*

$$\left| \frac{c(n)}{\log n} - \frac{c(m)}{\log m} \right| \leqslant 3\left(|b(m) - c(m)| + |b(n) - c(n)| \right)\left(\frac{1}{\log m} + \frac{1}{\log n} \right).$$ ∎

Now we may complete the proof of the theorem. From the last corollary it follows that if n_k is a sequence of natural numbers such that

$$b(n_k) - c(n_k) \to 0,$$

then $c(n_k) = O(\log n_k)$.

Since (2.13) implies the existence of such a sequence, we may suppose that $c(n)/\log n$ has an accumulation point, say a. If $\lim\limits_{r \to \infty} c(n_r)/\log n_r = a$, then from the Corollary to Lemma 2.7 we conclude that

$$\left| a - \frac{c(m)}{\log m} \right| \leqslant \frac{3|b(m) - c(m)|}{\log m}$$

holds for all m, and hence

$$|c(m) - a \log m| \leqslant 3|b(m) - c(m)|$$

and

$$|b(m) - a \log m| \leqslant 4|b(m) - c(m)|.$$

Since for a.e. m we have $|b(m) - c(m)| < \frac{1}{4}\epsilon$, the inequality $|b(m) - a \log m| \leqslant \epsilon$ holds for a.e. m. From this it follows that for a fixed m and $\epsilon > 0$ we may find an integer n relatively prime to m and satisfying

$$|b(n) - a \log n| < \epsilon, \qquad |b(mn) - a \log (mn)| < \epsilon.$$

But $b(mn) = b(m) + b(n)$ and eventually

$$|b(m) - a \log m| = |b(m) + b(n) - b(n) - a \log m|$$

$$\leqslant |b(mn) - a \log mn| + |a \log n - b(n)| < 2\epsilon$$

for any m and $\epsilon > 0$. Hence for every m we have $b(m) = a \log m$. ∎

Exercises

1. Give an example of a sequence of sets $A_1, A_2, \ldots$ pairwise disjoint and satisfying the conditions:

    (a) $d(A_i) > 0 \quad (i = 1, 2, \ldots)$,

    (b) $d(\bigcup_{i=1}^{\infty} A_i)$ exists and $\neq \sum_{i=1}^{\infty} d(A_i)$.

2. Find the density of the set of integers not divisible by any k-th power of an integer > 1.

3. Let $K(x)$ denote the number of square-free integers $\leqslant x$.

    (a) Prove that $K(x) = \sum_{n \leqslant x} \sum_{d^2 | n} \mu(d)$.

    (b) Strengthen Corollary to Theorem 2.18 by showing that

    $$K(x) = \frac{6}{\pi^2} x + O(\sqrt{x}).$$

4. Find $M(f)$ for $f(n) = \dfrac{\sigma_k(n)}{n^k}$.

5. Prove an analogy of Theorem 2.19 for the unitary convolution.

6. Can the assumption of positiveness and unboundedness of the function f in Theorem 2.21 be omitted?

CHAPTER III

PRIME NUMBERS

The fundamental purposes of this chapter are to prove the so-called prime number theorem giving an asymptotic formula for the number $\pi(x)$ of primes not exceeding x and to prove Dirichlet's theorem on primes in arithmetic progressions. We shall begin by proving some elementary results, and then discuss analytic methods which will be used to prove the fundamental theorems of this chapter.

§1. Čebyšev's theorem

1. In this section we shall prove an estimate for the function $\pi(x)$ discovered by P. Čebyšev in 1849. Different from the following sections of this chapter, we shall use here elementary methods exclusively.

We begin with the introduction of the function $\vartheta(x)$ which is closely related to prime numbers and the proof of a relation between the asymptotic behavior of $\vartheta(x)$ and $\pi(x)$.

The function $\vartheta(x)$ is defined by the formula

$$\vartheta(x) = \sum_{p \leqslant x} \log p = \log\left(\prod_{p \leqslant x} p \right) \ ,$$

and its relation with $\pi(x)$ is given by the following lemma:

Lemma 3.1. *For every positive ϵ and $x \geqslant 2$, the inequalities*

$$\left(1 - \epsilon\right)\left(\pi(x) - \pi(x^{1-\epsilon})\right) \ \leqslant \ \frac{\vartheta(x)}{\log x} \ \leqslant \ \pi(x)$$

hold.

Proof. The inequality on the right side follows from

$$\vartheta(x) = \sum_{p \leqslant x} \log p \leqslant \pi(x) \log x \ \ ,$$

and to prove the left side it is sufficient to note that

$$\vartheta(x) \geqslant \sum_{x^{1-\epsilon} \leqslant p \leqslant x} \log p \geqslant (1 - \epsilon)(\log x)\left(\pi(x) - \pi(x^{1-\epsilon})\right) . \quad \blacksquare$$

Corollary. *For $x \geqslant 2$ and for any positive ϵ, we have*

$$\frac{\vartheta(x)}{\log x} \leqslant \pi(x) \leqslant \frac{1}{1-\epsilon} \cdot \frac{\vartheta(x)}{\log x} + O(x^{1-\epsilon}) .$$

Proof. It suffices to apply the lemma and note that

$$\pi(x^{1-\epsilon}) \leqslant x^{1-\epsilon} . \quad \blacksquare$$

Now we can prove Čebyśev's theorem:

Theorem 3.1. *There exist positive constants c_1, c_2 such that*

$$c_1 \frac{x}{\log x} \leqslant \pi(x) \leqslant c_2 \frac{x}{\log x} .$$

Proof. In view of the Corollary to Lemma 3.1 it is sufficient to prove that for suitably chosen constants b_1, b_2 we have

(3.1) $$b_1 x \leqslant \vartheta(x) \leqslant b_2 x \qquad (x \geqslant 2) .$$

Let us prove the upper estimate of $\vartheta(x)$ in (3.1) by the method of P. Erdös and L. Kalmár. Evidently, it suffices to prove the inequality (3.1) for natural numbers x. We denote the product of all prime numbers not exceeding k by P_k, and hence $\vartheta(k) = \log P_k$. We shall prove, by induction, the inequality

(3.2) $$P_k < 4^k .$$

For $k = 2$ this is obvious. Suppose it is true for all $k < n$. If $n+1$ is an even number, then

$$P_{n+1} = P_n < 4^n < 4^{n+1} ,$$

and if $n+1$ is an odd number, say $n+1 = 2m + 1$, then let us write

$$P_{n+1} = \prod_{p \leqslant m+1} p \cdot \prod_{m+1 < p \leqslant 2m+1} p = P_{m+1} \cdot \prod_{m+1 < p \leqslant 2m+1} p .$$

Now we note that each of the prime numbers in the interval $[m + 2, 2m + 1]$ divides the numerator of the fraction $\binom{2m+1}{m} = \dfrac{(2m+1)!}{m!(m+1)!}$, but does not divide the denominator, so that

$$\prod_{m+1 < p \leqslant 2m+1} p \leqslant \binom{2m+1}{m} .$$

This gives

$$P_{n+1} \leqslant P_{m+1} \binom{2m+1}{m} \quad ,$$

and implies, on account of

$$\binom{3}{1} = 3 \quad \text{and} \quad \binom{2m+1}{m} : \binom{2m-1}{m-1} = \frac{2m(2m+1)}{m(m+1)} < 4 \quad ,$$

the inequality

$$\binom{2m+1}{m} \leqslant 4^m \quad ,$$

and hence

$$P_{n+1} < 4^{m+1} \, 4^m = 4^{2m+1} = 4^{n+1} \quad ,$$

which proves (3.2), and thus the upper estimate in (3.1) follows with $b_2 = 2 \log 2$.

The estimate on the left-hand side of the inequality (3.1) will be deduced from that on the right-hand side by the method of H. N. Shapiro [1]. To this end we note that taking logarithms of both sides of the equality

$$n! = \prod_{p \leqslant n} p^{\alpha_p(n!)}$$

and making use of Corollary 2 to Theorem 1.7, we obtain

$$\log n! = \sum_{p \leqslant n} \alpha_p(n!) \log p = \sum_{p \leqslant n} \sum_{k=1}^{\infty} \left[\frac{n}{p^k} \right] \log p$$

$$= \sum_{p \leqslant n} \left[\frac{n}{p} \right] \log p + O \left(n \sum_{p \leqslant n} \sum_{k=2}^{\infty} \frac{\log p}{p^k} \right)$$

$$= \sum_{p \leqslant n} \left[\frac{n}{p} \right] \log p + O \left(n \sum_{p \leqslant n} \frac{\log p}{p^2} \right)$$

$$= \sum_{p \leqslant n} \left[\frac{n}{p} \right] \log p + O(n) \quad ,$$

and since Stirling's formula gives $\log n! = n \log n + O(n)$, we obtain

(3.3) $$\sum_{p \leqslant n} \left[\frac{n}{p} \right] \log p = n \log n + O(n) \quad .$$

From this it follows that

$$n \sum_{p \leqslant n} \frac{\log p}{p} = n \log n + O(n) + O\left(\sum_{p \leqslant n} \log p \right)$$

$$= n \log n + O(n)$$

by virtue of that part of the theorem already proved, and so

(3.4) $$\sum_{p \leqslant n} \frac{\log p}{p} = \log n + R(n) \; ,$$

where $R(n) = O(1)$. If for a sufficiently large n, $|R(n)| \leqslant C$ holds, then for any $0 < a < 1$ and a sufficiently large n, we have

$$\sum_{an < p \leqslant n} \frac{\log p}{p} = \log \frac{1}{a} + R(n) - R(an) \geqslant \log \frac{1}{a} - 2C \; ,$$

and

$$\sum_{an < p \leqslant n} \frac{\log p}{p} \leqslant \frac{1}{an} \vartheta(n)$$

thus

$$\frac{\vartheta(n)}{an} \geqslant \log \frac{1}{a} - 2C \; .$$

If we take $a = \exp(-2C - 1)$, then $\vartheta(n) \geqslant an$, and this completes the proof of the theorem. ∎

2.   From the estimate (3.4) we shall now deduce an asymptotic formula for the sum of reciprocals of prime numbers not exceeding x. We have seen from Theorem 1.5 that the series $\sum 1/p$ is divergent, and the theorem below determines the rapidity of tending to infinity of its partial sums.

Theorem 3.2. *As x tends to ∞,*

$$\sum_{p \leqslant x} \frac{1}{p} = \log \log x + B + O\left(\frac{1}{\log x} \right),$$

where B is a constant.

Proof. Let us apply Theorem 2.3 with $a_n = \epsilon_n \log n/n$ (where ϵ_n is the characteristic function of the set of prime numbers) and $b_n = 1/\log n$. Then by (3.4) we have $A(x) = \log x + R(x)$, where $R(x) = O(1)$ and

$$\sum_{p \leqslant x} \frac{1}{p} = \sum_{m \leqslant x-1} (\log m + R(m)) \frac{\log\left(1 + \frac{1}{m}\right)}{\log m \cdot \log(1+m)} + \frac{\log x + R(x)}{\log x}$$

$$= \sum_{m \leqslant x-1} \frac{\log\left(1 + \frac{1}{m}\right)}{\log(1+m)} + \sum_{m \leqslant x-1} \frac{R(m) \log\left(1 + \frac{1}{m}\right)}{\log m \cdot \log(1+m)} + 1 + \frac{R(x)}{\log x} \quad .$$

Since $\log\left(1 + \frac{1}{m}\right) = \frac{1}{m} + S(m)$, where $\overset{\cdot}{S}(m) = O\left(\frac{1}{m^2}\right)$, from Corollary to Theorem 2.4 we obtain

$$\sum_{p \leqslant x} \frac{1}{p} = \sum_{m \leqslant x-1} \frac{1}{m \log(1+m)} + \sum_{m=1}^{\infty} \frac{S(m)}{\log(1+m)} + O\left(\frac{1}{x}\right)$$

$$+ \sum_{m=1}^{\infty} \frac{R(m) \log\left(1 + \frac{1}{m}\right)}{\log m \cdot \log(1+m)} + O\left(\frac{1}{\log x}\right)$$

$$= \log\log x + B + O\left(\frac{1}{\log x}\right),$$

where B is a constant.   ∎

Corollary 1. *As x tends to infinity, we have*

$$\sum_{n \leqslant x} \omega(n) = x \log\log x + Bx + O\left(\frac{x}{\log x}\right)$$

and

$$\sum_{n \leqslant x} \Omega(n) = x \log\log x + B_1 x + O\left(\frac{x}{\log x}\right) ,$$

where B is the constant appearing in the theorem and B_1 is also a constant.

Proof. First, we note that

$$\sum_{n \leqslant x} \omega(n) = \sum_{n \leqslant x} \sum_{p | n} 1 = \sum_{p \leqslant x} \sum_{\substack{n \leqslant x \\ p | n}} 1 = \sum_{p \leqslant x} \left[\frac{x}{p}\right]$$

$$= x \sum_{p \leqslant x} \frac{1}{p} + O\left(\pi(x)\right) = x \log\log x + Bx + O\left(\frac{x}{\log x}\right),$$

$$\sum_{n \leqslant x} \Omega(n) = \sum_{n \leqslant x} \sum_{p^k | n} 1 = \sum_{p^k \leqslant x} \sum_{\substack{n \leqslant x \\ p^k | n}} 1 = \sum_{p^k \leqslant x} \left[\frac{x}{p^k} \right]$$

$$= x \sum_{p^k \leqslant x} \frac{1}{p^k} + O\left(\frac{x}{\log x} \right) \quad,$$

for

$$\sum_{p^k \leqslant x} 1 = \pi(x) + \sum_{p \leqslant \sqrt{x}} \sum_{2 \leqslant k \leqslant \frac{\log x}{\log p}} 1 = O\left(\frac{x}{\log x} \right) + O\left(\log x \sum_{p \leqslant \sqrt{x}} \frac{1}{\log p} \right)$$

$$= O\left(\frac{x}{\log x} \right) + O(\sqrt{x} \log x) = O\left(\frac{x}{\log x} \right),$$

and hence

$$\sum_{n \leqslant x} \Omega(n) = x \sum_{p \leqslant x} \frac{1}{p} + O\left(\frac{x}{\log x} \right) + x \sum_{\substack{p^k \leqslant x \\ k \geqslant 2}} \frac{1}{p^k} \quad.$$

But

$$\sum_{\substack{p^k \leqslant x \\ k \geqslant 2}} \frac{1}{p^k} = \sum_{p \leqslant \sqrt{x}} \sum_{2 \leqslant k \leqslant \frac{\log x}{\log p}} \frac{1}{p^k}$$

$$= \sum_{p \leqslant \sqrt{x}} \frac{1}{p^2 \left(1 - \frac{1}{p} \right)} + O\left(\sum_{p \leqslant \sqrt{x}} \sum_{k > \frac{\log x}{\log p}} \frac{1}{p^k} \right) = C + O\left(\frac{1}{\sqrt{x}} \right)$$

for some constant C, and

$$\sum_{n \leqslant x} \Omega(n) = x \log \log x + Bx + Cx + O\left(\frac{x}{\log x} \right) \quad. \qquad \blacksquare$$

Corollary 2. *For* $x \geqslant 2$

$$\sum_{n \leqslant x} \omega^2(n) = x (\log \log x)^2 + O(x \log \log x) \quad.$$

Proof. Changing the order of summation in the sum

$$\sum_{n \leqslant x} \omega^2(n) = \sum_{n \leqslant x} \sum_{p | n} \sum_{q | n} 1$$

(where p, q denote prime numbers), we obtain

$$\sum_{n \leqslant x} \omega^2(n) = \sum_{\substack{p,q \leqslant n \\ p|n, q|n}} \sum_{n \leqslant x} 1 = \sum_{\substack{p \neq q \\ pq \leqslant x}} \left[\frac{x}{pq}\right] + \sum_{p \leqslant n} \left[\frac{x}{p}\right]$$

$$= \sum_{pq \leqslant x} \left[\frac{x}{pq}\right] - \sum_{p \leqslant x} \left(\left[\frac{x}{p^2}\right] - \left[\frac{x}{p}\right]\right)$$

$$= x \sum_{pq \leqslant x} \frac{1}{pq} - x \sum_{p \leqslant x} \left(\frac{1}{p^2} - \frac{1}{p}\right) + O\left(\sum_{pq \leqslant x} 1\right) + O\left(\pi(x)\right)$$

$$= x \sum_{pq \leqslant x} \frac{1}{pq} + O(x \log\log x).$$

Moreover,

$$(\log\log x)^2 + O(\log\log x) = \left(\sum_{p \leqslant \sqrt{x}} \frac{1}{p}\right)^2 \leqslant \sum_{pq \leqslant x} \frac{1}{pq}$$

$$\leqslant \left(\sum_{p \leqslant x} \frac{1}{p}\right)^2 = (\log\log x)^2 + O(\log\log x)$$

by Theorem 3.2, and this proves the assertion. ∎

Corollary 3 (G. H. Hardy, S. S. Ramanujan [1]). *The function* $\log\log n$ *is a normal order for the function* $\omega(n)$ *as well as for* $\Omega(n)$.

Proof. To begin with, let us note that by the Corollaries 1 and 2 we have

$$\sum_{n \leqslant x} (\omega(n) - \log\log x)^2 = \sum_{n \leqslant x} \omega^2(n) - 2\log\log x \sum_{n \leqslant x} \omega(n)$$

$$+ x(\log\log x)^2 = O(x \log\log x) \ ,$$

and therefore, denoting by $R(x)$ the number of integers not exceeding x for which the inequality

$$|\omega(n) - \log\log x| > \epsilon(\log\log x)$$

holds, we obtain

$$\epsilon^2 R(x) (\log\log x)^2 = O(x \log\log x) \ ,$$

or

$$R(x) = O\left(\frac{x}{\log\log x}\right) = o(x) \ .$$

Since for n in the interval $[x^{1/e}, x]$

$$\log\log x \geqslant \log\log n \geqslant \log\log x - 1 \ ,$$

the inequality

$$|\omega(n) - \log\log n| > \epsilon \log\log n$$

can hold only for $o(x)$ integers $n \leqslant x$, which proves the assertion in the case of $\omega(n)$. The case for the function $\Omega(n)$ reduces to the former, owing to the fact that by the equality

$$\sum_{n \leqslant x} (\Omega(n) - \omega(n)) = C_1 x + O\left(\frac{x}{\log x}\right),$$

derived from Corollary 1. The equality

$$\Omega(n) - \omega(n) > \epsilon \log \log x$$

can hold only for $o(x)$ integers $n \leqslant x$. ∎

Corollary 4. *If z is a complex number, $|z| < 2$, then*

$$\prod_{p \leqslant x} \left(1 + \frac{z}{p}\right) = A(z) (\log x)^z \left(1 + O\left(\frac{1}{\log x}\right)\right) \quad ,$$

where $A(z)$ is non-zero and does not depend on x.

Proof. Since $|z| < 2$, the series

$$\sum_{p} \sum_{k=2}^{\infty} \frac{(-1)^{k-1} z^k}{kp^k}$$

is convergent, being majorized by

$$\sum_{p} \left|\frac{z}{p}\right|^2 \left(1 - \left|\frac{z}{p}\right|\right)^{-1} \quad .$$

Therefore

$$\sum_{p \leqslant x} \sum_{k=2}^{\infty} \frac{(-1)^{k-1} z^k}{kp^k} = C(z) + O\left(\frac{|z|^2}{x}\right) = C(z) + O\left(\frac{1}{x}\right),$$

where $C(z)$ does not depend on x. Hence, using Theorem 3.2, we may write

$$\prod_{p \leqslant x} \left(1 + \frac{z}{p}\right) = \exp\left\{\sum_{p \leqslant x} \log\left(1 + \frac{z}{p}\right)\right\}$$

$$= \exp\left\{\sum_{p \leqslant x} \sum_{k=1}^{\infty} \frac{(-1)^{k-1} z^k}{kp^k}\right\}$$

$$= \exp\left\{z \sum_{p \leqslant x} \frac{1}{p} + \sum_{p \leqslant x} \sum_{k=2}^{\infty} \frac{(-1)^{k-1} z^k}{kp^k}\right\}$$

$$= \exp\left\{z \log \log x + Bz + O\left(\frac{1}{\log x}\right) + C(z) + O\left(\frac{1}{x}\right)\right\}$$

$$= A(z) (\log x)^z \left(1 + O\left(\frac{1}{\log x}\right)\right),$$

where $A(z) = \exp\left\{Bz + C(z)\right\}$. ∎

Corollary 5. *For some non-zero constants* A_1, A_2 *we have*

(i)
$$\prod_{p \leqslant x}\left(1 + \frac{1}{p}\right) = A_1 \log x + O(1) ,$$

(ii)
$$\prod_{p \leqslant x}\left(1 - \frac{1}{p}\right) = \frac{A_2}{\log x} + O\left(\frac{1}{\log^2 x}\right) .$$

Proof. It suffices to take $z = 1$ and $z = -1$ respectively in the preceding corollary. ∎

Exercises

1 (H. N. Shapiro [1]). Let $\{a_n\}$ be a sequence of non-negative real numbers satisfying the condition

$$\sum_{n \leqslant x}\left[\frac{x}{n}\right] a_n = x \log x + O(x) .$$

Prove that there exist positive constants a, b such that

$$ax \leqslant \sum_{n \leqslant x} a_n \leqslant bx .$$

2.   Prove that, if p_n denotes the n-th prime number, then for appropriate positive constants a and b,

$$an \log n \leqslant p_n \leqslant bn \log n .$$

3.   Prove that for $c > -1$ and $x \geqslant 2$, we have

$$a_1(c) \frac{x^{1+c}}{\log x} < \sum_{p \leqslant x} p^c < a_2(c) \frac{x^{1+c}}{\log x} ,$$

where $a_1(c) > 0$.

4 (E. Landau [2]). Show that for some constant C, we have

$$\varphi(n) \geqslant \frac{Cn}{\log \log n} .$$

5.   Investigate the convergence of the series

$$\sum_p \frac{1}{p(\log p)^\alpha} \qquad \text{for a given } \alpha.$$

6.   Show that for a sufficiently large n, at least one prime number lies between n and $3n$.

§2. Dirichlet series

1. In this section we shall study some series of functions which will be used in the study of arithmetical functions and in the theory of prime numbers. We have already come across such series in the preceding chapter. They are the so-called *Dirichlet series*

(after P. G. Lejeune Dirichlet who introduced them) and are of the form

$$(3.5) \qquad \sum_{n=1}^{\infty} a_n n^{-s} \quad ,$$

where a_n are complex numbers and s is also complex. We call the numbers a_n the *coefficients* of the series (3.5). Dirichlet series may be convergent at every point of the complex plane, e.g. the series

$$\sum_{n=1}^{\infty} \frac{1}{n! n^s} \quad ,$$

or may be divergent at every point, e.g. the series

$$\sum_{n=1}^{\infty} n! n^{-s} \quad ,$$

or may be convergent in some domains of the plane and divergent elsewhere. It turns out that the domain of convergence of a Dirichlet series consists of an open half-plane and a subset, which may be empty, of its boundary. This results from the following theorem:

Theorem 3.3.
(i)   *If the series* (3.5) *is convergent at a point* s_0, *then it is also convergent in the half-plane* $\mathrm{Re}\, s > \mathrm{Re}\, s_0$ *and the convergence is uniform in the domain* $\mathrm{Re}\, s > \mathrm{Re}\, s_0$, $\arg(s - s_0) \leqslant \vartheta < \frac{1}{2}\pi$.

(ii) *If the series* (3.5) *is absolutely convergent at a point* s_0, *then it is absolutely and uniformly convergent in the half-plane* $\mathrm{Re}\, s \geqslant \mathrm{Re}\, s_0$.

Proof. (i) Suppose that the series (3.5) is convergent at a point $s_0 = x_0 + iy_0$, take N so large that, under the notation

$$b_r = \sum_{n=N+1}^{r} a_n n^{-s_0}, \qquad b_N = 0 \, ,$$

the inequality $|b_r| < \epsilon \cos \vartheta$ $(r \geqslant N)$ holds and suppose that $s = x + iy$ satisfies $x \geqslant x_0$ and $\arg(s - s_0) \leqslant \vartheta$. Then using Theorem 2.3, we obtain for $M > N$

$$\sum_{n=1+N}^{M} a_n n^{-s} = \sum_{n=1+N}^{M} (b_n - b_{n-1}) n^{-(s-s_0)}$$

$$= b_M M^{-(s-s_0)} - \sum_{n=1+N}^{M-1} b_n \left((n+1)^{-(s-s_0)} - n^{-(s-s_0)} \right).$$

Since

$$\frac{|s - s_0|}{x - x_0} \leqslant \frac{1}{\cos \vartheta} \, ,$$

we have

$$|(n+1)^{-(s-s_0)} - n^{-(s-s_0)}| = |\exp\{-(s-s_0)\log(1+n)\} - \exp\{-(s-s_0)\log n\}|$$

$$= \left| (s-s_0) \int_{\log n}^{\log(1+n)} \exp\{-(s-s_0)z\}\,dz \right|$$

$$\leqslant |s-s_0| \int_{\log n}^{\log(1+n)} \exp\{-t(x-x_0)\}\,dt$$

$$= \frac{|s-s_0|}{x-x_0}\left(n^{-(x-x_0)} - (n+1)^{-(x-x_0)}\right)$$

$$\leqslant \frac{1}{\cos\vartheta}\left(n^{-(x-x_0)} - (n+1)^{-(x-x_0)}\right),$$

and so

$$\left| \sum_{n=1+N}^{M} a_n n^{-s} \right| \leqslant \epsilon\cos\vartheta M^{-(x-x_0)} + \epsilon \sum_{n=1+N}^{M} \left(n^{-(x-x_0)} - (n+1)^{-(x-x_0)}\right)$$

$$< \epsilon(N+1)^{-(x-x_0)} < \epsilon ,$$

which proves that the series (3.5) is uniformly convergent for $\mathrm{Re}\,s > \mathrm{Re}\,s_0$, $\arg(s-s_0) \leqslant \vartheta$. Since every point s satisfying $\mathrm{Re}\,s > \mathrm{Re}\,s_0$ lies in the angle $\arg(s-s_0) < \vartheta$ for a suitable $\vartheta < \frac{1}{2}\pi$, the series (3.5) is convergent in the open half-plane $\mathrm{Re}\,s > \mathrm{Re}\,s_0$, this proves (i).

The proof of (ii) is much simpler. Indeed, if the series (3.5) is absolutely convergent at the point s_0 and $\mathrm{Re}\,s \geqslant \mathrm{Re}\,s_0$, then $|a_n n^{-s}| \leqslant |a_n n^{-\mathrm{Re}\,s_0}|$, and therefore the series (3.5) is absolutely and uniformly convergent in the closed half-plane $\mathrm{Re}\,s \geqslant \mathrm{Re}\,s_0$. ∎

Corollary. *If the series (3.5) is convergent at a point s_0, then its sum is a holomorphic function in the open half-plane $\mathrm{Re}\,s > \mathrm{Re}\,s_0$.* ∎

The greatest lower bound τ of the numbers t such that for $\mathrm{Re}\,s > t$ the series (3.5) is convergent is called its *abscissa of convergence*, and the open half-plane $\mathrm{Re}\,s > \tau$ is called its *half-plane of convergence*.

Similarly, the greatest lower bound τ' of the numbers t such that for $\mathrm{Re}\,s > t$ the series is absolutely convergent is called the *abscissa of absolute convergence* of the series (3.5), and the half-plane $\mathrm{Re}\,s > \tau'$ is called its *half-plane of absolute convergence*.

We shall end this subsection by proving that a given holomorphic function can be represented at most in one way in the form of a Dirichlet series. Namely, we shall prove:

Theorem 3.4. *If, in some half-plane $\mathrm{Re}\,s > t$, the series $\sum_{n=1}^{\infty} a_n n^{-s}$ and $\sum_{n=1}^{\infty} b_n n^{-s}$ are convergent and represent the same holomorphic function $f(s)$ there,* *then for* $n = 1, 2, \ldots$ *the equality* $a_n = b_n$ *holds.*

Proof. Let m be the smallest index having the property that $a_m \neq b_m$. From the preceding theorem it follows that the series $\sum_{n=1}^{\infty} a_n n^{-x}$ and $\sum_{n=1}^{\infty} b_n n^{-x}$ are uniformly convergent for $x \geqslant x_0$. We can assert the same for the series

$$\sum_{n=m}^{\infty} a_n \left(\frac{n}{m}\right)^{-x} = \sum_{n=m}^{\infty} b_n \left(\frac{n}{m}\right)^{-x} \,,$$

therefore one can interchange the limit and the sum to obtain

$$a_m = \sum_{n=m}^{\infty} a_n \lim_{x \to \infty} \left(\frac{n}{m}\right)^{-x} = \lim_{x \to \infty} \sum_{n=m}^{\infty} a_n \left(\frac{n}{m}\right)^{-x}$$

$$= \lim_{x \to \infty} \sum_{n=m}^{\infty} b_n \left(\frac{n}{m}\right)^{-x} = \sum_{n=m}^{\infty} b_n \lim_{x \to \infty} \left(\frac{n}{m}\right)^{-x} = b_m \,,$$

which is a contradiction. ∎

In the sequel the following theorem of Landau concerning the singularities of Dirichlet series with non-negative coefficients will be useful:

Theorem 3.5. *If the coefficients* a_n *are non-negative numbers,* t *is the abscissa of absolute convergence of* (3.5) *and* $f(s)$ *is a holomorphic function representing the sum of* (3.5)*, then the point* $s = t$ *cannot be a regular point of* $f(s)$.

In other words, if $f(s)$ *is a function holomorphic in some region containing the open half-plane* $\mathrm{Re}\, s > t$*, where* $f(s)$ *is equal to the sum of* (3.5)*, then the point* $s = t$ *is a singular point of* $f(s)$.

Proof. Suppose the point $s = t$ is a regular point of the function $f(s)$. This means that if at any point $x_0 > t$ we expand $f(s)$ into a power series, then it is convergent in some neighborhood of the point t. Since

$$f^{(k)}(s) = \sum_{n=1}^{\infty} (-1)^k a_n \log^k n \cdot n^{-s} \qquad (k = 1, 2, \ldots) \,,$$

therefore the Taylor series of the function $f(s)$ at point $x_0 > t$ has the form

$$\sum_{k=0}^{\infty} \frac{(-1)^k (s - x_0)^k}{k!} \sum_{m=1}^{\infty} a_m m^{-x_0} \log^k m \,.$$

For some $x < t$ this series must be convergent, and is equal to

$$\sum_{k=0}^{\infty} \frac{(x_0 - x)^k}{k!} \sum_{m=1}^{\infty} a_m m^{-x_0} \log^k m \,.$$

Since this series has non-negative terms, we can exchange the order of summations without changing the value of the sum. This gives

$$\sum_{m=1}^{\infty} a_m m^{-x_0} \sum_{k=1}^{\infty} \frac{(x_0 - x)^k}{k!} \log^k m = \sum_{m=1}^{\infty} a_m m^{-x} \ ,$$

but on the right-hand side we have a divergent series. The obtained contradiction proves the theorem. ∎

2.   One of the most important Dirichlet series is the series

$$(3.6) \qquad\qquad \sum_{n=1}^{\infty} n^{-s}$$

whose abscissa of convergence coincides with that of absolute convergence and equals 1. In fact, for $s = 1$ this series is obviously divergent, and for $\text{Re}\, s > 1$ the convergence follows from the majorization of its terms by the terms of the convergent series

$$\sum_{n=1}^{\infty} n^{-\text{Re}\, s} \ .$$

We denote the sum of the series (3.6) by $\zeta(s)$. This function is called the *Riemann zeta-function* and is extremely important in the theory of primes about which we shall soon realize.

Its fundamental analytical properties are contained in the following theorem:

Theorem 3.6. *The function $\zeta(s)$ is a holomorphic function in the half-plane $\text{Re}\, s > 1$ and can be continued analytically to a meromorphic function on the whole plane. Its unique singularity is the point $s = 1$ at which it has a simple pole with residue 1. Furthermore, for $\text{Re}\, s > 1$ the equality*

$$(3.7) \qquad\qquad \zeta(s) = \prod_{p} (1 - p^{-s})^{-1}$$

holds (where p runs through all the primes).

Finally, for any complex s, we have the identity

$$(3.8) \qquad\qquad \zeta(s) = 2^s \pi^{s-1} \sin \frac{\pi s}{2} \ \Gamma(1-s) \ \zeta(1-s) \ .$$

(The equality (3.8) is called the fundamental equation of the zeta-function).

Proof. The holomorphy of the function $\zeta(s)$ in the half-plane $\text{Re}\, s > 1$ is a consequence of the Corollary to Theorem 3.3, and the equality (3.7) follows from the Corollary to Theorem 2.14. Let us now show that $\zeta(s)$ can be continued to a meromorphic function on the whole plane and satisfies the equality (3.8). To this end we shall use Theorem 2.4 (ii) in the case where $A = 1$, $x \in \mathcal{N}$ and $f(t) = t^{-s}$. This gives, for any complex s,

$$\sum_{n=1}^{x} n^{-s} = 1 + \int_{1}^{x} t^{-s}\, dt - s \int_{1}^{x} (t - [t])\, t^{-1-s}\, dt \ .$$

If now, we fix s satisfying $\operatorname{Re} s > 1$ and pass from x to infinity, we obtain

(3.9)
$$\zeta(s) = 1 + \frac{1}{s-1} - s \int_{1}^{\infty} \frac{t - [t]}{t^{1+s}}\, dt \ .$$

The integral appearing here is uniformly convergent for $\operatorname{Re} s > c$ if $c > 0$, and therefore the right-hand side of (3.9) defines a holomorphic function in the half-plane $\operatorname{Re} s > 0$ and gives an analytic continuation of $\zeta(s)$ there. The formula (3.9) also shows that $\zeta(s)$ has a simple pole at $s = 1$ with residue 1.

Noting that $\dfrac{1}{2} \displaystyle\int_{1}^{\infty} \frac{dt}{t^{1+s}} = \frac{1}{2^s}$, we see that the formula (3.9) implies the identity

(3.10)
$$\zeta(s) = \frac{1}{2} + \frac{1}{s-1} + s \int_{1}^{\infty} \frac{[t] - t + \frac{1}{2}}{t^{1+s}}\, dt$$

valid for $\operatorname{Re} s > 0$. Now note that the function on the right-hand side represents a function holomorphic for $\operatorname{Re} s > -1$ since for any $j \in \mathscr{Z}$ we have

$$\int_{j}^{j+1} ([t] - t + \tfrac{1}{2})\, dt = 0 \ ,$$

and hence for $A > 1$

$$\left| \int_{1}^{A} ([t] - t + \tfrac{1}{2})\, dt \right| \leq \left| \sum_{j=1}^{[A]-1} \int_{j}^{j+1} ([t] - t + \tfrac{1}{2})\, dt \right| + \left| \int_{[A]}^{A} ([t] - t + \tfrac{1}{2})\, dt \right| \leq \frac{1}{2} \ ;$$

and therefore

$$\int_{M}^{N} \frac{[t] - t + \frac{1}{2}}{t^{1+s}}\, dt = \left. \frac{\int_{1}^{x} ([t] - t + \frac{1}{2})\, dt}{x^{1+s}} \right|_{M}^{N} + (s+1) \int_{M}^{N} \frac{\int_{1}^{x} ([t] - t + \frac{1}{2})\, dt}{x^{s+2}}\, dx \ ,$$

which tends to zero as $M, N \to \infty$. Thus $\zeta(s)$ has a continuation to a meromorphic function in the half-plane $\operatorname{Re} s > -1$.

Finally, for $-1 < \operatorname{Re} s < 0$, we have

$$s \int_{0}^{1} \frac{[t] - t + \frac{1}{2}}{t^{1+s}}\, dt = \frac{1}{s-1} + \frac{1}{2}$$

and ultimately from (3.10), the equality

$$(3.11) \qquad \qquad \zeta(s) = s \int_0^\infty \frac{[t] - t + \frac{1}{2}}{t^{1+s}} \, dt \qquad (-1 < \operatorname{Re} s < 0)$$

follows.

Since for every non-integral t the equality

$$[t] - t + \tfrac{1}{2} = \sum_{n=1}^\infty \frac{\sin(2\pi nt)}{\pi n}$$

holds, and moreover the partial sums of the series appearing here are uniformly bounded, we can use Lebesgue's integration theorem to obtain for any $T > 0$ and $-1 < \operatorname{Re} s < 0$,

$$s \int_0^T \frac{[t] - t + \frac{1}{2}}{t^{1+s}} \, dt = \frac{s}{\pi} \sum_{n=1}^\infty \int_0^T \frac{\sin(2\pi nt)}{nt^{1+s}} \, dt$$

$$= \frac{s}{\pi} \sum_{n=1}^\infty \frac{(2\pi n)^s}{n} \int_0^{2\pi nT} \frac{\sin t}{t^{1+s}} \, dt$$

$$= \frac{s}{\pi} \sum_{n=1}^\infty \frac{(2\pi n)^s}{n} \int_0^\infty \frac{\sin t}{t^{1+s}} \, dt - \frac{s}{\pi} \sum_{n=1}^\infty \frac{1}{n} (2\pi n)^s \int_{2\pi nT}^\infty \frac{\sin t}{t^{1+s}} \, dt$$

$$= \frac{-s}{\pi} (2\pi)^s \Gamma(-s) \zeta(1-s) \sin \tfrac{1}{2}\pi s - \frac{s}{\pi} \sum_{n=1}^\infty \frac{1}{n} (2\pi n)^s \int_{2\pi nT}^\infty \frac{\sin t}{t^{1+s}} \, dt.$$

We shall show that the second term of the last expression tends to zero as T tends to infinity.

In fact for $a > 0$ we have

$$\int_a^\infty \frac{\sin t}{t^{1+s}} \, dt = \frac{-\cos t}{t^{1+s}} \Big|_a^\infty - (1+s) \int_a^\infty \frac{\cos t}{t^{s+2}} \, dt = O(a^{-1-\operatorname{Re} s}),$$

hence

$$\sum_{n=1}^\infty \frac{1}{n} (2\pi n)^s \int_{2\pi nT}^\infty \frac{\sin t}{t^{1+s}} \, dt = O\left(T^{-1-\operatorname{Re} s} \sum_{n=1}^\infty \frac{1}{2\pi n^2}\right) = O(T^{-1-\operatorname{Re} s}),$$

which tends to zero as T tends to infinity. This implies the equality (3.8) for $-1 < \operatorname{Re} s < 0$. Now note that its right-hand side is holomorphic for $\operatorname{Re} s < 0$, and so this equality gives an analytic continuation of $\zeta(s)$ over the whole plane with the exception of the point $s = 1$. This shows also that (3.8) holds for any $s \neq 0, 1$. ∎

Corollary. *For $s \neq 1$ the equality $\zeta(s) = \dfrac{A(s)}{s-1}$ holds, where $A(s)$ is an integral function and $A(1) = 1$.* ∎

B. Riemann perceived the significance of the zeta-function in the theory of numbers, in particular, for the problem of distribution of primes and devoted an important paper to its study (see B. Riemann [1]). He proved there Theorem 3.6, and stated a series of assertions, most of which have been subsequently proved. However, no proof has been found for the following statment which is presently called *Riemann's hypothesis*: *All the zeros of the zeta-function other than* $-2, -4, \ldots$ *lie on the line* $\operatorname{Re} s = \frac{1}{2}$. (From Theorem 3.6 it follows that they must lie symmetrically with respect to this line).

We shall now prove a considerably weaker result due to J. Hadamard and Ch. de la Vallée-Poussin (1896) concerning the zeros of the zeta-function, which will enable us to find an asymptotic formula for the number of primes in the interval $[2, x]$.

Theorem 3.7. *In the closed half-plane* $\operatorname{Re} s \geq 1$, $\zeta(s) \neq 0$.

Proof. For $\operatorname{Re} s > 1$ this follows from the formula (3.7) because its right-hand side does not vanish in this domain. In the case of $\operatorname{Re} s = 1$ we shall appeal to an auxiliary result which will also be used in the following section.

Lemma 3.2. *Let* $f(n)$ *be a completely multiplicative function satisfying the condition* $|f(n)| = 0, 1$. *Suppose that the functions*

$$F_1(s) = \sum_{n=1}^{\infty} f(n) n^{-s}, \qquad F_2(s) = \sum_{n=1}^{\infty} f^2(n) n^{-s}$$

(defined for $\operatorname{Re} s > 1$) *can be continued to functions holomorphic for* $\operatorname{Re} s \geq 1$, $s \neq 1$ *with the point* $s = 1$ *being either a regular point or a simple pole for both of them. Then in the half-plane* $\operatorname{Re} s \geq 1$ *we have* $F_1(s) \neq 0$ *for* $s \neq 1$. *If both the functions* F_1, F_2 *are holomorphic at the point* $s = 1$, *then* F_1 *does not vanish in the whole of the closed half-plane* $\operatorname{Re} s \geq 1$.

Proof. In the half-plane $\operatorname{Re} s > 1$ we have by virtue of Corollary 1 to Theorem 2.14 the identity

$$F_1(s) = \prod_p (1 - f(p) p^{-s})^{-1}, \qquad F_2(s) = \prod_p (1 - f^2(p) p^{-s})^{-1}$$

and utilizing the expansion of $\log \dfrac{1}{1-t}$, we may write

$$F_1(s) = \exp \sum_p \sum_{k=1}^{\infty} \frac{1}{k} \cdot \frac{f(p)^k}{p^{ks}}$$

and

$$F_2(s) = \exp \sum_p \sum_{k=1}^{\infty} \frac{1}{k} \cdot \frac{f(p)^{2k}}{p^{ks}}.$$

Taking $f(p) = 1$, we obtain

$$\zeta(s) = \exp \sum_{p} \sum_{k=1}^{\infty} \frac{1}{k} \cdot \frac{1}{p^{ks}} \ ,$$

hence it follows that for $u > 1$, $t \neq 0$ we have

$$\zeta^3(u) F_1^4(u + it) F_2(u + 2it)$$

$$= \exp\left\{ \sum_{p} \sum_{k=1}^{\infty} \frac{1}{k} \left\{ \frac{3}{p^{ku}} + \frac{4f(p)^k}{p^{k(u+it)}} + \frac{f(p)^{2k}}{p^{k(u+2it)}} \right\} \right\}$$

$$= \exp\left\{ \sum_{p} \sum_{k=1}^{\infty} \frac{1}{k \cdot p^{ku}} \{ 3 + 4f(p)^k p^{-kit} + (f(p)^k p^{-kit})^2 \} \right\}$$

and so

$$| \zeta^3(u) F_1^4(u + it) F_2(u + 2it) |$$

$$= \exp\left\{ \sum_{\substack{p \\ f(p) = 0}} \sum_{k=1}^{\infty} \frac{3}{kp^{ku}} + \sum_{\substack{p \\ |f(p)| = 1}} \frac{1}{kp^{ku}} \left\{ 3 + 4\cos\varphi + \cos 2\varphi \right\} \right\} \ ,$$

where $\varphi = \varphi(k, p, t)$ is a real number satisfying the equality $f(p)^k p^{-ikt} = \cos\varphi + i\sin\varphi$. Since

$$3 + 4\cos\varphi + \cos 2\varphi = 2(1 + \cos\varphi)^2 \geqslant 0 \ ,$$

it follows from the last equality that

(3.12) $| \zeta^3(u) F_1^4(u + it) F_2(u + 2it) | \geqslant 1 \ .$

   Now if $1 + it$ were a zero of the function $F_1(s)$ (where in the case when the functions F_1, F_2 are not both holomorphic at $s = 1$, we assume $t \neq 0$) then we have the following estimates as u tends to 1:

$$F_1(u + it) = O(u - 1), \qquad \zeta(u) = O\left(\frac{1}{u-1}\right), \qquad F_2(u + 2it) = O(1) \ ,$$

hence

$$| \zeta^3(u) F_1^4(u + it) F_2(u + 2it) | = O(u - 1) \ ,$$

which together with formula (3.12) leads to

$$1 = O(u - 1) \ ,$$

a clear contradiction.   ∎

Applying the lemma with $f(n) = 1$, we get the assertion of the theorem. ∎

Corollary 1. *For any $s \neq 1$ we have the equality*

$$\zeta(s) = \frac{A(s)}{s-1} \quad ,$$

where $A(s)$ is an entire function non-vanishing for $\mathrm{Re}\, s \geq 1$. ∎

Corollary 2. *For* $\mathrm{Re}\, s > 1$ *we have*

$$\zeta(s) = H(s)\exp\left(\sum_p \frac{1}{p^s}\right) \quad , \qquad \log \zeta(s) = \sum_p \frac{1}{p^s} + \log H(s) \quad ,$$

where $H(s)$ is a holomorphic and non-vanishing function for $\mathrm{Re}\, s > \frac{1}{2}$. *The function* $\log \zeta(s)$ *(where the branch of the logarithm is chosen so that it assumes real values for positive reals) is holomorphic for* $\mathrm{Re}\, s \geq 1$, $s \neq 1$.

Proof. From (3.7) we get

$$\zeta(s) = \prod_p \left(1 - \frac{1}{p^s}\right)^{-1} = \exp \sum_p \sum_{k=1}^{\infty} \frac{1}{kp^{ks}}$$

$$= \exp \sum_p \frac{1}{p^s} \cdot \exp \sum_{k=2}^{\infty} \frac{1}{k} \sum_p \frac{1}{p^{ks}} \quad .$$

The series $\displaystyle\sum_{k=2}^{\infty} \frac{1}{k} \sum_p \frac{1}{p^{ks}}$ is majorized by

$$\sum_p \sum_{k=2}^{\infty} \frac{1}{p^{k\,\mathrm{Re}\,s}} = \sum_p \frac{1}{p^{2\,\mathrm{Re}\,s}} \cdot \frac{1}{1 - \dfrac{1}{p^{\mathrm{Re}\,s}}} \leq B \cdot \sum_{n=1}^{\infty} \frac{1}{n^{2\,\mathrm{Re}\,s}}$$

which is uniformly convergent for $\mathrm{Re}\, s \geq \frac{1}{2} + \epsilon$ $(\epsilon > 0)$ and by Weierstrass' theorem, represents a holomorphic function. This gives the first part of the assertion and the remaining parts follow from Theorem 3.7. ∎

Exercises

1. Let $\displaystyle\sum_{n=1}^{\infty} a_n n^{-s}$ be a Dirichlet series divergent for $s = 0$, and let

$$s(k) = \sum_{n=1}^{k} a_n \quad .$$

Prove that its abscissa of convergence is equal to

$$\limsup_{k \to \infty} \frac{\log |s(k)|}{\log k} \quad .$$

2.   Prove a result analogous to the last exercise for the abscissa of absolute convergence.

3.   Show that the result of Exercise 1 remains true for the series convergent at the point $s = 0$ if the function $s(k)$ is replaced by

$$S(k) = \sum_{n=k}^{\infty} a_n .$$

Find the corresponding result for the abscissa of absolute convergence.

4.   Prove that if $\sum_{n \leqslant x} a_n = x + o(x)$, then the series $f(s) = \sum_{n=1}^{\infty} a_n n^{-s}$ is convergent for $\operatorname{Re} s > 1$, and

$$\lim_{s \to 1} (s - 1) f(s) = 1 .$$

5.   Prove that if $\sum_{n \leqslant x} \dfrac{a_n}{n} = \log x + o(\log x)$, then the series $f(s) = \sum_{n=1}^{\infty} a_n n^{-s}$ is convergent for $\operatorname{Re} s > 1$, and $\lim_{s \to 1} (s - 1) f(s) = 1$.

6.   Prove that if $f = g * h$ and for $\operatorname{Re} s > a$ the series $G(s) = \sum_{n=1}^{\infty} g(n) n^{-s}$, $H(s) = \sum_{n=1}^{\infty} h(n) n^{-s}$ are absolutely convergent, then the series $F(s) = \sum_{n=1}^{\infty} f(n) n^{-s}$ is also absolutely convergent there, and furthermore $F(s) = G(s) H(s)$.

7.   Show that the following formulas hold for $\operatorname{Re} s > 1$:

$$\zeta^2 (s) = \sum_{n=1}^{\infty} \frac{d(n)}{n^s} , \qquad \frac{1}{\zeta(s)} = \sum_{n=1}^{\infty} \frac{\mu(n)}{n^s} ,$$

$$\frac{\zeta'(s)}{\zeta(s)} = \sum_{n=1}^{\infty} \frac{-\Lambda(n)}{n^s} , \quad \text{where} \quad \Lambda(n) = \begin{cases} \log p, & \text{if} \quad n = p^k, \\ 0, & \text{if} \quad n \neq p^k, \end{cases}$$

$$\frac{\zeta(2s)}{\zeta(s)} = \sum_{n=1}^{\infty} \frac{(-1)^{\Omega(n)}}{n^s} , \qquad \frac{\zeta^2(s)}{\zeta(2s)} = \sum_{n=1}^{\infty} \frac{2^{\omega(n)}}{n^s} .$$

$$\frac{\zeta(s)}{\zeta(2s)} = \sum_{n=1}^{\infty} \frac{|\mu(n)|}{n^s} , \qquad \frac{\zeta^4(s)}{\zeta(2s)} = \sum_{n=1}^{\infty} \frac{d^2(n)}{n^s} .$$

§3. A Tauberian theorem

1. In order to utilize the known analytic properties of the zeta-function to number theory we shall now prove Ikehara-Delange's Tauberian theorem which allows us to deduce from the analytic properties of the sum of a Dirichlet series the behavior of its coefficients. This theorem enables us, among other things, to find an asymptotic formula

for the number of primes not exceeding x because the analytic properties of the corresponding Dirichlet series, that is those of the series

$$\sum_p p^{-s}$$

can be obtained from Corollary 2 to Theorem 3.7.

The proof of this Tauberian theorem uses some facts from the so-called "hard" analysis (among other things the Laplace transform) and the reader who is solely interested in the arithmetical applications can omit it without any harm, and read only the formulation of the theorem and corollaries following from it.

Theorem 3.8 (H. Delange [1]). *Let $b(t)$ be a measurable function, bounded on any finite interval and moreover let the integral*

$$G(s) = \int_0^\infty b(t) \exp(-st)\, dt$$

be convergent for $\operatorname{Re} s > 0$. *Let $a(t)$ be a non-negative and non-decreasing function, and a number a be chosen so that the integral*

$$f(s) = \int_0^\infty a(t) \exp(-st)\, dt$$

is convergent for $\operatorname{Re} s > a$. *Let us denote the difference $f(s) - G(s-a)$ by $F(s)$ and suppose that there exist a non-negative integer p and a function $c(u)$ which is continuous and non-decreasing in some interval $[0, l]$, positive in the interval $(0, l]$ and satisfies the following four conditions:*

(a) *The integral* $\displaystyle\int_0^l \log \frac{1}{c(u)}\, du$ *is convergent.*

(b) *As t tends to infinity, the expression*

$$t^{1+p} b(t) \int_0^l c(u) \exp(-tu)\, du$$

tends to 1.

(c) *The function $F(s)$ is holomorphic for* $\operatorname{Re} s \geqslant a$, $s \neq a$.

(d) *There exists a function $\psi(t)$ which is non-increasing in some interval $(0, c)$ such that both the integrals*

$$\int_0^c \psi(t)\, dt, \qquad \int_0^c \psi(t) \log \frac{1}{c(t)}\, dt$$

are convergent, and moreover for $\operatorname{Re} s > 0$ *we have*

$$F^{(p)}(a+s) = O\!\left(\frac{\psi(|s|)}{c(|s|)} \right) \qquad \text{as } s \to 0.$$

Then as t *tends to infinity, we have*

$$a(t) = (1 + o(1)) \exp(at) b(t) \quad .$$

Proof. For $\mathrm{Re}\, s > a$ we define the function $h(s)$ by the formula

$$h(s) = (-1)^{p+1} \int_0^l F^{(1+p)}(s+u) c(u)\, du \quad .$$

It is clearly holomorphic for $\mathrm{Re}\, s > a$, and integrating by parts, we obtain

$$h(s) = (-1)^{p+1} F^{(p)}(s+l) c(l) + (-1)^p F^{(p)}(s) c(0) + (-1)^p \cdot$$

$$\int_0^l F^{(p)}(s+u)\, dc(u) \quad ,$$

which shows that for y real and different from zero, the limit

(3.13) $\lim_{\epsilon \to 0+0} h(a + \epsilon + iy) = g(y)$

exists and the convergence is uniform in every closed interval not containing zero. This proves the continuity of the function $g(y)$ for $y \neq 0$.

Lemma 3.3. *For every* y *there exist a positive number* T *and a function* $\varphi_y(t)$ *defined in the interval* $[0, T]$, *non-negative and integrable in* $(0, T]$ *satisfying*

$$h(a + iy + s) = O(\varphi_y(|s|)) \quad .$$

Proof of the lemma. From conditions (c) and (d) it follows that for a suitable T and $|s| \leq T$ we can choose for every y a function $\psi_y(t)$ (which satisfies the same conditions as does the function ψ defined in condition (d)) such that

$$F^{(p)}(a + iy + s) = O\left(\frac{\psi_y(|s|)}{c(|s|)}\right) \quad .$$

In fact, for $y = 0$ this is just condition (d), and for $y \neq 0$ we can take $\psi_y(t) = 1$ in view of condition (c). Now fix y and choose a positive number $\eta < l$ so that with some M for $0 \leq r = |s| \leq 2\eta$ the estimate

$$|F^{(p)}(a + iy + s)| \leq \frac{M \psi_y(r)}{c(r)}$$

holds. Now let

$$I_1 = \int_0^\eta F^{(p+1)}(a + iy + s + u) c(u)\, du$$

and

$$I_2 = \int_\eta^l F^{(p+1)} (a + iy + s + u) c(u) \, du \quad .$$

Obviously we have $(-1)^{p+1} h(a + iy + s) = I_1 + I_2$, moreover, the integral I_2 is $O(1)$

as s tends to zero because $\int_\eta^l c(u) \, du = O(1)$ and the function $F^{(p+1)}(s)$ is bounded

in a neighborhood of zero as it is holomorphic there. In order to estimate the integral I_1 we transform it, integrating by parts, into the form

$$I_1 = F^{(p)} (a + iy + s + \eta) c(\eta) - F^{(p)} (a + iy + s) c(0)$$

$$- \int_0^\eta F^{(p)} (a + iy + s + u) \, dc(u) \quad .$$

If now $|s| \leqslant \eta$, then in view of the monotonicity of functions ψ_y and c we have the inequality for $0 \leqslant u \leqslant \eta$

$$|F^{(p)} (a + iy + s + u) c(u)| \leqslant M \frac{\psi_y(|s + u|) c(u)}{c(|s + u|)}$$

$$\leqslant \frac{M \psi_y(r) c(u)}{c(\sqrt{r^2 + u^2})} \leqslant M \psi_y(r) \quad ,$$

and therefore

$$I_1 \leqslant M \psi_y(r) \int_0^\eta \frac{dc(u)}{c(\sqrt{r^2 + u^2})}$$

But

$$\int_0^\eta \frac{dc(u)}{c(\sqrt{r^2 + u^2})} = \frac{c(\eta)}{c(\sqrt{r^2 + \eta^2})} - \frac{c(0)}{c(r)} + \int_0^\eta c(u) d\left(\frac{-1}{c(\sqrt{r^2 + u^2})}\right)$$

$$\leqslant 1 + \int_0^\eta c(\sqrt{r^2 + u^2}) d\left(\frac{-1}{c(\sqrt{r^2 + u^2})}\right) = 1 + \log \frac{c(\sqrt{r^2 + \eta^2})}{c(r)}$$

$$\leqslant 1 + \log\left(\frac{c(\eta\sqrt{5})}{c(r)}\right) \quad ,$$

and this gives

$$h(a + iy + s) = O(I_1 + I_2) = O(\varphi_y(|s|)) \quad ,$$

where

$$\varphi_y(t) = \begin{cases} \psi_y(t) \log \dfrac{1}{c(t)} & \text{if } c(0) = 0 \ , \\[2em] \psi_y(t) & \text{if } c(0) > 0 \ . \end{cases}$$

Therefore we find that the function $\varphi_y(t)$ is positive and non-increasing in the interval $[0, T]$ for some T, and furthermore it is integrable in this interval. Hence the lemma has been proved. ∎

Corollary 1. *If* $Y_1 < 0 < Y_2$ *and the function* $\vartheta(y)$ *is continuous in* $[Y_1, Y_2]$, *then*

$$\lim_{\epsilon \to 0+0} \int_{Y_1}^{Y_2} h(a + \epsilon + iy)\, \vartheta(y)\, dy = \int_{Y_1}^{Y_2} g(y)\, \vartheta(y)\, dy \ .$$

Proof. Follows at once from Lemma 3.3, Lebesgue's integration theorem and formula (3.13). ∎

Corollary 2. *If* λ *is a positive number and* $k(y)$ *is a function continuous in the interval* $[-2\lambda, 2\lambda]$, *then for all real* ξ

$$\lim_{\epsilon \to 0+0} \int_{-2\lambda}^{2\lambda} h(a + \epsilon + iy)\, k(y) \exp(i\xi y)\, dy = \int_{-2\lambda}^{2\lambda} g(y)\, k(y) \exp(i\xi y)\, dy \ .$$

Proof. This follows immediately from Corollary 1. ∎

Lemma 3.4. *If the functions* $k(y)$ *and* λ *satisfy the conditions of Corollary 2 to Lemma 3.3. and*

$$K(u) = \int_{-2\lambda}^{2\lambda} k(y) \exp(iuy)\, dy \in L_1(-\infty, \infty) \ ,$$

then with the notation

$$v(t) = \int_0^l c(u) \exp(-tu)\, du \ ,$$

the integral

$$\int_0^\infty \exp(-\epsilon t)\, K(\xi - t)\, t^{1+p}\, v(t) \exp(-at)\, a(t)\, dt$$

tends to

$$\int_{-2\lambda}^{2\lambda} g(y)\, k(y) \exp(i\xi y)\, dy + \int_0^\infty K(\xi - t)\, t^{1+p}\, v(t)\, b(t)\, dt$$

as ϵ *tends to zero.*

Proof. Note that for $\mathrm{Re}\,s > a$ we have

(3.14) $$h(s) = \int_0^\infty \exp(-st)\,t^{p+1}\,v(t)\,[a(t) - \exp(at)\,b(t)]\,dt$$

because we obtain

$$(-1)^{1+p}\,F^{(1+p)}(s) = \int_0^\infty \exp(-st)\,t^{1+p}\,[a(t) - \exp(at)\,b(t)]\,dt$$

by the $(p+1)$-times differentiation of

$$F(s) = \int_0^\infty \exp(-st)\,[a(t) - \exp(at)\,b(t)]\,dt \quad .$$

Observe that the integral (3.14) is uniformly convergent in the strip $|\mathrm{Im}\,s| \leqslant 2\lambda$.

Now we substitute $s = a + \epsilon + iy$ into (3.14), multiply both sides of the obtained equality by $k(y)\exp(i\xi y)$ and integrate with respect to y over the interval $[-2\lambda, 2\lambda]$. Then we get

$$\int_{-2\lambda}^{2\lambda} h(a + \epsilon + iy)\,k(y)\exp(i\xi y)\,dy$$

$$= \int_0^\infty \exp(-(a+\epsilon)t)\,K(\xi - t)\,t^{p+1}\,v(t)\,[a(t) - \exp(at)\,b(t)]\,dt \quad ,$$

whence, adding to both sides the integral,

$$\int_0^\infty \exp(-\epsilon t)\,K(\xi - t)\,t^{p+1}\,v(t)\,b(t)\,dt \quad ,$$

which is absolutely convergent, because $K(\xi - t)$ is bounded and $b(t)\,v(t)\,t^{p+1}$ tends to 1 as t tends to infinity, we get

$$\int_{-2\lambda}^{2\lambda} h(a + \epsilon + iy)\,k(y)\exp(i\xi y)\,dy + \int_0^\infty \exp(-\epsilon t)\,K(\xi - t)\,t^{p+1}\,v(t)\,b(t)\,dt$$

$$= \int_0^\infty \exp(-\epsilon t)\,K(\xi - t)\,t^{p+1}\,v(t)\exp(-at)\,a(t)\,dt \quad .$$

From the integrability of $K(u)$ we infer the absolute convergence of the integral

$$\int_0^\infty K(\xi - t)\,t^{p+1}\,v(t)\,b(t)\,dt \quad ,$$

and moreover,

$$\lim_{\epsilon \to 0} \int_0^\infty \exp(-\epsilon t)\,K(\xi - t)\,t^{p+1}\,v(t)\,b(t)\,dt = \int_0^\infty K(\xi - t)\,t^{p+1}\,v(t)\,b(t)\,dt \quad .$$

Therefore for all positive ϵ, the integral

$$\int_0^\infty \exp(-\epsilon t)\, K(\xi - t)\, t^{p+1}\, v(t)\exp(-at)a(t)\,dt$$

is convergent and tends to

$$\int_{-2\lambda}^{2\lambda} g(y)\,K(y)\exp(i\xi y)\,dy + \int_0^\infty K(\xi - t)\,t^{p+1}\,v(t)b(t)\,dt$$

as ϵ tends to zero. This completes the proof of the lemma. ∎

Lemma 3.5. *For every real* ξ, *the integral*

$$(3.15) \qquad A(\xi) = \int_0^\infty \frac{\sin^2(\lambda(t - \xi))}{\lambda(t - \xi)^2}\, t^{p+1}\, v(t)\exp(-at)a(t)\,dt$$

is convergent and tends to π *as* ξ *tends to infinity.*

Proof. We apply the preceding lemma to the function

$$k(y) = \tfrac{1}{2}\left(1 - \frac{|y|}{2\lambda}\right) .$$

Then

$$K(u) = \frac{\sin^2(\lambda u)}{\lambda u^2} .$$

Since the integrand in (3.15) is non-negative, this integral is convergent, and

$$\lim_{\epsilon \to 0} \int_0^\infty \exp(-\epsilon t)\, \frac{\sin^2(\lambda(t - \xi))}{\lambda(t - \xi)^2}\, t^{p+1}\, v(t)\exp(-at)a(t)\,dt$$

$$= \tfrac{1}{2}\int_{-2\lambda}^{2\lambda} g(y)\left(1 - \frac{|y|}{2\lambda}\right)\exp(i\xi y)\,dy + \int_0^\infty \frac{\sin^2\lambda(t - \xi)}{\cdot\lambda(t - \xi)^2}\, t^{p+1}\, v(t)b(t)\,dt$$

The first term of this sum tends to zero as ξ tends to infinity according to the Riemann-Lebesgue theorem, and the second term can be written in the form

$$\int_{-\infty}^\infty \frac{\sin^2\lambda u}{\lambda u^2}\, W_\xi(u)\,du ,$$

where

$$W_\xi(u) = \begin{cases} (\xi + u)^{p+1}\, v(\xi + u)\, b(\xi + u) & \text{when } u \geqslant -\xi , \\[2mm] 0 & \text{when } u < -\xi . \end{cases}$$

In view of the boundedness of the function $W_\xi(u)$ and $\lim\limits_{\xi \to \infty} W_\xi(u) = 1$, we obtain

$$\lim_{\xi \to \infty} \int_{-\infty}^{\infty} \frac{\sin^2 \lambda u}{\lambda u^2} \, W_\xi(u) \, du = \int_{-\infty}^{\infty} \frac{\sin^2 \lambda u}{\lambda u^2} \, du = \pi \quad,$$

which completes the proof of the lemma. ∎

Now we can complete the proof of the Tauberian theorem. Let

$$J(t) = t^{p+1} \, v(t) \exp(-at) a(t) \quad.$$

We must show that

(3.16) $$\lim_{t \to \infty} J(t) = 1 \quad.$$

The function $J(t)$ is clearly non-negative. We shall show that it is bounded for $t > 0$. To this end we choose any positive number h and any number $\xi > h$. Then for every $\lambda > 0$ we have

$$\int_0^{\infty} \frac{\sin^2 \lambda(t-\xi)}{\lambda(t-\xi)^2} \, J(t) \, dt \geq \int_{\xi-h}^{\xi+h} \frac{\sin^2 \lambda(t-\xi)}{\lambda(t-\xi)^2} \, J(t) \, dt \quad,$$

but for $t \in [\xi - h, \, \xi + h]$ we have

$$J(t) \geq (\xi - h)^{p+1} \, v(\xi + h) \exp(-(\xi + h) a) \, a(\xi - h)$$

$$= \; J(\xi - h) \exp(-2ah) \, \frac{v(\xi + h)}{v(\xi - h)} \quad.$$

therefore

$$\int_0^{\infty} \frac{\sin^2 \lambda(t-\xi)}{\lambda(t-\xi)^2} \, J(t) \, dt \geq J(\xi - h) \exp(-2ah) \, \frac{v(\xi + h)}{v(\xi - h)} \int_{\xi-h}^{\xi+h} \frac{\sin^2 \lambda(t-\xi)}{\lambda(t-\xi)^2} \, dt$$

$$\geq J(\xi - h) \exp(-2ah) \, \frac{v(\xi + h)}{v(\xi - h)} \int_{-\lambda h}^{\lambda h} \frac{\sin^2 u}{u^2} \, du \quad,$$

whence, in turn, we obtain

$$J(\xi - h) \leq \exp(2ah) \frac{v(\xi - h)}{v(\xi + h)} \cdot \frac{1}{\displaystyle\int_{-\lambda h}^{\lambda h} \frac{\sin^2 u \, du}{u^2}} \int_0^{\infty} \frac{\sin^2 \lambda(t - \xi)}{\lambda(t - \xi)^2} \, J(t) \, dt \quad.$$

Letting first ξ and next λ tend to infinity, we get

(3.17) $$\limsup_{t \to \infty} J(t) = \limsup_{\xi \to \infty} J(\xi - h) \leq 1 \quad,$$

and hence $J(t) = O(1)$.

Now if

$$M = \sup\left\{ J(t)\colon\ t \geqslant 0 \right\} \ ,$$

then for $\xi > h > 0$ and $\lambda > 0$ we have

(3.18) $$\int_0^\infty \frac{\sin^2 \lambda(t-\xi)}{\lambda(t-\xi)^2} J(t)\,dt \leqslant M \int_0^{\xi-h} \frac{\sin^2 \lambda(t-\xi)}{\lambda(t-\xi)^2}\,dt$$

$$+ \int_{\xi-h}^{\xi+h} \frac{\sin^2 \lambda(t-\xi)}{\lambda(t-\xi)^2} J(t)\,dt + M \int_{\xi+h}^\infty \frac{\sin^2 \lambda(t-\xi)}{\lambda(t-\xi)^2}\,dt \ .$$

The sum of the first and the third terms on the right-hand side does not exceed

$$2M \int_{\lambda h}^\infty \frac{\sin^2 u}{u^2}\,du \ ,$$

and we estimate the second term, by repeating the idea used in the proof of (3.17). For t in the interval $[\xi - h, \xi + h]$ we have

$$J(t) \leqslant (\xi + h)^{p+1}\, v(\xi - h)\exp(-(\xi - h)a)\, a(\xi + h)$$

$$= J(\xi + h)\exp(2ah)\,\frac{v(\xi - h)}{v(\xi + h)} \ ,$$

and so the second term does not exceed

$$J(\xi + h)\exp(2ah)\,\frac{v(\xi - h)}{v(\xi + h)} \int_{\xi-h}^{\xi+h} \frac{\sin^2 \lambda(t-\xi)}{\lambda(t-\xi)^2}\,dt$$

$$= J(\xi + h)\exp(2ah)\,\frac{v(\xi - h)}{v(\xi + h)} \int_{-\lambda h}^{\lambda h} \frac{\sin^2 u}{u^2}\,du \ .$$

Altogether, the right-hand side of (3.18) is at most equal to

$$2M \int_{\lambda h}^\infty \frac{\sin^2 u}{u^2}\,du + J(\xi + h)\exp(2ah)\,\frac{v(\xi - h)}{v(\xi + h)} \int_{-\lambda h}^{\lambda h} \frac{\sin^2 u}{u^2}\,du$$

which implies

$$J(\xi + h) \geqslant \exp(-2ah)\,\frac{v(\xi + h)}{v(\xi - h)} \cdot \frac{1}{\displaystyle\int_{-\lambda h}^{\lambda h} \frac{\sin^2 u}{u^2}\,du} \left(\int_0^\infty \frac{\sin^2 \lambda(t-\xi)}{\lambda(t-\xi)^2} J(t)\,dt \right.$$

$$\left. - 2M \int_{\lambda h}^\infty \frac{\sin^2 u}{u^2}\,du \right) \ .$$

From this it follows that

$$\liminf_{t \to \infty} J(t) = \liminf_{\xi \to \infty} J(\xi + h) \geqslant \frac{\exp(-2ah)}{\displaystyle\int_{-\lambda h}^{\lambda h} \frac{\sin^2 u}{u^2}\, du} \left(\pi - 2M \int_{\lambda h}^{\infty} \frac{\sin^2 u}{u^2}\, du \right)$$

and analogously to the proof of formula (3.17), we conclude that $\liminf_{t \to \infty} J(t) \geqslant 1$.

Therefore $\lim_{t \to \infty} J(t) = 1$. ∎

2.   From the general theorem of Ikehara-Delange we shall now deduce some special theorems which have direct applications to some arithmetical problems. The applications will be stated in the next section.

In the following theorems and corollaries, we shall always make the many-valued function

$$(s - a)^w = \exp(w \log(s - a)) \qquad (s - \text{complex}) \quad,$$

single-valued by choosing that branch of $\log(s - a)$ which takes real values for real $s > a$.

Theorem 3.9. *Let w be a positive number. If $a(t)$ is a non-negative and non-decreasing function defined for $t \geqslant 0$, and the integral*

$$f(s) = \int_0^{\infty} \exp(-st)\, a(t)\, dt$$

is convergent in the half-plane $\operatorname{Re} s > a$, *where we have*

$$f(s) = g(s)\, (s - a)^{-w} + h(s) \quad,$$

(where the functions $g(s)$ and $h(s)$ are holomorphic in the half-plane $\operatorname{Re} s \geqslant a$*), and furthermore $g(a) \neq 0$, then*

$$\lim_{t \to \infty} a(t) \exp(-at)\, t^{1-w} = g(a) \cdot \Gamma(w)^{-1} \quad.$$

Proof. Let us take

$$b(t) = \begin{cases} 0 & \text{for } 0 \leqslant t < 1 \ , \\[2ex] \dfrac{g(a)\, t^{w-1}}{\Gamma(w)} & \text{for } t \geqslant 1 \end{cases}$$

and evaluate

$$G(s) = \int_0^{\infty} \exp(-st)\, b(t)\, dt = \frac{g(a)}{\Gamma(w)} \int_1^{\infty} t^{w-1} \exp(-st)\, dt \ .$$

Since for $w > 0$ we have

$$\int_0^\infty \exp(-st)\, t^{w-1}\, dt = \Gamma(w)\, s^{-w} \quad,$$

it follows that

$$G(s) = g(a)\, s^{-w} - g(a)\,\Gamma(w)^{-1}\int_0^1 \exp(-st)\, t^{w-1}\, dt = g(a)\, s^{-w} + H(s) \quad,$$

where $H(s)$ is an entire function. From this we conclude that

$$F(s) = f(s) - G(s-a) = \frac{g(s) - g(a)}{(s-a)^w} + H_1(s) = \frac{g_1(s)}{(s-a)^{w-1}} + H_2(s) \quad,$$

where $H_1(s)$ and $H_2(s)$ are entire, and $g_1(s)$ is holomorphic for $\operatorname{Re} s \geqslant a$.

The First Case: $w \geqslant 1$.

We take $p = 0$, $l = \frac{1}{2}$, $c(u) = u^{w-1}/g(a)$ and $\psi(u) = 1$ (note that $g(a)$ is positive because the non-negativity of $a(t)$ implies that $f(s) \geqslant 0$ for all real $s > a$). Let us show that the assumptions of Theorem 3.8 are satisfied. The conditions (a), (c) and (d) in this case are clear, and in order to check condition (b) we shall evaluate the integral

$$\int_0^{\frac{1}{2}} \exp(-tu)\, c(u)\, du = g(a)^{-1}\int_0^{\frac{1}{2}} \exp(-tu)\, u^{w-1}\, du \quad.$$

This evaluation makes use of the following lemma which will also be useful subsequently:

Lemma 3.6. *If $\rho \geqslant 0$, and m is a non-negative integer, then for t tending to infinity, we have*

$$\int_0^{\frac{1}{2}} \exp(-tu)\, u^\rho \left(\log\frac{1}{u}\right)^{-m} du = (\Gamma(\rho+1) + o(1))\, t^{-\rho-1}\,(\log t)^{-m} \quad.$$

Proof of the lemma. For $t > 4$ we shall make the substitution $u = x/t$. Then we may write

$$\int_0^{\frac{1}{2}} \exp(-tu)\, u^\rho \left(\log\frac{1}{u}\right)^{-m} = t^{-\rho-1}\int_0^{t^{\frac{1}{2}}} \exp(-x)\, x^\rho \left(\log\frac{t}{x}\right)^{-m} dx$$

$$+ t^{-\rho-1}\int_{t^{\frac{1}{2}}}^{\frac{1}{2}t} \exp(-x)\, x^\rho \left(\log\frac{t}{x}\right)^{-m} dx \quad.$$

We estimate the second of those integrals appearing here from above by

$$\left(\tfrac{1}{2}t - t^{\frac{1}{2}}\right) t^{-1}\left(\frac{t}{2}\right)^\rho (\log 2)^{-m} \exp(-\sqrt{t}) = o\left(t^{-\rho-1}(\log t)^{-m}\right) \quad,$$

and we rewrite the first in the form

$$t^{-\rho-1} (\log t)^{-m} \int_0^\infty w_t(x)\,dx \quad ,$$

where

$$w_t(x) = \begin{cases} \exp(-x)x^\rho \left(\dfrac{\log t}{\log t/x} \right)^m & \text{for } 0 < x \leqslant t^{\frac{1}{2}} \quad , \\[3ex] 0 & \text{for } x > t^{\frac{1}{2}} \quad . \end{cases}$$

In view of $0 < w_t(x) \leqslant 2^m \exp(-x)x^\rho$ and

$$\lim_{t \to \infty} w_t(x) = \exp(-x)x^\rho$$

(uniformly in each compact subinterval of $(0, \infty)$), we have

$$\lim_{t \to \infty} \int_0^\infty w_t(x)\,dx = \int_0^\infty \exp(-x)x^\rho\,dx = \Gamma(1 + \rho) \quad . \quad \blacksquare$$

We can therefore apply the Tauberian theorem to obtain the assertion of our theorem in the case $w \geqslant 1$.

The Second Case: $0 < w < 1$.

We take $p = 1$, $l = \frac{1}{2}$ and

$$c(u) = u^w \, w^{-1} \, g(a)^{-1} \quad .$$

($g(a)$ is positive by the same reasoning as in the first case). Here, it also suffices to check that condition (b) is satisfied. For this purpose, it suffices to note that from Lemma 3.6 it follows that

$$\int_0^{\frac{1}{2}} c(u)\exp(-tu)\,du = g(a)^{-1}\, w^{-1} \int_0^{\frac{1}{2}} u^w \exp(-tu)\,du = \left(\frac{\Gamma(w)}{g(a)} + o(1) \right) t^{-w-1} \quad .$$

Applying the Tauberian theorem, we obtain our assertion. $\quad \blacksquare$

Corollary. *Let the Dirichlet series* $f(s) = \displaystyle\sum_{n=1}^\infty a_n n^{-s}$ *with non-negative coefficients be convergent for* $\mathrm{Re}\,s > a > 0$. *Assume that in its domain of convergence,*

$$f(s) = g(s)\,(s - a)^{-w} + h(s) \quad ,$$

holds, where $g(s), h(s)$ *are holomorphic functions in the closed half-plane* $\mathrm{Re}\,s \geqslant a$, $g(a) \neq 0$, *and* $w > 0$.

Then for x tending to infinity, we have

$$\sum_{n \leqslant x} a_n = \left(\frac{g(a)}{a\Gamma(w)} + o(1)\right) x^a (\log x)^{w-1} \quad .$$

Proof. We shall make use of the lemma which allows us to write the Dirichlet series in the form of an integral:

Lemma 3.7. *If a_n is an arbitrary sequence of complex numbers and $a(t)$ is defined, for non-negative real t, by*

$$a(t) = \sum_{n \leqslant \exp t} a_n \quad ,$$

then for each complex s such that the series $\sum_{n=1}^{\infty} a_n n^{-s}$ is convergent, we have

$$\sum_{n=1}^{\infty} a_n n^{-s} = s \int_0^{\infty} \exp(-st) a(t) \, dt \quad .$$

The proof of the lemma follows from the chain of identities:

$$\sum_{n=1}^{\infty} a_n n^{-s} = a_1 + \sum_{n=2}^{\infty} a_n n^{-s} = a_1 + \sum_{n=2}^{\infty} [a(\log n) - a(\log(n-1))] n^{-s}$$

$$= \sum_{n=1}^{\infty} \frac{a(\log n)}{n^s} - \sum_{n=1}^{\infty} \frac{a(\log n)}{(n+1)^s} = \sum_{n=1}^{\infty} a(\log n)\left(\frac{1}{n^s} - \frac{1}{(n+1)^s}\right)$$

$$= \sum_{n=1}^{\infty} a(\log n) \cdot s \int_{\log n}^{\log(n+1)} \exp(-st) \, dt$$

$$= \sum_{n=1}^{\infty} s \int_{\log n}^{\log(n+1)} \exp(-st) a(t) \, dt = s \int_0^{\infty} \exp(-st) a(t) \, dt. \quad \blacksquare$$

The Corollary follows immediately from the lemma and the preceding theorem because $a(t)$ is clearly non-negative and non-decreasing. ∎

Theorem 3.10. *Let w be a non-negative number. If $a(t)$ is a non-negative and non-decreasing function defined for $t \geqslant 0$ and the integral*

$$f(s) = \int_0^{\infty} \exp(-st) a(t) \, dt$$

is convergent for $\operatorname{Re} s > a$, and, in this half-plane we may write

$$f(s) = (s-a)^{-w} \sum_{j=0}^{q} g_j(s) \log^j \frac{1}{s-a} + h(s) \quad ,$$

where $q \geqslant 1$, the functions g_j and h are holomorphic for $\operatorname{Re} s \geqslant a$, $g_q(a) \neq 0$, and *in the case* $w = 0$, *we have* $g_0(s) = 0$, *then for a positive* w *we have*

$$\lim_{t \to \infty} a(t) \exp(-at) t^{1-w} (\log t)^{-q} = \frac{g_q(a)}{\Gamma(w)} \quad ,$$

and for $w = 0$,

$$\lim_{t \to \infty} a(t) \exp(-at) t (\log t)^{1-q} = q g_q(a) \quad .$$

Proof. We shall apply the Tauberian theorem by taking

$$b(t) = \sum_{j=0}^{q} g_j(a) b_{w,j}(t) \quad ,$$

where for $\langle w, j \rangle \neq \langle 0, 0 \rangle$, we have

$$(3.19) \quad b_{w,j}(t) = \begin{cases} 0 & \text{when } 0 < t < 1, \\[2ex] t^{w-1} \sum_{i=0}^{j} \binom{j}{i} \log^{j-i} t \, \dfrac{d^i}{dw^i}\left(\dfrac{1}{\Gamma(w)}\right) , & \text{when } t \geqslant 1. \end{cases}$$

To check the conditions of the Tauberian theorem we shall need a lemma:

Lemma 3.8. *If* $b_{w,j}(t)$ *is defined by formula* (3.19), *then for* $\operatorname{Re} s > 0$ *we have*

$$\int_0^\infty b_{w,j}(t) \exp(-st) \, dt = \frac{\left(\log \dfrac{1}{s}\right)^j}{s^w} + H(s) \quad ,$$

where $H(s) = H_{w,j}(s)$ *is an entire function.*

Proof of the lemma. Let us begin by considering the case $j = 0$. Since

$$b_{w,0}(t) = \begin{cases} 0 & \text{for } 0 < t < 1 , \\[2ex] \Gamma(w)^{-1} t^{w-1} & \text{for } t \geqslant 1 , \end{cases}$$

we have (recall that for $j = 0$, we have $w > 0$):

$$\Gamma(w) \int_0^\infty \exp(-st) b_{w,0}(t) \, dt = \int_1^\infty \exp(-st) t^{w-1} \, dt$$

$$= \Gamma(w) s^{-w} - \int_0^1 \exp(-st) t^{w-1} \, dt \quad ,$$

and the last integral obviously defines an entire function.

Now we shall consider the case of a positive j. Let

$$d_{w,j}(t) = \frac{d^j}{dw^j} (t^{w-1} \Gamma^{-1}(w)) \quad,$$

note that for $t \geqslant 1$, we have $d_{w,j}(t) = b_{w,j}(t)$, and that

$$\frac{d}{dt}(d_{w,j}(t)) = d_{w-1,j}(t) \quad.$$

For positive w and s, we have the equality

$$\frac{1}{\Gamma(w)} \int_0^\infty \exp(-st)\, t^{w-1}\, dt = s^{-w}$$

and, differentiating this equality j-times with respect to w, we obtain

$$\int_0^\infty \exp(-st)\, d_{w,j}(t)\, dt = s^{-w} \log^j \frac{1}{s} \quad.$$

Since both sides are holomorphic for $\mathrm{Re}\, s > 0$, this equality is valid for any complex number s in this half-plane by analytic continuation. Hence we may write for $\mathrm{Re}\, s > 0$:

$$\int_0^\infty \exp(-st)\, b_{w,j}(t)\, dt = \int_0^\infty \exp(-st)\, d_{w,j}(t)\, dt - \int_0^1 \exp(-st)\, d_{w,j}(t)\, dt$$

$$= s^{-w} \log^j \frac{1}{s} + H(s) \quad,$$

where $H(s)$ is entire.

In the case $w = 0$, integrating by parts, we obtain for $\mathrm{Re}\, s > 0$

$$\int_0^\infty \exp(-st)\, b_{0,j}(t)\, dt = \int_1^\infty \exp(-st)\, d_{0,j}(t)\, dt$$

$$= -\exp(-s)\, d_{1,j}(1) + s \int_1^\infty \exp(-st)\, d_{1,j}(t)\, dt \quad.$$

Since

$$\int_1^\infty \exp(-st)\, d_{1,j}(t)\, dt = \int_0^\infty \exp(-st)\, d_{1,j}(t)\, dt - \int_0^1 \exp(-st)\, d_{1,j}(t)\, dt$$

$$= s^{-1} \log^j \frac{1}{s} - \int_0^1 \exp(-st)\, d_{1,j}(t)\, dt \quad,$$

we obtain

$$\int_0^\infty \exp(-st)\, b_{0,j}(t)\, dt = \log^j \frac{1}{s} + H(s) \quad,$$

where $H(s)$ is an entire function.   ∎

Returning to the proof of the theorem, we note that as t tends to infinity, we have

$$b(t) = \begin{cases} \left(g_q(a)\,\Gamma(w)^{-1} + o(1)\right)t^{w-1}\log^q t \ , & \text{when } w > 0 \ , \\[2ex] \left(qg_{\stackrel{}{q}}(a) + o(1)\right)t^{-1}\,(\log t)^{q-1} \ , & \text{when } w = 0 \ , \end{cases}$$

so that the assertion of our theorem will follow from the Tauberian theorem once we check its conditions.

In view of Lemma 3.8 we have

$$G(s) = s^{-w} \sum_{j=0}^{q} g_j(a)\log^j \frac{1}{s} + H(s) \ ,$$

where $H(s)$ is some entire function, and this tells us that condition (c) is satisfied.

In the case $w \geq 1$ we see that $g_q(a)$ must be a positive number. Here we take $p = 0$, $l = \frac{1}{2}$ and

$$c(u) = \begin{cases} g_q(a)^{-1}\,u^{w-1}\log^{-q}\dfrac{1}{u} & \text{for } 0 < u \leq \frac{1}{2} \ , \\[2ex] 0 & \text{for } u = 0 \ . \end{cases}$$

This function is continuous and non-decreasing in $[0, \frac{1}{2}]$ and is positive for $u \neq 0$. The condition (a) is clearly satisfied, and from Lemma 3.6 it follows that condition (b) is also satisfied. Therefore it remains to check (d), but this follows at once from the behavior of the function $F(s)$ in a neighborhood of the point a.

In the case $0 < w < 1$, we see again that $g_q(a)$ is positive. Now we take $p = 1$, $l = \frac{1}{2}$ and

$$c(u) = \begin{cases} u^w\,w^{-1}\log^{-q}\dfrac{1}{u} & \text{for } 0 < u \leq \frac{1}{2} \ , \\[2ex] 0 & \text{for } u = 0 \ . \end{cases}$$

We can check the conditions of the Tauberian theorem in the same way as in the preceding case.

There remains to consider the case $w = 0$. We use the same method here, taking $p = 1$, $l = \frac{1}{2}$ and

$$c(u) = \begin{cases} \dfrac{1}{qg_q(a)}\log^{1-q}\dfrac{1}{u} & \text{for } 0 < u \leq \frac{1}{2} \ , \\[2ex] 0 & \text{for } q > 1, \ u = 0 \ , \\[2ex] 1 & \text{for } q = 1, \ u = 0. \end{cases}$$

Having checked that the conditions of the Tauberian theorem are satisfied, we may apply it, and this gives us our assertion. ∎

Corollary. *Let the Dirichlet series* $f(s) = \sum\limits_{n=1}^{\infty} a_n n^{-s}$ *with non-negative coefficients be convergent for* $\operatorname{Re} s > a > 0$, *and moreover, in its domain of convergence let the equality*

$$f(s) = (s-a)^{-w} \sum_{j=0}^{q} g_j(s) \log^j \frac{1}{s-a} + h(s)$$

be satisfied where $q \geqslant 1$, $w \geqslant 0$, *the functions* $g_j(s), h(s)$ *are holomorphic for* $\operatorname{Re} s \geqslant a$ *and* $g_q(a) \neq 0$. *Then, as* x *tends to infinity, we have*

$$\sum_{n \leqslant x} a_n = \begin{cases} \left(\dfrac{1}{a} g_q(a) \, \Gamma^{-1}(w) + o(1) \right) x^a (\log x)^{w-1} \, (\log\log x)^q \; , & \text{if } w > 0 \; , \\[4mm] \left(\dfrac{q g_q(a)}{a} + o(1) \right) x^a (\log x)^{-1} \, (\log\log x)^{q-1} \; , & \text{if } w = 0 \; . \end{cases}$$

Proof. Follows from the preceding theorem and Lemma 3.7. ∎

§4. The prime number theorem. Dirichlet's theorem.

1. We now apply the method which we have got acquainted with in §3 to the theory of prime numbers. In particular, we shall find an asymptotic formula for the number of primes not exceeding x as $x \to \infty$, investigate the asymptotic behavior of the numbers having the given number of prime divisors etc. The Tauberian theorem allows us to find asymptotic formulas, but it does not give any estimate of the remainders. For that purpose, other methods, for example the method of complex integration, are used, which, however, goes beyond the extent of this book. The interested reader can refer to K. Prachar [1] or W. J. Ellison [2].

We shall begin with the prime number theorem:

Theorem 3.11 (J. Hadamard [1], C. de la Vallée Poussin [1]).

(i) *If* $\operatorname{Re} s > 1$, *then*

$$\sum_p \frac{1}{p^s} = \log \frac{1}{s-1} + g(s) \, ,$$

where $g(s)$ *is a function holomorphic for* $\operatorname{Re} s \geqslant 1$.

(ii) *If* $\pi(x)$ *denotes the number of primes not exceeding* x, *then as* $x \to \infty$,

$$\pi(x) = \frac{(1 + o(1))x}{\log x} \; .$$

Proof. By Corollary 2 to Theorem 3.7, we have for $\mathrm{Re}\,s > 1$

$$\sum_p p^{-s} = \log \zeta(s) + g(s) \ ,$$

where $g(s)$ is a holomorphic function for $\mathrm{Re}\,s \geqslant 1$. Next, from Corollary 1 to the same theorem, it follows that

$$\log \zeta(s) = \log \frac{1}{s-1} + g_1(s) \ ,$$

where $g_1(s)$ is also holomorphic for $\mathrm{Re}\,s \geqslant 1$. We thus see that

$$\sum_p p^{-s} = \log \frac{1}{s-1} + g_2(s) \ ,$$

where $g_2(s)$ is holomorphic for $\mathrm{Re}\,s \geqslant 1$. This gives (i), and applying Corollary to Theorem 3.10, we get

$$\pi(x) = \sum_{p \leqslant x} 1 = (1 + o(1)) \frac{x}{\log x} \ . \quad \blacksquare$$

By more complicated methods, it is possible to prove that

$$\pi(x) = \mathrm{li}\,x + O\left(x \exp \left\{ \frac{-C \log^a x}{(\log \log x)^b} \right\} \right) \ ,$$

where $\mathrm{li}\,x = \displaystyle\int_2^\infty \frac{dt}{\log t} + O(1)$, $a = \frac{3}{5}$, $b = \frac{1}{5}$, $C > 0$ (see, e.g. A. Walfisz [2]).

The reader can easily check the relation using integration by parts:

$$\mathrm{li}\,x = \frac{x}{\log x} + \frac{x}{\log^2 x} + \ldots + \frac{(n-1)!\,x}{\log^n x} + O\left(\frac{x}{\log^{n+1} x} \right),$$

so that the result given above implies the estimate

$$\pi(x) = \sum_{k=1}^n \frac{(k-1)!\,x}{\log^k x} + O\left(\frac{x}{\log^{n+1} x} \right)$$

for all natural numbers n.

We shall now study the problem of distribution of natural numbers with a given number of distinct prime divisors. Let $\pi_k(x)$ denote the number of integers $n \leqslant x$ for which $\omega(n) = k$. For the function $\pi_k(x)$, we shall prove the following theorem:

Theorem 3.12. *For a fixed k we have, as x tends to infinity*

$$\pi_k(x) = \left(\frac{1}{(k-1)!} + o(1) \right) \frac{x(\log \log x)^{k-1}}{\log x} \ .$$

Since $\pi_1(x) = \pi(x)$, this theorem is a generalization of Theorem 3.11.

Proof. For $|z| \leqslant \frac{1}{2}$ and $\mathrm{Re}\,s > 1$, let us consider the function

$$(3.20) \qquad F(s, z) = \sum_{n=1}^{\infty} \frac{z^{\omega(n)}}{n^s} = \sum_{k=1}^{\infty} z^k \sum_{\omega(n)=k} \frac{1}{n^s} \quad .$$

Since $\omega(n)$ is an additive function, $z^{\omega(n)}$ is multiplicative and from Corollary to Theorem 2.14 we obtain

$$F(s, z) = \prod_p \left(1 + \frac{z^{\omega(p)}}{p^s} + \frac{z^{\omega(p^2)}}{p^{2s}} + \dots\right)$$

$$= \prod_p \left(1 + \frac{z}{p^s} + \frac{z}{p^{2s}} + \dots\right) = \prod_p \left(1 + \frac{z}{p^s} \cdot \frac{1}{1 - p^{-s}}\right)$$

$$= \prod_p \left(1 + \frac{z}{p^s - 1}\right)$$

In the region under consideration, we have the inequality $|z/(p^s - 1)| < 1$, since none of the factors $1 + \dfrac{z}{p^s - 1}$ vanishes and so we have $F(s, z) \neq 0$. Hence we can write

$$F(s, z) = \exp \sum_p \log\left(1 + \frac{z}{p^s - 1}\right) = \exp\left(\sum_p \sum_{j=1}^{\infty} \frac{(-1)^{j+1}}{j} \cdot \frac{z^j}{(p^s - 1)^j}\right)$$

$$= \exp\left\{z \sum_p \frac{1}{p^s - 1}\right\} \cdot \exp\left\{\sum_p \sum_{j=2}^{\infty} \frac{(-1)^{j+1}}{j} \cdot \frac{z^j}{(p^s - 1)^j}\right\} ,$$

where $\log$ denotes that branch of the logarithm which is real for positive reals.

Functions

$$h_j(s) = \frac{(-1)^{j+1}}{j} \sum_p \frac{1}{(p^s - 1)^j}$$

are holomorphic for $j = 2, 3, \dots$ and $\mathrm{Re}\,s \geqslant 1$, and the inequality

$$|h_j(s)| \leqslant B$$

holds for a suitable positive constant B.

Furthermore

$$\sum_p \frac{1}{p^s - 1} = \sum_p \frac{1}{p^s} + \sum_p \frac{1}{p^s(p^s - 1)} = \log \frac{1}{s - 1} + h_1(s)$$

(by Corollary 2 to Theorem 3.7), where $h_1(s)$ is holomorphic for $\operatorname{Re} s \geqslant 1$, and hence

$$F(s, z) = \exp\left\{ z \log \frac{1}{s-1} \right\} \cdot \exp\left\{ \sum_{j=1}^{\infty} h_j(s) z^j \right\}$$

$$= \sum_{k=0}^{\infty} \frac{\log^k \frac{1}{s-1} \cdot z^k}{k!} \sum_{l=0}^{\infty} \frac{1}{l!} \left(\sum_{j=1}^{\infty} h_j(s) z^j \right)^l$$

$$= \sum_{k=0}^{\infty} \frac{\log^k \frac{1}{s-1} \cdot z^k}{k!} \sum_{l=0}^{\infty} g_l(s) z^l \ ,$$

where $g_l(s)$ are holomorphic for $\operatorname{Re} s \geqslant 1$, and $g_0(s) = 1$. From this, we get

$$\sum_{k=0}^{\infty} z^k \sum_{\omega(n)=k} \frac{1}{n^s} = F(s, z) = \sum_{k=0}^{\infty} \left(\sum_{n=0}^{k} g_{k-n}(s) \log^n \frac{1}{s-1} \cdot \frac{1}{n!} \right) z^k$$

and comparing the coefficients of z^k, we obtain

$$\sum_{\omega(n)=k} \frac{1}{n^s} = \sum_{n=0}^{k} \frac{1}{n!} g_{k-n}(s) \log^n \frac{1}{s-1} \ .$$

Applying Corollary to Theorem 3.10 we get the assertion of our theorem. ∎

The same method can be applied to other related problems, for example, to the determination of the number of integers $n \leqslant x$ for which $\Omega(n) = k$, or also to those for which $\Omega(n) - \omega(n) = k$. The reader will find a collection of such problems at the end of this section.

2. As we have already mentioned, to determine whether a given function of a natural argument represents infinitely many primes is a difficult problem and is solved only in a few cases. We shall now be concerned with one of the already solved problems of this type and prove *Dirichlet's theorem* which tells us that every arithmetic progression with common difference relatively prime to the initial term represents infinitely many prime numbers. We shall also find an asymptotic expression for the number of primes lying in such a progression and not exceeding x. It happens that among such progressions with a given difference, prime numbers are distributed equally.

For the proof of Dirichlet's theorem we require characters of the group $G(N)$. Let us begin with a more general definition of characters of a finite Abelian group.

Let G be a finite Abelian group. Every homomorphism $G \to T$, where T denotes the multiplicative group of complex numbers with absolute value 1, is called a *character* of G. If φ_1, φ_2 are characters of the group G and $\psi(g) = \varphi_1(g)\varphi_2(g)$ $(g \in G)$, then ψ is also a character which is called the *product of characters* φ_1, φ_2.

With such multiplication the set of all characters of G forms a group. In fact, the character $\varphi(g) = 1$ is the neutral element, and $\psi(g) = \overline{\varphi(g)} = \varphi(g^{-1})$ is the inverse element of φ. We denote the group of characters by $\hat{G}$. It is often called the *dual group* of G.

It is also possible to define characters of an infinite group in which a topology is defined. The theory of characters of topological groups is a very extensive field of mathematics having a wide range of applications. Within the framework of this theory, it is possible to deduce (as J. Tate did in 1950) in a unified way functional equations for a wide class of Dirichlet series having an arithmetic meaning. This class contains in particular the Riemann zeta-function and the L-functions, with which we shall meet in this section.

We shall be interested mainly in the characters of the group $G(N)$, that is the group of residues (mod N) relatively prime to N. Corollary 3 to Theorem 1.18 shows that the complete description of characters of $G(N)$ follows from the following theorem:

Theorem 3.13. *If* $G = \bigoplus\limits_{i=1}^{n} G_i$, *where each of the groups* G_i *is cyclic,* G_i *has* r_i *elements and* g_i *is a generator of* G_i $(i = 1, \ldots, n)$, *then the map*

$$\varphi : G \to T$$

given by

(3.21)
$$\varphi(\langle g_1^{k_1}, \ldots, g_n^{k_n} \rangle) = \prod_{j=1}^{n} \exp\left\{ \frac{2\pi i k_j l_j}{r_j} \right\} \quad ,$$

where $0 \leqslant l_i \leqslant r_i - 1$ *are fixed, and* $k_i = 0, 1, \ldots, r_i - 1$, *is a character of* G *and every character of* G *is obtained in this way. In particular, the group* $\hat{G}$ *has as many elements as* G.

Proof. Let us begin with the case $n = 1$, where G is a cyclic group, say $G = C_r$ and let g be its generator. The formula (3.21) in this case takes the form

(3.22) $\quad \varphi(g^k) = \exp\left\{ \dfrac{2\pi i k l}{r} \right\} \quad (0 \leqslant k \leqslant r-1, \ 0 \leqslant l \leqslant r-1 \text{ fixed}) \quad .$

Since from $0 \leqslant k, k' \leqslant r - 1$, it follows that

$$\varphi(g^k)\varphi(g^{k'}) = \exp\left\{ \frac{2\pi i}{r}(k + k') l \right\} = \exp\left\{ \frac{2\pi i k_0 l}{r} \right\} \quad ,$$

where $k_0 \equiv k + k' \pmod{r}$, $0 \leqslant k_0 \leqslant r - 1$, and on the other hand, we have $g^{k+k'} = g^{k_0}$, so

$$\varphi(g^{k+k'}) = \varphi(g^{k_0}) = \exp\left\{ \frac{2\pi i k_0 l}{r} \right\} \quad ,$$

hence (3.21) actually defines a certain character of the group G.

Now let φ be a character of G and put $z = \varphi(g)$. Then

$$z^r = \varphi(g^r) = \varphi(1) = 1 \quad ,$$

and so for some l satisfying $0 \leqslant l \leqslant r - 1$, we have $z = \exp(2\pi i l/r)$. But then $\varphi(g^k) = z^k = \exp(2\pi i k l/r)$ holds for every k and hence φ is of the form given by (3.22).

In the general case we shall need the following lemma:

Lemma 3.19. *If G, H are finite Abelian groups and $K = G \oplus H$, then $\hat{K} \simeq \hat{G} \oplus \hat{H}$, and every character of the group K can be expressed uniquely in the form*

$$(3.23) \qquad \varphi(\langle g, h \rangle) = \varphi_1(g)\,\varphi_2(h) \quad ,$$

where $g \in G$, $h \in H$, $\varphi_1 \in \hat{G}$, $\varphi_2 \in \hat{H}$.

Proof. By formula (3.23), a map

$$\Phi: \hat{G} \oplus \hat{H} \to \hat{K}$$

is defined. Both parts of the lemma will be proved if we can show that Φ is an isomorphism. If $a = \langle \varphi_1, \varphi_2 \rangle$, $b = \langle \psi_1, \psi_2 \rangle \in \hat{G} \oplus \hat{H}$ and $c = \langle g, h \rangle \in K$, then

$$\Phi_{ab}(c) = \varphi_1(g)\,\psi_1(g)\,\varphi_2(h)\,\psi_2(h) = \Phi_a(c)\,\Phi_b(c) \quad ,$$

and so Φ is a homomorphism. If $\langle \varphi_1, \varphi_2 \rangle$ lies in the kernel of Φ, then for any $g \in G$, $h \in H$ we have $\varphi_1(g)\,\varphi_2(h) = 1$ and taking here successively $g = 1$ and $h = 1$, we obtain $\varphi_1 = 1$ and $\varphi_2 = 1$. Therefore Φ is a monomorphism and it remains to show that it is surjective. For this purpose let φ be an arbitrary character of the group K and let us denote by q_1, q_2 the injections of G and H into K respectively given by the formulas

$$q_1(g) = \langle g, 1 \rangle \quad , \qquad q_2(g) = \langle 1, h \rangle \quad .$$

Finally, let

$$\varphi_i = \varphi \circ q_i \qquad (i = 1, 2) \quad .$$

Then

$$\varphi_1 \in \hat{G}, \qquad \varphi_2 \in \hat{H} \quad ,$$

and for $\psi = \Phi_{\varphi_1 \varphi_2}$, we have

$$\psi(\langle g, h \rangle) = \psi(\langle g, 1 \rangle) \cdot \psi(\langle 1, h \rangle) = \varphi_1(g)\,\varphi_2(h)$$

$$= \varphi(q_1(g))\,\varphi(q_2(h))$$

$$= \varphi(\langle g, 1 \rangle)\,\varphi(\langle 1, h \rangle) = \varphi(\langle g, h \rangle) \quad ,$$

and hence $\varphi = \psi$, that is, φ lies in the image of Φ. ∎

Corollary. *If* $G_1, \ldots, G_n$ *are finite Abelian groups, and* K *is their direct sum, then*

$$\hat{K} \simeq \bigoplus_{i=1}^{n} \hat{G}_i \quad ,$$

and each character φ *of the group* K *can be expressed uniquely in the form*

$$\varphi(\langle g_1, \ldots, g_n \rangle) = \prod_{i=1}^{n} \varphi_i(g_i) \qquad (\varphi_i \in \hat{G}_i, \; g_i \in G_i) \; .$$

Proof. Follows from the preceding lemma by induction. ∎

Now it is easy to complete the proof of the theorem. In fact, it suffices to apply the above corollary to the case when all the groups are cyclic and utilize the already proven part of the theorem. ∎

Corollary 1. *The group* $G(N)$ *has exactly* $\varphi(N)$ *distinct characters.*

Proof. In view of Corollary 3 to Theorem 1.18, $G(N)$ is a direct sum of cyclic groups, hence we may apply the preceding theorem. ∎

Corollary 2. *If* G *is a finite group expressed as a direct sum of cyclic groups, then for any of its element* $g \neq 1$, *there exists a character* $\varphi \in G$ *satisfying* $\varphi(g) \neq 1$.

Proof. If $g = \langle g_1^{k_1}, \ldots, g_r^{k_r} \rangle \neq 1$ $(0 \leq k_i < r_i)$, then at least one of the indices, e.g. k_t, is different from zero. If now φ is a character which written in the form (3.21) has

$$l_j = \begin{cases} 0 & \text{for } j \neq t \; , \\ \\ 1 & \text{for } j = t \; , \end{cases}$$

then

$$\varphi(g) = \exp\left\{\frac{2\pi i k_t}{n_t}\right\} \neq 1 \; . \quad \blacksquare$$

Corollary 3. *If* G *is a finite group expressed as a direct sum of cyclic groups, then*

$$(3.24) \qquad \sum_{\chi \in \hat{G}} \chi(g) = \begin{cases} |G| & \text{when } g = 1 \; , \\ \\ 0 & \text{when } g \neq 1 \; , \end{cases}$$

and

$$(3.25) \qquad \sum_{g \in G} \chi(g) = \begin{cases} |G| & \text{when } \chi = 1 \; , \\ \\ 0 & \text{when } \chi \neq 1 \; . \end{cases}$$

Proof. We begin with the proof of (3.24). If $g \neq 1$, then by Corollary 2 there exists a character χ such that $\chi(g) \neq 1$. Note that if χ runs through all the characters of G, so does $\varphi\chi$, and hence

$$\sum_{\chi \in \hat{G}} \chi(g) = \sum_{\chi \in \hat{G}} \varphi\chi(g) = \sum_{\chi \in \hat{G}} \varphi(g)\chi(g) = \varphi(g) \sum_{\chi \in \hat{G}} \chi(g) \quad ;$$

thus $\sum_{\chi \in \hat{G}} \chi(g) = 0$. The case $g = 1$ is trivial because then for every character χ, we have $\chi(g) = 1$.

The proof of the second equality is exactly the same. If $\chi = 1$, then all the terms are equal to 1 and the assertion is obvious, and if $\chi \neq 1$, then there exists an element $g_0 \in G$ such that $\chi(g_0) \neq 1$. In exactly the same way as in the proof of the preceding equality, we now obtain

$$\sum_{g \in G} \chi(g) = \sum_{g \in G} \chi(gg_0) = \chi(g_0) \sum_{g \in G} \chi(g)$$

and

$$\sum_{g \in G} \chi(g) = 0 \quad . \quad \blacksquare$$

Corollary 4. *If G is a finite group which is a direct sum of cyclic groups, and f is a complex-valued function defined on G, then for any element g_0 of G we have*

$$f(g_0) = \frac{1}{|G|} \sum_{\chi \in \hat{G}} c_\chi \overline{\chi(g_0)} \quad ,$$

where

$$c_\chi = \sum_{g \in G} f(g) \chi(g) \quad .$$

Proof. A direct calculation gives

$$\sum_{\chi \in \hat{G}} c_\chi \overline{\chi(g_0)} = \sum_{\chi \in \hat{G}} \sum_{g \in G} f(g) \chi(g) \overline{\chi(g_0)} = \sum_{\chi \in \hat{G}} \sum_{g \in G} f(g) \chi(gg_0^{-1})$$

$$= \sum_{g \in G} f(g) \sum_{\chi \in \hat{G}} \chi(gg_0^{-1}) \quad .$$

The inner sum vanishes by (3.24) except in the case when $gg_0^{-1} = 1$, i.e. $g = g_0$, and so

$$\sum_{\chi \in \hat{G}} c_\chi \overline{\chi(g_0)} = f(g_0) \cdot |G| \quad . \quad \blacksquare$$

Dirichlet did not use the notion of the character of the group $G(N)$ for the proof of his theorem; instead he considered certain arithmetical functions closely related

Number Theory

to the characters. We call these functions Dirichlet's characters nowadays, and they are defined in the following way:

If χ is a character of the group $G(k)$, then the function given by the formula

$$f(n) = \begin{cases} \chi(n \bmod k) & \text{when } (k, n) = 1 \ , \\ 0 & \text{when } (k, n) \neq 1 \end{cases}$$

is called the *Dirichlet's character* determined by the character χ.

We also say that f is a character (mod k). Dirichlet's character is denoted traditionally by the same letter as the character of $G(k)$ determining it. The Dirichlet character determined by the character $\chi(g) = 1$ on $G(k)$ is called the *principal character* (mod k) and is usually denoted by χ_0.

The following theorem gives the fundamental properties of Dirichlet's characters and allows us to define them directly, i.e. without the notion of a character of the group $G(k)$:

Theorem 3.14.

(i) *If f is a character* (mod k), *then*

    (a) $f(x + k) = f(x)$ *for all* x,

    (b) $f(n) = 0$ *iff* $(k, n) \neq 1$,·

    (c) $f(xy) = f(x) f(y)$ *holds for all integers* x, y.

(ii) *If f is a function satisfying the conditions* (a), (b) *and* (c) *for some* k, *then* f *is a character* (mod k).

(iii) *If χ is a character* (mod k), *then*

$$\sum_{n=1}^{k} \chi(n) = \begin{cases} \varphi(k) & \text{for } \chi = \chi_0 \ , \\ 0 & \text{for } \chi \neq \chi_0 \ . \end{cases}$$

(iv) *For any natural number k, we have*

$$\sum_{\chi(\bmod k)} \chi(n) = \begin{cases} \varphi(k) & \text{when } n \equiv 1 \ (\text{mod } k) \ , \\ 0 & \text{when } n \not\equiv 1 \ (\text{mod } k) \ , \end{cases}$$

where χ in the above sum runs through all characters (mod k).

Proof. Part (i) follows immediately from the definition of Dirichlet's character. If, on the other hand, a function f satisfies conditions (a), (b) and (c), then its value at the point n depends only on the residue $n (\text{mod } k)$, and hence the function defined on $G(k)$ by

$$\chi(n \ (\text{mod } k)) = f(n) \qquad (n, k) = 1 \ ,$$

is, by condition (c), a homomorphism of this group into the multiplicative group of complex numbers. Moreover, for any $g \in G(k)$

$$1 = \chi(1) = \chi(g^{\varphi(k)}) = \chi(g)^{\varphi(k)} \quad,$$

and so $\chi(g)$ is a root of unity. Therefore $|\chi(g)| = 1$, that is χ is a character of $G(k)$. It remains to observe that the Dirichlet character determined by χ coincides with f. Thus we have proved (ii). Parts (iii) and (iv) follow immediately from Corollary 3 to Theorem 3.13. ∎

Corollary 1. *If p is an odd prime, then Legendre's symbol $\left(\dfrac{a}{p}\right)$ is a Dirichlet character* (mod p). *In the case $p = 2$, the symbol $\left(\dfrac{a}{2}\right)$ is a Dirichlet character* (mod 8).

Proof. In the case $p \neq 2$, it suffices to make use of Theorem 1.25 and part (ii) of the preceding theorem, and in the case $p = 2$, verification of the conditions (a), (b) and (c) is immediate. ∎

The Dirichlet characters appearing in the above corollary take on real values. We shall call every such character a *real character* or a *quadratic character* since each of them satisfies $f^2 = 1$.

Corollary 2. *If χ is a Dirichlet character* (mod k), *different from the principal character χ_0, then for any natural number N we have*

$$\left| \sum_{n=1}^{N} \chi(n) \right| \leqslant k \quad.$$

Proof. If we write N in the form $N = kq + r$, with $q \geqslant 0$, $1 \leqslant r \leqslant k$, then we obtain

$$\left| \sum_{n=1}^{N} \chi(n) \right| = \left| \sum_{j=0}^{q-1} \sum_{n=jk+1}^{(j+1)k} \chi(n) + \sum_{n=qk+1}^{qk+r} \chi(n) \right|$$

$$= \left| q \sum_{n=1}^{k} \chi(n) + \sum_{n=qk+1}^{qk+r} \chi(n) \right| = \left| \sum_{n=qk+1}^{qk+r} \chi(n) \right| \leqslant r \leqslant k. \quad ∎$$

3. The idea of the proof of Dirichlet's theorem is simple — one considers the series

$$(3.26) \qquad F_{k,l}(x) = \sum_{p \equiv l(\text{mod } k)} p^{-x}$$

convergent for $x > 1$ and shows that its sum can be expressed in the form

$$(3.27) \qquad F_{k,l}(x) = \sum_{r=1}^{\varphi(k)} c_r g_r(x) \quad,$$

where c_r's are some complex numbers different from zero, and $g_r(x)$ are functions defined for $x > 1$ with the property that

(3.28) $\lim\limits_{x \to 1} g_1(x) = \infty$, $\lim\limits_{x \to 1} g_r(x) = a_r$ $(r \neq 1)$,

where a_r's are some complex numbers. From this it follows that

$$\lim_{x \to 1} \left| F_{k,l}(x) \right| = \infty ,$$

and hence infinitely many terms must appear in the sum (3.26), which proves the existence of infinitely many primes of the form $kn + l$.

In order to obtain an asymptotic formula, we must consider, instead of the series (3.26), the Dirichlet series

$$\sum_{p \equiv l (\mathrm{mod}\ k)} p^{-s} ,$$

but this change has no influence on the main idea of the proof. The characters which we have introduced in the preceding discussion are used for finding identities (3.27), and the most difficult point of the proof is to show that the second of the relations (3.28) holds for the function $g_r(x)$.

Before we shall be able to realize the idea described above, we must prove a series of auxiliary results of analytic type.

Let χ be any Dirichlet character. The formula

(3.29) $L(s, \chi) = \sum\limits_{n=1}^{\infty} \chi(n) n^{-s}$

defines a function holomorphic for $\mathrm{Re}\, s > 1$. In fact, the series appearing here is majorized by one defining $\zeta(\mathrm{Re}\, s)$. The functions $L(s, \chi)$ are called *Dirichlet's L-functions*. It happens that except for some trivial cases, the series (3.29) is convergent for $\mathrm{Re}\, s > 0$.

Theorem 3.15.

(i)   *If χ is not the principal character, then the series (3.29) is convergent in $\mathrm{Re}\, s > 0$ and defines a holomorphic function here.*

(ii)   *If χ is any Dirichlet character, then for $\mathrm{Re}\, s > 1$ the identity*

$$L(s, \chi) = \prod_{p} (1 - \chi(p) p^{-s})^{-1}$$

holds.

Proof. The assertion of part (i) follows from Corollary 2 to Theorem 3.14 and Corollary 2 to Theorem 2.3. Part (ii) follows from the complete multiplicativity of $\chi(n)$ together with Corollary 1 to Theorem 2.14.   ∎

Corollary 1. *In $\mathrm{Re}\, s > 1$ we have $L(s, \chi) \neq 0$.*

Proof. This is true since none of the factors $(1 - \chi(p) p^{-s})^{-1}$ vanishes there.   ∎

Corollary 2. *If* χ_0 *is the principal character* (mod k), *then for* $\text{Re}\, s > 1$, *we have*

$$L(s, \chi_0) = \zeta(s) \cdot \prod_{p \mid k} (1 - p^{-s}) \ .$$

Proof. It suffices to note that $\chi_0(p) = 1$ for $p \nmid k$, $\chi_0(p) = 0$ for $p \mid k$ and use the preceding corollary and (3.7). ∎

Corollary 3. *If* χ_0 *is the principal character* (mod k), *then* $L(s, \chi_0)$ *is a meromorphic function in the whole plane, and its sole singularity is a simple pole at* $s = 1$. ∎

Now let us prove an analogue of Theorem 3.7:

Theorem 3.16. *None of the functions* $L(s, \chi)$ *has zeros in the closed half-plane* $\text{Re}\, s \geqslant 1$.

Proof. We already know (Corollary 1 to Theorem 3.15) that none of the Dirichlet's L-functions vanishes for $\text{Re}\, s > 1$. To prove the same fact for the line $\text{Re}\, s = 1$, we shall consider two cases according to whether the character χ is real or not. The case of a non-real character is simple. In fact, it is sufficient to note that the function $L(s, \chi^2)$ is then Dirichlet's L-function determined by a non-principal character and apply Lemma 3.2.

In the case of a real character the assertion for $s \neq 1$ follows also from Lemma 3.2, and when $s = 1$, we shall turn to Theorem 3.5 and the following lemma:

Lemma 3.10. *If* χ *is a real Dirichlet character, then in the half-plane* $\text{Re}\, s > 1$ *we have the equality*

$$\zeta(s) L(s, \chi) = \sum_{n=1}^{\infty} a_n n^{-s} \ ,$$

where the coefficients a_n *are non-negative and* $a_{n^2} \geqslant 1$ *for* $n = 1, 2, \ldots$.

Proof of the lemma. For $\text{Re}\, s > 1$ the equality (3.7) and Theorem 3.15 imply the equality

$$\zeta(s) L(s, \chi) = \prod_p \left(1 - \frac{1}{p^s}\right)^{-1} (1 - \chi(p) p^{-s})^{-1} = \sum_{n=1}^{\infty} a_n n^{-s} \ ,$$

where

(3.30) $$a_{mn} = a_m a_n$$

provided that $(m, n) = 1$.

Now let us determine the values a_{p^k}:

If $\chi(p) = 0$, then

$$\left(1 - \frac{1}{p^s}\right)^{-1} \left(1 - \frac{\chi(p)}{p^s}\right)^{-1} = \left(1 - \frac{1}{p^s}\right)^{-1} = 1 + p^{-s} + p^{-2s} + \ldots,$$

and so $a_{p^k} = 1$.

   If $\chi(p) = 1$, then

$$(1 - p^{-s})^{-1}\,(1 - \chi(p)\,p^{-s})^{-1} = \left(1 - \frac{1}{p^s}\right)^{-2} = 1 + 2p^{-s} + 3p^{-2s} + \ldots\,,$$

and hence $a_{p^k} = 1 + k$.

   If finally $\chi(p) = -1$, then

$$(1 - p^{-s})^{-1}\,(1 - \chi(p)\,p^{-s})^{-1} = (1 - p^{-2s})^{-1} = 1 + p^{-2s} + p^{-4s} + \ldots\,,$$

and so

$$a_{p^k} = \begin{cases} 1 & \text{for } k \text{ even}, \\[2mm] 0 & \text{for } k \text{ odd}. \end{cases}$$

   Now the assertion of the lemma follows from the equality (3.30).   ■

Corollary. *The series*

$$\sum_{n=1}^{\infty} a_n\, n^{-s}$$

is divergent for $s = \frac{1}{2}$.

   Proof. For a natural number N, we have

$$\sum_{n=1}^{N} a_n\, n^{-\frac{1}{2}} \geqslant \sum_{k^2 \leqslant N} a_{k^2}\, k^{-1} \geqslant \sum_{k^2 \leqslant N} \frac{1}{k}\,,$$

hence the assertion of the corollary follows from the divergence of the harmonic series.   ■

   Now we can complete the proof of the theorem in the case of a real character and $s = 1$. Suppose that $L(1, \chi) = 0$. Since the Dirichlet series representing the function $f(s) = \zeta(s)\,L(s, \chi)$ has non-negative coefficients, $f(s)$ has, by Theorem 3.5, a singular point at $s = x$, where x is the abscissa of convergence of this series. From the Corollary to Lemma 3.15 it follows that $x \geqslant \frac{1}{2}$, and hence $f(s)$ has a singular point in the half-plane $\mathrm{Re}\,s \geqslant \frac{1}{2}$. However, by virtue of Theorems 3.6 and 3.15, $f(s)$ is holomorphic for $\mathrm{Re}\,s > 0$ as the pole $s = 1$ of the zeta-function cancels out with a zero of $L(s, \chi)$ at this point. This contradiction shows that the function $L(s, \chi)$ cannot vanish at the point $s = 1$.   ■

   Corollary 1. *If we choose that branch of the logarithm which is real for positive arguments, then the function* $\log L(s, \chi)$ *is, for the non-principal character* χ, *a holomorphic function for* $\mathrm{Re}\,s \geqslant 1$, *and for the principal character* χ_0, *a holomorphic function for* $\mathrm{Re}\,s \geqslant 1$, $s \neq 1$.

   Proof. It suffices to make us of Corollary 3 to Theorem 3.15 and Theorem 3.16.   ■

Finally we can start realizing the program sketched at the beginning of this subsection. We shall begin by expressing the sum

$$\sum_{p \equiv l(\mathrm{mod}\ k)} p^{-s}$$

in terms of Dirichlet's L-functions, and for this purpose we use the following lemma:

Lemma 3.11. *If* χ *is an arbitrary Dirichlet character, then for* $\mathrm{Re}\,s > 1$ *we have*

$$\sum_{p} \chi(p)p^{-s} = \log L(s, \chi) + g(s, \chi) \ ,$$

where $g(s, \chi)$ *is holomorphic for* $\mathrm{Re}\,s > \frac{1}{2}$ *and bounded for* $\mathrm{Re}\,s \geq \frac{1}{2} + \delta$, δ *being any positive number.*

Proof. From Theorem 3.15 (ii) it follows that for $\mathrm{Re}\,s > 1$ we have

$$\log L(s, \chi) = \sum_{p} \log(1 - \chi(p)p^{-s})^{-1} = \sum_{p} \sum_{k=1}^{\infty} \frac{1}{k} (\chi(p)p^{-s})^{k}$$

$$= \sum_{p} \chi(p)p^{-s} + \sum_{p} \sum_{k=2}^{\infty} \frac{1}{k} (\chi(p)p^{-s})^{k} \ .$$

We denote the second term by $-g(s, \chi)$. Since the series appearing here is majorized for $\mathrm{Re}\,s \geq \frac{1}{2} + \delta$ by the series

$$\sum_{p} \sum_{k=2}^{\infty} \frac{1}{p^{k\,\mathrm{Re}\,s}} = \sum_{p} \frac{1}{p^{2\,\mathrm{Re}\,s}} \cdot \frac{1}{1 - \dfrac{1}{p^{\mathrm{Re}\,s}}} \leqslant \frac{1}{1 - 2^{-1/2}} \sum_{p} p^{-1-2\delta} \ ,$$

$g(s, \chi)$ is holomorphic and bounded there. ∎

Corollary 1. *If* χ *is not the principal character, then there exists a function* $g(s)$ *holomorphic for* $\mathrm{Re}\,s \geq 1$ *such that*

$$\sum_{p} \chi(p)p^{-s} = g(s) \qquad (\mathrm{Re}\,s > 1) \ .$$

Proof. Follows from the lemma and from Corollary to Theorem 3.16. ∎

Corollary 2. *If* χ_0 *is the principal character, then for* $\mathrm{Re}\,s > 1$ *we have*

$$\sum_{p} \chi_0(p)p^{-s} = \log \frac{1}{s-1} + g(s) \ ,$$

where $g(s)$ *is holomorphic for* $\mathrm{Re}\,s \geq 1$.

Proof. Follows from Corollary 2 to Theorem 3.15, Lemma 3.11 and Theorem 3.9 (i). ∎

Theorem 3.17 (Dirichlet).

(i) *If* $(k, l) = 1$, *then for* $\mathrm{Re}\,s > 1$, *we have*

$$\sum_{p \equiv l(\text{mod } k)} \frac{1}{p^s} = \frac{1}{\varphi(k)} \log \frac{1}{s-1} + g(s) \ ,$$

where $g(s)$ *is holomorphic for* $\text{Re } s \geqslant 1$.

(ii) *If* $(k, l) = 1$, *then there exist infinitely many prime numbers* p *satisfying*
$p \equiv l(\text{mod } k)$.

(iii) *If* $(k, l) = 1$ *and*

$$\pi(x; k, l) = \sum_{\substack{p \leqslant x \\ p \equiv l(\text{mod } k)}} 1 \ ,$$

then we have, as x *tends to infinity,*

$$\pi(x; k, l) = \frac{1}{\varphi(k)} + o(1) \ \frac{x}{\log x} \ .$$

Proof. Note that from (i) it is easy to deduce the remaining parts of the theorem. In fact, if the set of prime numbers of the form $kx + l$ is finite, then the function

$$\sum_{p \equiv l(\text{mod } k)} p^{-s}$$

must be entire, contrary to (i). Finally (iii) follows from the Corollary to Theorem 3.10. Therefore it remains to prove (i), and this follows from the following identity:

$$(3.31) \qquad \sum_{p \equiv l(\text{mod } k)} p^{-s} = \varphi(k)^{-1} \sum_{\chi(\text{mod } k)} \overline{\chi(l)} \sum_{p} \chi(p) p^{-s}$$

which is valid for $\text{Re } s > 1$. The proof of this identity follows from Corollary 4 to Theorem 3.13, where it is necessary to put $G = G(k)$, $g_0 = l(\text{mod } k)$ and $g = r(\text{mod } k)$

$$f(g) = \sum_{p \equiv r(\text{mod } k)} \frac{1}{p^s} \ .$$

In fact, this corollary gives us

$$\sum_{p \equiv l(\text{mod } k)} \frac{1}{p^s} = \frac{1}{\varphi(k)} \sum_{\chi(\text{mod } k)} \overline{\chi(l)} \sum_{r(\text{mod } k)} \chi(r) \sum_{p \equiv r(\text{mod } k)} \frac{1}{p^s}$$

$$= \frac{1}{\varphi(k)} \sum_{\chi(\text{mod } k)} \overline{\chi(l)} \sum_{p} \frac{\chi(p)}{p^s} \ ,$$

and hence (3.31). Therefore

$$\sum_{p \equiv l(\text{mod } k)} \frac{1}{p^s} = \frac{1}{\varphi(k)} \sum_{p} \frac{\chi_0(p)}{p^s} + \frac{1}{\varphi(k)} \sum_{\chi \neq \chi_0} \overline{\chi(l)} \sum_{p} \frac{\chi(p)}{p^s}$$

and the result follows by applying Corollaries 1 and 2 to Lemma 3.11. ∎

In part (iii) of the theorem proven above, we have no estimate of the remainder. By means of complex integration (cf. Walfisz [2]) it can be proved that for a fixed k, we have as x tends to infinity

$$\pi(x;k,l) = \frac{1}{\varphi(k)} \text{li} x + O(x \exp(-C \log^{3/5} x(\log\log x)^{-1/5}))$$

where $C > 0$.

To find an estimate uniform in k in this formula, that is with the constant implied by Landau's symbol independent of k, is an interesting problem. It turns out that it depends in a most substantial way the existence of the zeros of $L(s, \chi)$ for real characters $\chi(\text{mod } k)$ on the segment $(0, 1)$. If such a function has a zero at the point b, then, as A. Page [1] and E. C. Titchmarsh [1] have shown, the estimate

$$\pi(x;k,l) = \frac{\text{li} x}{\varphi(k)} + O\left(\frac{x^b}{\varphi(k)}\right) + O(x \exp(-C\sqrt{\log x}))$$

holds for all k not exceeding $\exp(C_1 \sqrt{\log x})$, where C, C_1 and the constant in the symbol $O(\cdot)$ do not depend on k. For smaller intervals this estimate can be refined. Thus, for example, C. L. Siegel [1] has shown that for $k \leqslant \log^a x$ we have

$$\pi(x;k,l) = \frac{\text{li} x}{\varphi(k)} + O(x \exp(-C\sqrt{\log x}))$$

where the implied constant depends upon a, and the number a can be chosen arbitrarily.

The reader can find the proofs of theorems cited here in Prachar's book [1].

In certain problems the upper estimate of the number $\pi(x;k, l)$ uniform with respect to k is important. Here various sieve methods are used. We shall study the simplest sieves in Chapter IV.

4.   As an application of Dirichlet's theorem we shall now prove a result concerning Legendre's symbol which can be formulated as a theorem concerning finite fields:

Theorem 3.18. *If a is a natural number such that for prime numbers $p \nmid a$ we have*

$$(3.32) \qquad \left(\frac{a}{p}\right) = 1,$$

then a is a square of a natural number.

This theorem can be expressed as follows: *If we denote by f_p the canonical epimorphism of $\mathscr{Z}$ onto $\mathscr{Z}/p\mathscr{Z} = GF(p)$, and for all p, $f_p(a)$ is a square in $GF(p)$, then a is a square in $\mathscr{Z}$.*

Proof. Suppose a is not a square in $\mathscr{Z}$ and that for all prime numbers p not dividing a, we have the equality (3.32). We write a in the form

$$a = p_0 p_1 \cdots p_t Q^2 \ ,$$

where $p_1, \ldots, p_t$ are distinct odd prime numbers,

$$p_0 = 2 \quad \text{if } 2 \nmid \alpha_2(a), \qquad p_0 = 1 \quad \text{if } 2 \mid \alpha_2(a)$$

and Q is a natural number. Note that the condition $t \geqslant 1$ must hold, otherwise, we would have $a = 2Q^2$ and by Theorem 1.30, for every prime number p not dividing $2Q$, we would have

$$(-1)^{(p^2-1)/8} \left(\frac{2}{p} \right) = \left(\frac{2Q^2}{p} \right) = \left(\frac{a}{p} \right) = 1 \ ,$$

that is, every sufficiently large prime number p would satisfy the congruence

$$p^2 \equiv 1 \ (\text{mod } 16) \ ,$$

and hence there would exist only finitely many prime numbers congruent to 3 (mod 16), contrary to Dirichlet's theorem. By the Corollary to Theorem 1.18 there exists a solution X of the system of congruences:

$$x \equiv 1 \ (\text{mod } 16) \ ,$$

$$x \equiv 1 \ (\text{mod } p_i) \qquad \text{for } i = 1, \ldots, t-1 \ ,$$

$$x \equiv b \ (\text{mod } p_t) \ ,$$

where b is a fixed quadratic non-residue (mod p_t). Now let p be a prime congruent to X (mod $16 p_1 \cdots p_t$) and greater than a. Then we have

$$\left(\frac{p_0}{p} \right) = 1$$

and

$$1 = \left(\frac{a}{p} \right) = \left(\frac{p_0}{p} \right) \cdots \left(\frac{p_t}{p} \right) = \left(\frac{p_1}{p} \right) \cdots \left(\frac{p_t}{p} \right) \ .$$

But Theorem 1.29 implies

$$\left(\frac{p_i}{p} \right) = \left(\frac{p}{p_i} \right) (-1)^{\frac{1}{2}(p-1)\frac{1}{2}(p_i-1)} = \left(\frac{p}{p_i} \right) \quad (i = 1, \ldots, t) \ ,$$

and hence

$$1 = \left(\frac{p}{p_1} \right) \cdots \left(\frac{p}{p_t} \right) = \left(\frac{1}{p_1} \right) \cdots \left(\frac{1}{p_{t-1}} \right) \left(\frac{p}{p_t} \right) = -1 \ .$$

The obtained contradiction proves the validity of our assertion. ∎

Exercises

1. Prove that for $k = 0, 1, 2, \ldots$ we have

$$\sum_{\substack{n \leqslant x \\ \Omega(n) - \omega(n) = k}} 1 = (C_k + o(1))\, x \, , \qquad C_k > 0 \, .$$

2. Show that for all r, s we have

$$\sum_{\substack{n \leqslant x \\ \omega(n) \equiv s \,(\mathrm{mod}\; r)}} 1 = \left(\frac{1}{r} + o(1) \right) x \, .$$

Hint. Consider the function $\displaystyle\sum_{n=1}^{\infty} \frac{z^{\omega(n)}}{n^s}$ with $z = \exp(2\pi i / r)$.

3. Show that if $(r, s) = 1$, then for $m = 0, 1, \ldots$ we have

$$\sum_{\substack{n \leqslant x \\ \omega(n) = m \\ n \equiv s \,(\mathrm{mod}\; r)}} 1 = \left(\frac{1}{\varphi(r)\, (m-1)!} + o(1) \right) \frac{x\, (\log\log x)^{m-1}}{\log x} \, .$$

4. Prove that for $(r, s) = 1$ the equality

$$\sum_{\substack{p \leqslant x \\ p \equiv s \,(\mathrm{mod}\; r)}} \frac{1}{p} = \frac{1}{\varphi(r)} \log\log x + O(1)$$

holds.

5. By considering the function $\displaystyle\sum_{n=1}^{\infty} \frac{\chi(\varphi(n))}{n^s}$, where χ is a character (mod r), find an

asymptotic formula for

$$\sum_{\substack{n \leqslant x \\ \varphi(n) \equiv s \,(\mathrm{mod}\; r)}} 1 \qquad \text{in the case } (r, s) = 1, \quad (6, r) = 1 \, .$$

CHAPTER IV

SIEVE METHODS

§1. Eratosthenes' sieve

1. Certain elementary methods in number theory are called *sieve methods* which serve to estimate the cardinalities of various sets defined by the use of multiplicative properties. Examples of such sets are the set of prime numbers lying in some interval, the set of prime numbers of a given form lying in some interval, or the set of integers of a given form having a given number of prime factors. Sieve methods often lead to results which cannot be obtained by analytic methods. In this chapter we shall study three sieve methods — Eratosthenes' sieve, Selberg's sieve and the so-called large sieve. We shall confine ourselves, however, to the estimations from above.

Eratosthenes' sieve in its simplest form is used to find prime numbers and can be stated as follows:

Suppose we have already found all prime numbers in the interval $[2, x^{\frac{1}{2}}]$ and wish to find prime numbers in $[x^{\frac{1}{2}}, x]$. Let us write down all natural numbers in $[1, x]$, and next omit those among them which are divisible by some prime number not exceeding $x^{\frac{1}{2}}$. The remaining are the prime numbers. In this primitive form, Eratosthenes' sieve does not lead to any non-trivial information about $\pi(x)$. In fact, the number of integers which remain after the above process is clearly $\pi(x) - \pi(\sqrt{x})$. The number of natural numbers in $[2, x]$ is $[x] - 1$. If p is a prime number $\leqslant x^{\frac{1}{2}}$, then all its multiples were omitted and so we have made $\displaystyle\sum_{p \leqslant \sqrt{x}} \left[\frac{x}{p}\right]$ cancellations. Here. however, those numbers which are divisible by two prime numbers $\leqslant \sqrt{x}$ have been cancelled twice, and if we cancelled each number divisible by two prime numbers exactly once, then the number of cancellations would be

$$\sum_{p \leqslant \sqrt{x}} \left[\frac{x}{p} \right] - \sum_{\substack{p_1 \leqslant \sqrt{x} \\ p_2 \leqslant \sqrt{x} \\ p_1 \neq p_2}} \left[\frac{x}{p_1 p_2} \right].$$

This time, however, those numbers which are divisible by three prime numbers p_1, p_2, p_3 are left uncancelled because such a number is counted thrice in the first term and also in the second term. Taking this into account, we obtain

$$\sum_{p \leqslant \sqrt{x}} \left[\frac{x}{p} \right] - \sum_{\substack{p_1, p_2 \leqslant \sqrt{x} \\ p_1 \neq p_2}} \left[\frac{x}{p_1 p_2} \right] + \sum_{\substack{p_1, p_2, p_3 \leqslant \sqrt{x} \\ \text{distinct}}} \left[\frac{x}{p_1 p_2 p_3} \right]$$

cancellations, but this time the numbers divisible by four prime numbers have been cancelled too many times and we must again add a term to the number of cancellations. It is not difficult to note (and we shall give a formal proof in Theorem 4.1) that we get, in the end

$$\sum_{s \geqslant 1} (-1)^{s+1} \sum_{\substack{p_{i_1}, \ldots, p_{i_s} \leqslant \sqrt{x} \\ \text{distinct}}} \left[\frac{x}{p_{i_1} \cdots p_{i_s}} \right]$$

cancelled numbers, so that

$$\pi(x) = \pi(\sqrt{x}) + \sum_{s=1}^{\pi(x)} (-1)^s \sum_{\substack{p_{i_1}, \ldots, p_{i_s} \leqslant \sqrt{x} \\ \text{distinct}}} \left[\frac{x}{p_{i_1} \cdots p_{i_s}} \right].$$

If we now replace $\left[\dfrac{x}{p_{i_1} \cdots p_{i_s}} \right]$ by $\dfrac{x}{p_{i_1} \cdots p_{i_s}}$, then in each of these terms we commit an error of order $O(1)$. Thus we obtain some main term and the error term of order $O(2^{\pi(\sqrt{x})})$, which is not useful at all, for it does not even give the trivial estimate $\pi(x) \leqslant x$.

Now we shall show that a modification of that procedure can lead to a non-trivial, although weaker than that obtained in the preceding chapter, estimate for the function $\pi(x)$. Here we shall turn to the following result which is a basis of most of the existing sieve methods:

Theorem 4.1. *Let* $A_1, \ldots, A_r$ *be subsets of a set* A, $B = A \setminus \bigcup_{i=1}^{r} A_i$, *and let* $f(x)$ *be any complex-valued function defined on* A. *For* $j \leqslant r$ *we put*

$$T_j = \sum_{x \in A} f(x) + \sum_{s=1}^{j} (-1)^s \sum_{\{i_1, \ldots, i_s\} \subset \{1, \ldots, r\}} \sum_{x \in A_{i_1} \cap \ldots \cap A_{i_s}} f(x).$$

Then

$$\sum_{x \in B} f(x) = T_r,$$

and, if moreover $f(x)$ is non-negative, then for all odd $k \leqslant r$ and all even $l \leqslant r$ we have

$$T_k \leqslant \sum_{x \in B} f(x) \leqslant T_l.$$

Proof. If g is a characteristic function of the set B, and f_i that of A_i, then for any $x \in A$ we have the equality

(4.1) $$g(x) = 1 + \sum_{s=1}^{r} (-1)^s \sum_{\{i_1, \ldots, i_s\} \subset \{1, \ldots, r\}} f_{i_1}(x) \ldots f_{i_s}(x).$$

Indeed, if x lies in B, then both sides of this equality are equal to one. If, on the contrary, x does not lie in B, and belongs to exactly m sets A_i, say, to $A_{j_1}, \ldots, A_{j_m}$, then the left-hand side is equal to zero, and the right is equal to

$$1 + \sum_{s=1}^{m} (-1)^s \sum_{\{i_1, \ldots, i_s\} \subset \{j_1, \ldots, j_m\}} 1 = 1 + \sum_{s=1}^{m} (-1)^s \binom{m}{s} = (1-1)^m = 0.$$

Multiplying both sides of (4.1) by $f(x)$ and summing over all $x \in A$, we obtain $T_r = \sum_{x \in B} f(x)$. Now let $k \leqslant r$ be odd, $l \leqslant r$ be even, and let $g(x)$ and $f_i(x)$ have the above meaning. We prove the inequality

(4.2) $$1 + \sum_{s=1}^{k} (-1)^s \sum_{\{i_1, \ldots, i_s\} \subset \{1, \ldots, r\}} f_{i_1}(x) \ldots f_{i_s}(x) \leqslant g(x)$$

$$\leqslant 1 + \sum_{s=1}^{l} (-1)^s \sum_{\{i_1, \ldots, i_s\} \subset \{1, \ldots, r\}} f_{i_1}(x) \ldots f_{i_s}(x).$$

In the case $x \in B$, it is trivial, and if x belongs to exactly m sets A_i, say, to $A_{j_1}, \ldots, A_{j_m}$, then its extremal terms are respectively equal to

$$\sum_{s=0}^{k} (-1)^s \binom{m}{s} \quad \text{and} \quad \sum_{s=0}^{l} (-1)^s \binom{m}{s}.$$

We now prove the following lemma.

Lemma 4.1. *If* $0 \leqslant k \leqslant m$, *then*

$$\sum_{s=0}^{k} (-1)^s \binom{m}{s} \geqslant 0 \quad \textit{for } k \textit{ even}$$

and

$$\sum_{s=0}^{k} (-1)^s \binom{m}{s} \leqslant 0 \quad \text{for } k \text{ odd}.$$

Proof. For $k \leqslant [\frac{1}{2}m]$ the assertion follows immediately from the inequality

$$\binom{m}{0} \leqslant \binom{m}{1} \leqslant \ldots \leqslant \binom{m}{[\frac{1}{2}m]}. \text{ If } k > [\frac{1}{2}m], \text{ then}$$

$$\sum_{s=0}^{k} (-1)^s \binom{m}{s} = - \sum_{s=1+k}^{m} (-1)^s \binom{m}{s} = (-1)^{m+1} \sum_{s=0}^{m-k-1} (-1)^s \binom{m}{s},$$

and by $m - k - 1 \leqslant [\frac{1}{2}m]$ the sign of the right-hand side of the equality obtained is already known, and an easy verification leads to the assertion. ■

From the lemma, the inequality (4.2) follows immediately, and we may now obtain the assertion in the second part of the theorem by multiplying (4.2) by $f(x)$ and summing the inequalities obtained. ■

Corollary 1. *Under the conditions of Theorem 4.1 we have*

$$|B| = |A| + \sum_{s=1}^{r} (-1)^s \sum_{\{i_1, \ldots, i_s\} \subset \{1, \ldots, r\}}{}' |A_{i_1} \cap \ldots \cap A_{i_s}|.$$

Proof. Apply the theorem with $f(x) = 1$. ■ ■

Corollary 2. *Let $D = \{p_1, \ldots, p_n\}$ be a set of distinct prime numbers and let A be a given finite set of integers. Denote by S the number of elements of A which are not divisible by any of p_i's, and by S_d the number of elements of A divisible by d. Then we have*

(4.3) $$S = \sum_{d \mid p_1 \ldots p_n} \mu(d) S_d,$$

and furthermore for $r = 1, 2, 3, \ldots, [\frac{1}{2}n]$ we have

(4.4) $$\sum_{\substack{d \mid p_1 \ldots p_n \\ \omega(d) \leqslant 2r-1}} \mu(d) S_d \leqslant S \leqslant \sum_{\substack{d \mid p_1 \ldots p_n \\ \omega(d) \leqslant 2r}} \mu(d) S_d.$$

The inequality (4.4) is usually called the Eratosthenes' sieve.

Proof. We apply the theorem proved above by taking for A_i the set of elements of A divisible by p_i and putting $f(x) = 1$. Then for $d = p_{i_1} \ldots p_{i_s}$

$$S_d = S_{p_{i_1} \ldots p_{i_s}} = |A_{i_1} \cap \ldots \cap A_{i_s}|, \quad s = \omega(d)$$

and $\mu(p_{i_1} \ldots p_{i_s}) = (-1)^s$, so that

$$T_j = \sum_{\substack{d \mid p_1 \ldots p_n \\ \omega(d) \leqslant j}} \mu(d) S_d$$

and the assertion follows directly from the theorem. ∎

2.   Now we shall give two results which follow from the result of Eratosthenes' sieve formulated in the inequality (4.4). The first of them implies an estimate for $\pi(x)$ weaker than that obtained in the preceding chapter, and the second is concerned with the number of twin primes $p \leqslant x$, i.e. such that $p+2$ is also a prime like 3, 5, 7, 11, 17, ... It is not known whether there are infinitely many of them.

Theorem 4.2. *Let* $u = u(x)$ *satisfy the condition*

$$2^{u(x)} \log u(x) < x$$

for all x.

If $N(x) = N_u(x)$ *is the number of integers* $n \leqslant x$ *none of whose prime divisors is less than* $u(x)$, *then* $N_u(x) = O\left(\dfrac{x}{\log u(x)}\right)$.

Proof. In Corollary 2 to Theorem 4.1 we take D to be the set of all prime numbers $\leqslant u(x)$, and let A be the set of all integers in the interval $[2, x]$. Then $n = \pi(u(x))$, $S = N_u(x)$ and $S_d = \left[\dfrac{x}{d}\right]$, so that from formula (4.3) we obtain

$$N_u(x) = \sum_{d \mid p_1 \ldots p_n}{}' \mu(d)\left[\frac{x}{d}\right] = x \sum_{d \mid p_1 \ldots p_n}{}' \frac{\mu(d)}{d} + O\left(2^{u(x)}\right)$$

$$= x \prod_{p \leqslant u(x)} \left(1 - \frac{1}{p}\right) + O\left(2^{u(x)}\right) = O\left(\frac{x}{\log u(x)}\right) + O\left(2^{u(x)}\right) = O\left(\frac{x}{\log u(x)}\right),$$

where $D = \{p_1, \ldots, p_n\}$. ∎

Corollary 1. *The number of integers* $n \leqslant x$, *all of whose prime divisors are at least equal to* $c \log x$, *where* c *is any number less than* $1/\log 2$, *is equal to* $O(x/\log\log x)$.

Proof. Apply the theorem to $c \log x$. ∎

Corollary 2. *For* $x \geqslant 3$

$$\pi(x) \ll \frac{x}{\log\log x}.$$

Proof. Since $\pi(x) - \pi(\log x) \leqslant N_u(x)$ for $u = \log x$, we may apply the preceding corollary. ∎

The following result is remarkably more difficult and was first proved by Viggo Brun in 1919.

Theorem 4.3. *Let* a *be an even integer and* $B_a(x)$ *denote the number of primes* $p \leqslant x$ *for which* $|p + a|$ *is also a prime. Then we have*

$$B_a(x) = O\left(x(\log\log x)^2 \log^{-2} x\right).$$

Proof. We shall apply Corollary 2 to Theorem 4.1 in the following situation: $A = \{|n(n+a)|: n \leqslant x\}$, D is the set of primes not exceeding z, where $z = z(x)$ satisfies $a < z(x) < x^{\frac{1}{2}}$ and will be fixed later. Since clearly $B_a(x) \leqslant S + O(z)$ (where S has a fixed meaning in Corollary 2 to Theorem 4.1), we have for even $k \leqslant \pi(x)$, by virtue of (4.4), the inequality

$$(4.5) \qquad\qquad B_a(x) \leqslant \sum_{\substack{m|P \\ \omega(m) \leqslant k}} \mu(m) S_m + O(z),$$

where P denotes the product of all primes not exceeding z, and $S_m = S_m(x; a)$ is the number of $n \leqslant x$ satisfying the congruence

$$(4.6) \qquad\qquad n(n+a) \equiv 0 \pmod{m}.$$

Now we turn to the estimation of S_m in the case $\mu(m) \neq 0$, for we are interested only in this case. We prove the asymptotic equality

$$(4.7) \qquad\qquad S_m(x; a) = \frac{d(m_1)}{m} x + O\big(d(m_1)\big),$$

where $m_1 = m/(a, m)$, and the constant implied by the O-symbol does not exceed 1 in absolute value.

First let $(a, m) = 1$. Then, of course we also have

$$(n, n+a, m) = 1.$$

Under this condition the congruence (4.6) is satisfied iff there exist a decomposition $m = m_1 m_2$, $(m_1, m_2) = 1$ such that $n \equiv 0 \pmod{m_1}$ and $n+a \equiv 0 \pmod{m_2}$. Indeed, the sufficiency of this condition is immediate, and in order to prove its necessity suppose that $m = p_1 \ldots p_t$ divides $n(n+a)$. None of the p_i's divide n and $n+a$ simultaneously, so that the set $\{p_1, \ldots, p_t\}$ is a disjoint union of two sets P_1 and P_2, where

$$n \equiv 0 \left(\bmod \prod_{p \in P_1} p\right)$$

and

$$n+a \equiv 0 \left(\bmod \prod_{p \in P_2} p\right).$$

Hence we can put $m_1 = \prod_{p \in P_1} p$ and $m_2 = \prod_{p \in P_2} p$. Note also that the factors m_1 and m_2 are uniquely determined by a solution n of the congruence (4.6). Since in our case, $(m_1, m_2) = 1$, the system

$$n \equiv 0 \pmod{m_1}, \qquad n+a \equiv 0 \pmod{m_2}$$

has precisely one solution (mod m), and hence the congruence (4.6) has exactly $d(m)$
solutions under the assumption $(a, m) = 1$. This gives for some $0 \leqslant \vartheta \leqslant 1$,

$$S_m = \left[\frac{x}{m}\right] d(m) + \vartheta d(m) = \frac{x}{m} d(m) + \left(\vartheta - \left\{\frac{x}{m}\right\}\right) d(m) = \frac{x}{m} d(m) + \vartheta_1 d(m),$$

where $|\vartheta_1| \leqslant 1$. This proves the validity of (4.7) in the case $(a, m) = 1$. If
$(a, m) = \delta > 1$ and we write $a = \delta a_1$, $m = \delta m_1$ $((a_1, m_1) = 1)$, then the congruence
(4.6) takes the form

$$n(n + a) \equiv 0 (\text{mod } m_1 \delta) \quad .$$

If n is its solution, then clearly δ divides n and writing $n = \delta n_1$, we see that

$$S_m(x; a) = \sum_{\substack{n_1 \leqslant x/\delta \\ n_1(n_1 + a_1) \equiv 0 (\text{mod } m_1)}} 1 = S_{m_1}\left(\frac{x}{\delta}; a_1\right),$$

but $(a_1, m_1) = 1$, and using the part of formula (4.7) already proved we obtain

$$S_m(x; a) = \frac{d(m_1)}{m_1} \cdot \frac{x}{\delta} + \vartheta_2 d(m_1) = \frac{d(m_1)}{m} x + \vartheta_2 d(m_1),$$

where $|\vartheta_2| \leqslant 1$. Hence formula (4.7) is true in general. Formulas (4.5) and (4.7) now
lead to

$$B_a(x) \leqslant x \sum_{\substack{m|P \\ \omega(m) \leqslant k}} \mu(m) \frac{d(m_1)}{m} + O\left(\sum_{\substack{m|P \\ \omega(m) \leqslant k}} d(m)\right) + O(z).$$

In view of the multiplicativity of the function $f(m) = \dfrac{\mu(m) d(m_1)}{m}$, we may
write by Corollary 6 to Theorem 2.7,

$$\sum_{m|P} \mu(m) \frac{d(m_1)}{m} = \prod_{\substack{p \leqslant z \\ p \nmid a}} \left(1 - \frac{2}{p}\right) \prod_{p|a} \left(1 - \frac{1}{p}\right) = \frac{\varphi(a)}{a} \prod_{\substack{p \leqslant z \\ p \nmid a}} \left(1 - \frac{2}{p}\right),$$

and so

$$B_a(x) \leqslant x \frac{\varphi(a)}{a} \prod_{\substack{p \leqslant z \\ p \nmid a}} \left(1 - \frac{2}{p}\right) + O(z) + O\left(x \sum_{\substack{m|P \\ \omega(m) > k}} \frac{d(m)}{m}\right) + O\left(\sum_{\substack{m|P \\ \omega(m) \leqslant k}} d(m)\right).$$

We estimate the last two terms of this sum. Changing the order of summation, we obtain

$$\sum_{\substack{m|P \\ \omega(m)>k}} \frac{d(m)}{m} = \sum_{j=k+1}^{\pi(z)} \sum_{\substack{m|P \\ \omega(m)=j}} \frac{d(m)}{m} = \sum_{j=k+1}^{\pi(z)} 2^j \sum_{\substack{m|P \\ \omega(m)=j}} \frac{1}{m} = \sum_{j=1+k}^{\pi(z)} 2^j R_j,$$

where

$$R_j = \sum_{\substack{m|P \\ \omega(m)=j}} \frac{1}{m} = \sum_{p_1 < \dots < p_j \leq z} \frac{1}{p_1 \cdots p_j}.$$

From Theorem 3.2 follows the estimate $R_1 \leq \log\log z + C$ with a constant C, and since

$$\left(\sum_{p \leq z} \frac{1}{p}\right)^j = \sum_{p_1,\dots,p_j \leq z} \frac{1}{p_1 \cdots p_j} \geq j!\, R_j,$$

we have

$$\sum_{\substack{m|P \\ \omega(m)>k}} \frac{d(m)}{m} \leq \sum_{j=1+k}^{\pi(z)} \frac{2^j}{j!} (\log\log z + C)^j \leq \sum_{j=1+k}^{\pi(z)} \left(\frac{2e}{j}\right)^j (\log\log z + C)^j.$$

If now we take $k = 4[e \log\log z + eC] + 4$, then from the above estimate it follows that

$$\sum_{\substack{m|P \\ \omega(m)>k}} \frac{d(m)}{m} \leq \sum_{j=1+k}^{\infty} \left(\frac{k}{2j}\right)^j \ll \left(\frac{k}{2(k+1)}\right)^{k+1} \ll \frac{1}{2^{k+1}}$$

$$\ll \exp\{(\log 2)(-4e \log\log z - 4eC - 4)\} \ll (\log z)^{-4e \log 2}$$

Moreover

$$\sum_{\substack{m|P \\ \omega(m) \leq k}} d(m) \leq \sum_{j=0}^{k} \sum_{\substack{m|P \\ \omega(m)=j}} d(m) = \sum_{j=0}^{k} 2^j \binom{\pi(z)}{j}$$

$$\leq \sum_{j=0}^{k} \frac{2^j}{j!} \pi(z)^j \leq \pi(z)^k \exp 2 \ll \frac{z^k}{\log^k z},$$

and since we have on the basis of Corollary 4 to Theorem 3.2

$$\prod_{\substack{p \leq z \\ p \nmid a}} \left(1 - \frac{2}{p}\right) = O(\log^{-2} z),$$

we get ultimately

$$B_a(x) \ll \frac{x}{\log^2 z} + \frac{x}{\log^{4e \log 2} z} + \frac{z^k}{\log^k z} + z \quad .$$

Now, taking $z = x^{1/k}$, we obtain

$$B_a(x) = O\left(\frac{k^2 x}{\log^2 x}\right) = O\left(\frac{x \log \log^2 x}{\log^2 x}\right)$$

as asserted.　■

Corollary. *If B is the set of all the primes p such that $p + 2$ is also a prime, then*

$$\sum_{\substack{p \in B \\ p \leqslant x}} 1 \ll \frac{x (\log \log x)^2}{\log^2 x}$$

and the series

$$\sum_{p \in B} \frac{1}{p}$$

is convergent.

Proof. The first part is an immediate consequence of the theorem. To prove the second, we put

$$\varepsilon_n = \begin{cases} 1, & \text{if} \quad n \in B, \\ 0, & \text{if} \quad n \notin B. \end{cases}$$

Using Theorem 2.3 we obtain

$$\sum_{N \geqslant p \in B} \frac{1}{p} = \sum_{n \leqslant N} \frac{\varepsilon_n}{n} = \sum_{n \leqslant N} \frac{1}{n(n+1)} \sum_{k \leqslant n} \varepsilon_k + \frac{1}{N} \sum_{k \leqslant N} \varepsilon_k \ll$$

$$\ll \sum_{n \leqslant N} \frac{(\log \log n)^2}{n \log^2 n} + \frac{(\log \log N)^2}{\log^2 N} = O(1). \quad ■$$

Remark. As proved by J. Bohman [1], the value of the series $\sum\limits_{p \in B} 1/p$ is equal to 1.70215 ± 0.00001. As we have already mentioned, we do not know whether the set B is finite or not. At present the greatest known pair of primes differing by 2 is the pair $76 \cdot 3^{139} \pm 1$ found by H. C. Williams and C. R. Zarnke [1] in 1972.[*]

---

[*] Added in proof by the translator: The greatest known twin primes are $1,159,142,985 \times 2^{2304} \pm 1$ found by A. O. L. Atkin and N. W. Rickert, *Notices Amer. Math. Soc.* **26** (1979), A-373.

Let

$$a = \liminf_{n} \frac{p_{n+1} - p_n}{\log p_n},$$

where p_n denotes the n-th successive prime. If the set B is infinite, then clearly $a = 0$. The best known estimate of the number a is given by M. N. Huxley [1], viz.

$$a \leqslant \tfrac{1}{4} + \tfrac{1}{6}\pi = 0.4463\ldots \quad^{\dagger}.$$

A related problem is that of determining the lower bound b of the numbers r such that for a sufficiently large x at least one prime lies in the interval $[x, x+x^r]$. From the prime number theorem follows the estimate $b \leqslant 1$, and D. R. Heath-Brown and H. Iwaniec [1] has given the best known estimate, viz.

$$b \leqslant \tfrac{1}{2} + \tfrac{1}{20} \quad.$$

It is conjectured that $b = 0$, and more precisely (*Cramér's hypothesis*) that with a suitable constant C at least one prime number lies in the interval $[x, x+C\log^2 x]$.‡

Exercises

1.   Apply Eratosthenes' sieve to estimate from above the number of primes $\leqslant x$ of the form $n^2 + 1$.

2.   Find the estimate from above of the number of primes $p \leqslant x$ for which $p + 2$ and $p + 6$ are also primes.

3.   Prove that if $\Phi(x, y)$ is the number of integers $\leqslant x$ not having prime divisors $\leqslant y$, then

$$\Phi(x, y) = x \prod_{p < y} \left(1 - \frac{1}{p}\right) + O\left(2^{\pi(y)}\right).$$

§2. Selberg's sieve

1.   The method found by A. Selberg in 1945, which we shall present in this section, is more complicated than Eratosthenes' sieve, but relies, as in the case of Eratosthenes' sieve, upon formula (4.3). Its idea is to replace in this formula the Möbius function by another one which is easier to handle. As we shall see, this method is again completely elementary; however, in its application to concrete problems the necessity arises using the analytic methods particularly in the case of lower bounds. Anyway, we shall not deal with such estimations in this book, and refer the reader to the book by

---

†M. N. Huxley has improved this estimate of his own by proving that $a \leqslant 0.442540\ldots$, *Mathematika* **24** (1977), 142-152.

‡For recent developments, see e.g. D. R. Heath-Brown, *J. London Math. Soc.* (2) **26** (1982), 385-396.

H. Halberstam and K. F. Roth [1]. A complete account of sieve theory is contained in the book by H. Halberstam and H. E. Richert, *Sieve Methods*, Academic Press (1974).

After presenting the method, we shall apply it to two problems: We shall estimate the number of primes $p \leqslant x$ in an arithmetic progression, and also estimate the number of solutions of the equation $p_1 + p_2 = N$, where N is given and p_1, p_2 are primes. This estimate will enable us to prove in the next chapter, Schnirelman's theorem on the representation of natural numbers as a sum of primes.

Let $D = \{p_1, \ldots, p_n\}$ be a set of primes, P their product, and let A be a set of integers with N elements. As in the preceding subsection, we denote by S the number of elements of A not having divisors in D, and by S_d the number of elements of A divisible by a given number d. Assume that the set A satisfies the following condition:

(*)   *There exists a multiplicative function* $f(d)$ *such that if* $d|P$, *then*

$$(4.8) \qquad S_d = \frac{f(d)}{d} N + R(d) \quad,$$

where $|R(d)| \leqslant f(d)$ *and* $d > f(d) \geqslant 1$.

If $r(d)$ is any function satisfying the condition

$$(4.9) \qquad \sum_{d|m} \mu(d) \leqslant \sum_{d|m} r(d)$$

(that is, $r(1) \geqslant 1$, $\sum_{d|m} r(d) \geqslant 0$ for $m \neq 1$), then by (4.3) we have

$$S = \sum_{d|P} \mu(d) S_d = \sum_{d|P} \mu(d) \sum_{\substack{a \in A \\ d|a}} 1 = \sum_{a \in A} \sum_{d|(P,a)} \mu(d) \leqslant \sum_{a \in A} \sum_{d|(P,a)} r(d)$$

$$= \sum_{d|P} r(d) \sum_{\substack{a \in A \\ d|a}} 1 = \sum_{d|P} r(d) S_d .$$

Using formula (4.8) we can rewrite the last inequality in the form:

$$(4.10) \qquad S \leqslant \sum_{d|P} r(d) \frac{f(d)}{d} N + \sum_{d|P} r(d) R(d)$$

$$= N \sum_{d|P} \frac{r(d) f(d)}{d} + O\left(\sum_{d|P} |r(d) R(d)| \right) \quad,$$

and we shall try to choose the function $r(d)$ so that the right-hand side of the last formula becomes as small as possible.

Let us prove now a lemma estimating S from above.

Lemma 4.2. *If* $\lambda(d)$ *is any real number for* $d > 1$, $d \mid P$, $\lambda(1) = 1$, *and the function* $g(n)$ *is defined by the formula*

$$g = \mu * h,$$

where $h(n) = n/f(n)$, *then*

$$S \leqslant N \sum_{d \mid P} g(d) \left(\sum_{\substack{\delta \mid P \\ d \mid \delta}} \frac{\lambda(\delta)}{h(\delta)} \right)^2 + O\left(\sum_{d_1, d_2 \mid P} |\lambda(d_1)\, \lambda(d_2)\, R([d_1,\, d_2])| \right).$$

Proof. Note first that the condition $f(d) < d$ implies $h(d) > 1$, and hence the right-hand side of the inequality has a meaning. Moreover, for a prime p we have $g(p) = h(p) - 1 > 0$, and so for square-free d, we have $g(d) \neq 0$.

Let

$$r(d) = \sum_{[d_1, d_2] = d} \lambda(d_1)\, \lambda(d_2).$$

Then $r(1) = 1$, and moreover

$$\sum_{d \mid n} r(d) = \sum_{d \mid n} \sum_{[d_1, d_2] = d} \lambda(d_1)\, \lambda(d_2)$$

$$= \sum_{d_1, d_2 \mid n} \lambda(d_1)\, \lambda(d_2) = \left(\sum_{d \mid n} \lambda(d) \right)^2 \geqslant 0,$$

so that the condition (4.9) is satisfied.

Such a choice of the function $r(d)$ does not limit the generality of our consideration, because if the condition (4.9) is to hold, then we take for $\lambda(d)$ a function satisfying

$$r * 1 = (\lambda * 1)^2.$$

From the inequality (4.10) we obtain

$$S \leqslant N \sum_{d \mid P} \frac{f(d)}{d} \sum_{[d_1, d_2] = d} \lambda(d_1)\, \lambda(d_2) + O\left(\sum_{d \mid P} |R(d)| \sum_{[d_1, d_2] = d} |\lambda(d_1)\, \lambda(d_2)| \right).$$

Now note that for square-free d_1, d_2 the equality

$$f([d_1, d_2]) = \frac{f(d_1)\, f(d_2)}{f((d_1, d_2))}$$

holds and hence

$$\sum_{d|P} \frac{f(d)}{d} \sum_{[d_1,d_2]=d} \lambda(d_1)\lambda(d_2) = \sum_{d_1,d_2|P} \frac{\lambda(d_1)\lambda(d_2)}{[d_1,d_2]} f([d_1,d_2])$$

$$= \sum_{d_1,d_2|P} \frac{\lambda(d_1)\lambda(d_2)(d_1,d_2)}{d_1 d_2} \cdot \frac{f(d_1)f(d_2)}{f((d_1,d_2))}$$

$$= \sum_{d_1,d_2|P} \frac{\lambda(d_1)\lambda(d_2)}{h(d_1)h(d_2)} h((d_1,d_2))$$

$$= \sum_{d_1,d_2|P} \sum_{\delta|(d_1,d_2)} g(\delta) \frac{\lambda(d_1)\lambda(d_2)}{h(d_1)h(d_2)}$$

$$= \sum_{\delta|P} g(\delta) \left(\sum_{\delta|d|P} \frac{\lambda(d)}{h(d)} \right)^2 .$$

Finally

$$\sum_{d|P} |R(d)| \sum_{[d_1,d_2]=d} |\lambda(d_1)\lambda(d_2)| = \sum_{d_1,d_2|P} |\lambda(d_1)\lambda(d_2) R([d_1,d_2])|$$

and we obtain the assertion of the lemma. ∎

The following lemma allows us to choose $\lambda(d)$ so that the main term given by Lemma 4.2 is as small as possible.

Lemma 4.3. *Let $z \geqslant 2$ be a given number. If g and h are as in the preceding lemma, then the expression*

$$(4.11) \qquad \sum_{d|P} g(d) \left(\sum_{\substack{\delta|P \\ d|\delta}} \frac{\lambda(\delta)}{h(\delta)} \right)^2$$

attains, under the conditions $\lambda(1) = 1$, $\lambda(d) = 0$ for $d > z$, its minimum equal to

$$\left(\sum_{z \geqslant d|P} \frac{1}{g(d)} \right)^{-1} \text{ for}$$

$$(4.12) \qquad \lambda(d) = \frac{\mu(d)}{\sum_{m \leqslant z} \frac{1}{g(m)}} \prod_{p|d} \left(1 - \frac{f(p)}{p} \right)^{-1} \sum_{\substack{m \leqslant z/d \\ (m,d)=1 \\ m|P}} \frac{1}{g(m)} .$$

Proof. Make the linear transformation of the variables $\lambda(d)$ $(d \leqslant z)$ onto $x_d (d \leqslant z)$ by the formula

$$x_d = \sum_{d|\delta|P} \lambda(\delta) h^{-1}(\delta) \qquad (d \leqslant z).$$

Let us show that it is non-singular. For convenience we put $x_d = 0$ for $d > z$. For a fixed d dividing P we have

$$\sum_{n|d|P} x_d \mu\left(\frac{d}{n}\right) = \sum_{n|d|P} \mu\left(\frac{d}{n}\right) \sum_{d|\delta|P} \lambda(\delta) h^{-1}(\delta) = \sum_{\delta|P} \lambda(\delta) h^{-1}(\delta) \sum_{n|d|\delta} \mu\left(\frac{\delta}{n}\right),$$

and since

$$\sum_{n|d|\delta} \mu\left(\frac{d}{n}\right) = \sum_{r|\delta/n} \mu(r) = \begin{cases} 1 & \text{for} \quad \delta = n, \\ 0 & \text{for} \quad \delta \neq n, \end{cases}$$

we have

$$\sum_{n|d|P} x_d \mu\left(\frac{d}{n}\right) = \lambda(n) h(n)^{-1},$$

so that

$$\lambda(n) = h(n) \sum_{n|d|P} x_d \mu\left(\frac{d}{n}\right).$$

We see that our transformation can be inverted, thus it is non-singular.

Now the condition $\lambda(1) = 1$ takes the form

(4.13)
$$\sum_{z \geqslant d|P} x_d \mu(d) = 1$$

(because $h(1) = 1$ by multiplicativity of h).

We have to determine the point at which the quadratic form

$$\sum_{\substack{d|P \\ d \leqslant z}} g(d) x_d^2$$

assumes the minimum and (4.13) is satisfied.

Let

$$Q = \sum_{\substack{\delta|P \\ \delta \leqslant z}} \frac{1}{g(\delta)}.$$

Considering the identity

$$\sum_{\substack{d|P \\ d \leqslant z}} g(d) x_d^2 = \sum_{z \geqslant d|P} \frac{1}{g(d)} \left(g(d) x_d - \frac{\mu(d)}{Q}\right)^2 + \frac{1}{Q},$$

we see that our form attains its minimum equal to $\dfrac{1}{Q}$ for

$$x_d = \frac{1}{Q} \cdot \frac{\mu(d)}{g(d)},$$

where the condition (4.13) is satisfied. Indeed, in our case we have

$$\sum_{z \geqslant d \mid P} x_d \mu(d) = \sum_{z \geqslant d \mid P} \frac{\mu^2(d)}{g(d)} \cdot \frac{1}{\sum_{z \geqslant \delta \mid P} g^{-1}(\delta)}.$$

Returning to the variables $\lambda(d)$, we see that the expression (4.11) attains its minimum for

$$\lambda(d) = h(d) \sum_{\substack{\delta \leqslant z \\ d \mid \delta \mid P}} \mu\left(\frac{\delta}{d}\right) \frac{\mu(\delta)}{g(\delta)} \cdot \frac{1}{Q} = \frac{\mu(d) h(d)}{Q} \sum_{\substack{d \mid \delta \mid P \\ \delta \leqslant z}} \frac{1}{g(\delta)}$$

$$= \frac{\mu(d) h(d)}{Q} \sum_{\substack{r \leqslant z/d \\ r \mid P/d}} g^{-1}(rd) = \frac{\mu(d) h(d)}{Q g(d)} \sum_{\substack{r \leqslant z/d \\ r \mid P \\ (r, d) = 1}} g^{-1}(r),$$

since the conditions $r \mid P$, $(r, d) = 1$ are equivalent to the condition $r \mid P/d$. Since for square-free d we have

$$h(d) g^{-1}(d) = \prod_{p \mid d} h(p) g^{-1}(p) = \prod_{p \mid d} \left(1 - \frac{1}{h(p)}\right)^{-1} = \prod_{p \mid d} \left(1 - \frac{f(p)}{p}\right)^{-1},$$

we obtain the form of $\lambda(d)$ given in the assertion. ∎

After these introductory results we now prove a fundamental theorem concerning Selberg's sieve:

Theorem 4.4. *Let D be a finite set of primes and P be the product of them. Let A be a set of natural numbers consisting of N elements and satisfying the condition* (∗) *and let S be the number of elements of A not having divisors in D not exceeding a given number $z \geqslant 2$. If for $p \mid P$ we have $f(p) > 1$,*

$$g(n) = \sum_{d \mid n} \mu(n/d) \, d/f(d)$$

and

$$Q_z = \sum_{\substack{d \mid P \\ d \leqslant z}} g^{-1}(d),$$

and the numbers $\lambda(d)$ are defined by formula (4.12), then

$$S \leqslant \frac{N}{Q_z} + z^2 \prod_{\substack{p \in D \\ p \leqslant z}} \left(1 - \frac{f(p)}{p}\right)^{-2}.$$

Proof. Let us begin by estimating

$$R = \sum_{d_1, d_2 \mid P} |\lambda(d_1) \lambda(d_2)| \, |R([d_1, d_2])|,$$

where $\lambda(d)$ are given by formula (4.12) for $d \leq z$, and are zero for $d > z$, and $R(m)$ is defined by (4.8) and satisfies the inequality $|R(m)| \leq f(m)$.

In view of the assumption $f(d) > 1$ (for $d|P$) we obtain

$$R \leq \sum_{d_1, d_2 | P} |\lambda(d_1)\,\lambda(d_2)| \frac{f(d_1)f(d_2)}{f((d_1, d_2))}$$

$$\leq \sum_{d_1, d_2 | P} |\lambda(d_1)\,\lambda(d_2)| f(d_1) f(d_2) = \left(\sum_{\substack{d|P \\ d \leq z}} |\lambda(d)| f(d) \right)^2,$$

but formula (4.12) gives

$$|\lambda(d)| \leq \prod_{p|d} \left(1 - \frac{f(p)}{p} \right)^{-1} = \frac{d}{f(d)g(d)},$$

and so

$$R \leq \left(\sum_{z \geq d | P} dg^{-1}(d) \right)^2 \leq z^2 \left(\sum_{z \geq d | P} g^{-1}(d) \right)^2$$

$$\leq z^2 \prod_{z \geq p \in D} (1 + g^{-1}(p)) = z^2 \prod_{z \geq p \in D} \left(1 - \frac{f(p)}{p} \right)^{-1}.$$

From Lemmas 4.2 and 4.3 we obtain the assertion of the theorem. ∎

2. As an application we shall prove

Theorem 4.5. *There exists a constant* C *such that if* $k < x$, *then for any* l $((k, l) = 1)$ *the inequality*

$$\pi(x; k, l) \leq C \; \frac{x}{\varphi(k) \log \dfrac{x}{k}}$$

holds.

Proof. In Theorem 4.4 we take for D the set of all primes not dividing k and not exceeding $x^{\frac{1}{2}}$, for A the set of integers of the form $kn + l$ not exceeding x, with n natural. Then $N = [x/k]$, and

$$S = \pi(x; k, l) - \pi(\sqrt{x}; k, l) = \pi(x; k, l) + O(x^{\frac{1}{2}}) .$$

To determine the function $f(n)$, we note that if $(d, k) = 1$, then the congruence $ky + l \equiv 0 \pmod{d}$ has a solution which is determined uniquely $\pmod{d}$. Thus

$$S_d = \sum_{\substack{n \leq N \\ kn+l \equiv 0 (\mathrm{mod}\, d)}} 1 = \left[\frac{N}{d} \right] + \vartheta,$$

where $\vartheta = 0$ or 1, so that $S_d = N/d + R_d$, $|R_d| \leqslant 1$ and we see that $f(d) = 1$. Therefore

$$g(n) = \sum_{d|n} \mu\left(\frac{n}{d}\right) d = \varphi(n)$$

and

$$Q_z = \sum_{z \geqslant d|P} \varphi^{-1}(d).$$

In the sequel we shall suppose that N is sufficiently large because if

$$N = \left[\frac{x}{k}\right] \leqslant B \text{ with a constant } B > 1, \text{ then}$$

$$\pi(x; k, l) \ll \frac{x}{k} \ll \frac{x}{\varphi(k)\log B} \ll \frac{x}{\varphi(k)\log\frac{x}{k}},$$

and so the assertion becomes trivial.

Taking $z = N^{\frac{1}{2}} \log^{-2} N$ and using Theorem 4.4 we obtain

$$\pi(x; k, l) \leqslant \frac{N}{\sum\limits_{z \geqslant d|P} \varphi^{-1}(d)} + z^2 \sum_{\substack{p \leqslant z \\ p \nmid k}} \left(1 - \frac{1}{p}\right)^{-2} + O(x^{\frac{1}{2}}).$$

We now note that

$$\sum_{z \geqslant d|P} \varphi^{-1}(d) = \sum_{\substack{d \leqslant z \\ (k, d) = 1}} \frac{\mu^2(d)}{d} \prod_{p|d}\left(1 - \frac{1}{p}\right)^{-1}$$

$$= \sum_{\substack{d \leqslant z \\ (k, d) = 1}} \mu^2(d) \prod_{p|d}\left(\frac{1}{p} + \frac{1}{p^2} + \frac{1}{p^3} + \dots\right) \geqslant \sum_{\substack{n \leqslant z \\ (k, n) = 1}} \frac{1}{n},$$

and, if $p_1, \dots, p_r$ are all the prime divisors of k, then

$$\sum_{\substack{n \leqslant z \\ (k, n) = 1}} \frac{1}{n} = \sum_{\substack{n \leqslant z \\ (n, p_1 \dots p_{r-1}) = 1}} \frac{1}{n} - \sum_{\substack{n \leqslant z \\ p_r|n \\ (n, p_1, \dots, p_{r-1}) = 1}} \frac{1}{n}$$

$$= \sum_{\substack{n \leqslant z \\ (n, p_1 \dots p_{r-1}) = 1}} \frac{1}{n} - \frac{1}{p_r} \sum_{\substack{n \leqslant z/p_r \\ (n, p_1 \dots p_{r-1}) = 1}} \frac{1}{n} \geqslant \left(1 - \frac{1}{p_r}\right) \sum_{\substack{n \leqslant z \\ (n, p_1 \dots p_{r-1}) = 1}} \frac{1}{n}.$$

Repeating this procedure, we arrive at

$$\sum_{z \geqslant d|P} \varphi^{-1}(d) \geqslant \prod_{p|k}\left(1-\frac{1}{p}\right)\sum_{n \leqslant z}\frac{1}{n} \geqslant C\frac{\varphi(k)}{k}\log z,$$

where $C > 0$ is a constant independent of k and z. Moreover, on the basis of Corollary 5 to Theorem 3.2 we have

$$\prod_{\substack{p \leqslant z \\ p \nmid k}}\left(1-\frac{1}{p}\right)^{-2} \leqslant C_1 \log^2 z$$

with a suitable constant C_1, and hence we obtain

$$\pi(x;k,l) \leqslant \frac{kN}{C\varphi(k)\log z} + C_1 z^2 \log^2 z + O(x^{\frac{1}{2}}).$$

Since for a sufficiently large N we have

$$\log z = \tfrac{1}{2}\log N - z \log\log N > C_2 \log N$$

with a suitable constant C_2, it follows that

$$\pi(x;k,l) \leqslant \frac{kN}{CC_2\varphi(k)\log N} + C_1\frac{N}{\log^2 N} + O(\sqrt{x}).$$

In view of

$$N \log^{-2} N \leqslant \frac{kN}{\varphi(k)\log N},$$

we now obtain (with a suitable constant C_3)

$$\pi(x;k,l) \leqslant \frac{C_3 kN}{\varphi(k)\log N} + O(\sqrt{x}) \leqslant \frac{Cx}{\varphi(k)\log\dfrac{x}{k}},$$

where C is a constant independent of x and k. ∎

Corollary (Brun-Titchmarsh theorem). *If ϵ is a positive number, then there exists a positive constant $C = C(\epsilon)$ such that for all $k \leqslant x^{1-\epsilon}$ and $(k, l) = 1$ we have*

$$\pi(x;k,l) \leqslant C\frac{x}{\varphi(k)\log x}. \qquad ■$$

The second application of Selberg's theorem will be the estimation of the number $p(n)$ of representations of an even number as the sum of two primes. Namely, we shall prove

Theorem 4.6. *There exists a constant C such that*

$$p(a) = \sum_{\substack{p_1+p_2=a \\ p_i-\text{prime}}} 1 \leqslant \frac{Ca}{\log^2 a} \prod_{p|a}\left(1+\frac{2}{p}\right).$$

Proof. Let a be an even integer. In Theorem 4.4 we put $A = \{n(n-a): n = 1, 2, \ldots, a\}$, $N = a$, $D =$ the set of primes not dividing a and not exceeding $a^{\frac{1}{2}}$. Then, under the notation of Theorem 4.4 we have

$$p(a) = S + O(a^{\frac{1}{2}}) + O(\omega(a)) = S + O(a^{\frac{1}{2}}).$$

Formula (4.7) shows that for m dividing $P = \prod_{p\in D} p$ we have

$$S_m = \frac{d\left(\dfrac{m}{(a,\,m)}\right)}{m} N + R(m),$$

where

$$|R(m)| \leqslant d\left(\frac{m}{(a,\,m)}\right),$$

hence

$$f(m) = d\left(\frac{m}{(a,\,m)}\right) = d(m),$$

since from $m|P$ follows $(a, m) = 1$. For n dividing P we have

$$g(n) = \sum_{m|n} \mu\left(\frac{n}{m}\right)\frac{m}{d(m)} = \mu(n)\prod_{p|n}\left(1-\frac{p}{2}\right)$$

$$= \prod_{p|n}\left(\frac{p}{2}-1\right) = n2^{-\omega(n)}\prod_{p|n}\left(1-\frac{2}{p}\right),$$

and so

$$Q_z = \sum_{\substack{n|P \\ n\leqslant z}} \frac{2^{\omega(n)}}{n} \prod_{p|n}\left(1-\frac{2}{p}\right)^{-1},$$

and from Theorem 4.4, we obtain immediately the estimate

(4.14) $$S \leqslant \frac{N}{Q_z} + z^2 \prod_{\substack{p\leqslant a^{\frac{1}{2}} \\ p\leqslant z \\ p\nmid a}}\left(1-\frac{2}{p}\right)^{-2}.$$

Hence the whole question has been reduced to giving an estimation of Q_z and the optimal choice of the number z . We shall proceed as in the proof of the preceding theorem. If $a = p_1 \ldots p_r$, and $z < a^{\frac{1}{2}}$, then

$$Q_z \geqslant \sum_{\substack{n \leqslant z \\ (n,a)=1}} \frac{\mu^2(n) \, 2^{\omega(n)}}{n} = \sum_{\substack{n \leqslant z \\ (n,p_1 \ldots p_{r-1})=1}} \frac{\mu^2(n) \, 2^{\omega(n)}}{n} - \frac{1}{p_r} \sum_{\substack{n \leqslant z/p_r \\ (n,p_1 \ldots p_{r-1})=1}} \frac{\mu^2(np_r) \, 2^{\omega(np_r)}}{n}$$

$$\geqslant \left(1 - \frac{2}{p_r}\right) \sum_{\substack{n \leqslant z \\ (n,p_1 \ldots p_{r-1})=1}} \frac{\mu^2(n) \, 2^{\omega(n)}}{n}$$

and proceeding further in the same way, we arrive finally at

$$Q_z \geqslant \prod_{p|a} \left(1 - \frac{2}{p}\right) \sum_{n \leqslant z} \frac{\mu^2(n) \, 2^{\omega(n)}}{n}.$$

We may now write

$$\sum_{n \leqslant z} \frac{\mu^2(n) \, 2^{\omega(n)}}{n} = \sum_{n \leqslant z} \frac{\mu^2(n) \, d(n)}{n} = \sum_{mn \leqslant z} \frac{\mu^2(mn)}{mn}$$

$$\geqslant \sum_{m,n \leqslant z^{\frac{1}{2}}} \frac{\mu^2(mn)}{mn} = \sum_{m \leqslant z^{\frac{1}{2}}} \frac{\mu^2(m)}{m} \sum_{\substack{n \leqslant z^{\frac{1}{2}} \\ (m,n)=1}} \frac{\mu^2(n)}{n}.$$

Moreover, note that

$$\sum_{\substack{n \leqslant z^{\frac{1}{2}} \\ (m,n)=1}} \frac{\mu^2(n)}{n} \geqslant \sum_{\substack{n \leqslant z^{\frac{1}{2}} \\ (m,n)=1}} \frac{1}{n} - \sum_p \sum_{\substack{n \leqslant z^{\frac{1}{2}} \\ (m,n)=1 \\ p^2|n}} \frac{1}{n} \geqslant \sum_{\substack{n \leqslant z^{\frac{1}{2}} \\ (m,n)=1}} \frac{1}{n} - \sum_p \frac{1}{p^2} \sum_{\substack{n \leqslant z^{\frac{1}{2}} \\ (m,n)=1}} \frac{1}{n}$$

$$= \left(1 - \sum_p \frac{1}{p^2}\right) \sum_{\substack{n \leqslant z^{\frac{1}{2}} \\ (m,n)=1}} \frac{1}{n} \geqslant C \sum_{\substack{n \leqslant z^{\frac{1}{2}} \\ (m,n)=1}} \frac{1}{n},$$

where $C = 1 - \sum_p \dfrac{1}{p^2} \geqslant 1 - (\frac{1}{6}\pi^2 - 1) > 0$. Hence

$$\sum_{\substack{n \leqslant z^{\frac{1}{2}} \\ (m,n)=1}} \frac{\mu^2(n)}{n} \geqslant \tfrac{1}{2}C \prod_{p|m} \left(1 - \frac{1}{p}\right) \log z = \frac{\varphi(m)}{m} C_1 \log z$$

and

$$Q_z \geqslant C_1 \prod_{p|a} \left(1 - \frac{2}{p}\right) \sum_{m \leqslant z^{\frac{1}{2}}} \frac{\mu^2(m)}{m} \cdot \frac{\varphi(m)}{m} \cdot \log z.$$

Finally

$$\sum_{m \leqslant z^{\frac{1}{2}}} \frac{\mu^2(m)\varphi(m)}{m^2} = \sum_{d \leqslant z^{\frac{1}{2}}} \frac{\mu(d)}{d} \sum_{d|m \leqslant z^{\frac{1}{2}}} \frac{\mu^2(m)}{m} = \sum_{d \leqslant z^{\frac{1}{2}}} \frac{\mu^2(d)}{d^2} \sum_{\substack{m \leqslant z^{\frac{1}{2}}/d \\ (m,d)=1}} \frac{\mu^2(m)}{m}$$

$$\geqslant \sum_{d \leqslant z^{\frac{1}{2}}} \frac{\mu^2(d)}{d^2} \cdot \frac{\varphi(d)}{d} C_1 \log\left(\frac{z}{d^2}\right)$$

$$\geqslant C_1 \log z \sum_{d \leqslant z^{\frac{1}{2}}} \frac{\mu^2(d)\varphi(d)}{d^3} - 2C_1 \sum_{d \leqslant z^{\frac{1}{2}}} \frac{\mu^2(d)\varphi(d)\log d}{d^3}$$

$$\geqslant C_2 \log z - C_3$$

and

$$Q_z \gg \prod_{p|a}\left(1 - \frac{2}{p}\right)\log^2 z .$$

Since

$$\prod_{\substack{p \leqslant a^{\frac{1}{2}} \\ p \leqslant z \\ p \nmid a}}\left(1 - \frac{2}{p}\right)^{-2} \leqslant \prod_{p \leqslant a^{\frac{1}{2}}}\left(1 - \frac{2}{p}\right)^{-2} \ll \log^4 a$$

and

$$\prod_{p|a}\left(1 - \frac{2}{p}\right) = \frac{\prod_{p|a}\left(1 - \frac{4}{p^2}\right)}{\prod_{p|a}\left(1 + \frac{2}{p}\right)} \gg \prod_{p|a}\left(1 + \frac{2}{p}\right)^{-1},$$

we conclude from (4.14) that

$$S \ll \frac{a}{\log^2 z} \prod_{p|a}\left(1 + \frac{2}{p}\right) + z^2 \log^4 a .$$

Now, taking $z = a^{\frac{1}{2}} \log^{-3} a$, we obtain

$$S \ll \frac{a}{\log^2 a} \prod_{p|a}\left(1 + \frac{2}{p}\right) + \frac{a}{\log^2 a} . \qquad \blacksquare$$

In Chapter V the following result will be useful:

Corollary. *As* $x \to \infty$ *we have*

$$\sum_{n \leqslant x} p^2(n) = O\left(\frac{x^3}{\log^4 x}\right).$$

Proof.

$$\sum_{n \leqslant x} p^2(n) \ll \frac{x^2}{\log^4 x} \sum_{n \leqslant x} \prod_{p|n}\left(1+\frac{2}{p}\right)^2,$$

but

$$\left(1+\frac{2}{p}\right)^2 = 1+\frac{4}{p}+\frac{4}{p^2} \leqslant 1+\frac{8}{p},$$

so

$$\sum_{n \leqslant x} \prod_{p|n}\left(1+\frac{2}{p}\right)^2 \leqslant \sum_{n \leqslant x} \prod_{p|n}\left(1+\frac{8}{p}\right) = \sum_{n \leqslant x} \sum_{d|n} \mu^2(d)\frac{8^{\omega(d)}}{d}$$

$$= \sum_{d \leqslant x} \mu^2(d)\frac{8^{\omega(d)}}{d}\left[\frac{x}{d}\right] \leqslant x \sum_{d \leqslant x} \frac{\mu^2(d)8^{\omega(d)}}{d^2} \leqslant x \prod_{p \leqslant x}\left(1+\frac{8}{p^2}\right) \ll x,$$

for the series $\displaystyle\sum_p \frac{8}{p^2}$ is convergent. ∎

Exercises

1.   Prove that using Selberg's sieve one can remove the factor $(\log \log x)^2$ in Theorem 4.3.

2.   Show that if k is an even integer, then the number of primes $p \leqslant x$ for which $kp + 1$ is also a prime does not exceed

$$c\frac{x}{\log^2 x} \prod_{p|k}\left(1-\frac{1}{p}\right)^{-1},$$

where c does not depend upon k and x (P. Erdös [1]).

3.   Prove that the number of primes of the form $n^2 + 1$ with $n \leqslant x$ is equal to $O\left(\dfrac{x}{\log x}\right)$.

4.   Find an estimate from above for the number of $n \leqslant x$ such that all the numbers

$$a_1 n + b_1, \; a_2 n + b_2, \; \ldots, \; a_r n + b_r,$$

are primes. Here $a_1, \ldots, a_r, b_1, \ldots, b_r$ are given integers.

§3. The large sieve

1. In 1941 Ju. V. Linnik obtained a result concerning the distribution of terms of any sequence of natural numbers in arithmetic progressions whose common difference are prime numbers. This result enabled him to prove important results in the theory of distribution of quadratic non-residues and Dirichlet L-function. Later many analogous results were found, proofs were simplified and a series of interesting applications were obtained. This whole circle of methods is the *large sieve*, because in the sieves of Eratosthenes and of Selberg, the integers divisible by primes from a given set one sieves out, while in the large sieve all integers are sieved out which lie in $\Phi(p)$ given residue classes (mod p). Here it has become usual to call the case $\Phi(p) = O(1)$ the small sieve, and the general case, the large sieve. Thus the sieves of Eratosthenes and of Selberg are small sieves, for in those cases, $\Phi(p) = 0, 1$.

In this section we shall prove by the method of P. X. Gallagher, the result first obtained by E. Bombieri [1] in 1965 by considerably much difficult methods, which leads easily to the results of Linnik, and also enables us to obtain many interesting applications. As an example we shall prove Gallagher's result on primitive roots.

2.   In our considerations, the following theorem will be fundamental:

Theorem 4.7 (E. Bombieri). *Let* $M \geqslant 0$, $N \geqslant 1$ *be integers and* $a_{M+1}, \cdots, a_{M+N}$ *be complex numbers. If*

$$S(t) = \sum_{n=1+M}^{N+M} a_n \exp(2\pi i n t), \quad Z = \sum_{n=1+M}^{N+M} |a_n|^2,$$

then for any $Q \geqslant 2$ *the estimate*

$$\sum_{q \leqslant Q} \sum_{\substack{1 \leqslant a \leqslant q \\ (a,q)=1}} \left| S\left(\frac{a}{q}\right) \right|^2 \leqslant (Q^2 + \pi N) Z.$$

holds.

Remark. Recently, H. L. Montgomery and R. C. Vaughan [1] have shown that in the last inequality, the factor π can be omitted.

Proof. If $F(t)$ is any complex-valued function, periodic with period 1, having the continuous first derivative, then for any x, y we have obviously

$$F(x) = F(y) - \int_x^y F'(t)\, dt,$$

and so

$$|F(x)| \leqslant |F(y)| + \left| \int_x^y |F'(t)|\, dt \right|.$$

We apply this with $x = a/q$. Integrate with respect to y the obtained inequality

$$\left| F\left(\frac{a}{q}\right) \right| \leqslant |F(y)| + \left| \int_{a/q}^{y} |F'(t)| \, dt \right|$$

over the interval $I_{a,q} = (a/q - 1/2Q^2, \, a/q + 1/2Q^2)$ of length $1/Q^2$. This gives

$$Q^{-2}\left| F\left(\frac{a}{q}\right) \right| \leqslant \int_{I_{a,q}} |F(y)| \, dy + \int_{I_{a,q}} dy \left| \int_{a/q}^{y} |F'(t)| \, dt \right|,$$

but in view of

$$\int_{I_{a,q}} dy \left| \int_{a/q}^{y} |F'(t)| \, dt \right| = \int_{a/q}^{a/q+\frac{1}{2}Q^2} dy \left\{ \left| \int_{a/q}^{y} |F'(t)| \, dt \right| + \left| \int_{2a/q-y}^{a/q} |F'(t)| \, dt \right| \right\}$$

$$\leqslant \int_{a/q}^{a/q+\frac{1}{2}Q^2} dy \int_{I_{a,q}} |F'(t)| \, dt = \tfrac{1}{2} Q^{-2} \int_{I_{a,q}} |F'(t)| \, dt$$

we obtain finally

$$\left| F\left(\frac{a}{q}\right) \right| \leqslant Q^2 \int_{I_{a,q}} |F(y)| \, dy + \tfrac{1}{2} \int_{I_{a,q}} |F'(t)| \, dt.$$

Denote by $J_{a,q}$ the set $\{\{x\}: \, x \in I_{a,q}\} \subset [0,1)$. By using $F(x+1) = F(x)$ we may rewrite the obtained inequality as follows:

$$(4.15) \qquad \left| F\left(\frac{a}{q}\right) \right| \leqslant Q^2 \int_{J_{a,q}} |F(y)| \, dy + \tfrac{1}{2} \int_{J_{a,q}} |F'(t)| \, dt.$$

Now we note that if $1 \leqslant a \leqslant q$, $(a, q) = 1$ and $q \leqslant Q$, then the sets $J_{a,q}$ are disjoint. Indeed, if for example $J_{a,q}$ and $J_{b,r}$ contain a common point, say x, then there exist y in $I_{a,q}$ and z in $I_{b,r}$ for which $\{y\} = \{z\} = x$, and hence the difference $y - z$ is integral. If $y = z$, then from

$$(4.16) \qquad \left| y - \frac{a}{q} \right| < \frac{1}{2Q^2}, \qquad \left| z - \frac{b}{r} \right| < \frac{1}{2Q^2}$$

we obtain

$$\frac{1}{Q^2} \leqslant \frac{|ar - bq|}{rq} < \frac{1}{Q^2},$$

which is impossible. If, on the contrary, $|y - z| \geqslant 1$, then the inequalities (4.16) lead to

$$1 \leqslant |y-z| \leqslant \left|y - \frac{a}{q}\right| + \left|z - \frac{b}{r}\right| + \left|\frac{a}{q} - \frac{b}{r}\right| < \frac{1}{Q^2} + \left|\frac{a}{q} - \frac{b}{r}\right| = \frac{1}{Q^2} + \frac{|ar - bq|}{qr},$$

and so

$$Q^{-2} > 1 - \frac{|ar - bq|}{qr}.$$

On the other hand, we have

$$\frac{|ar - bq|}{qr} = \left|\frac{a}{q} - \frac{b}{r}\right| < 1,$$

and so

$$1 - \frac{|ar - bq|}{qr} > 0 \quad \text{and} \quad 1 - \frac{|ar - bq|}{qr} = \frac{qr - |ar - bq|}{qr} \geqslant \frac{1}{Q^2}.$$

Finally we obtain

$$Q^{-2} > 1 - \frac{|ar - bq|}{qr} \geqslant Q^{-2},$$

which is impossible.

Using the disjointness of the sets $J_{a,q}$ we get from the inequality (4.15),

$$(4.17) \qquad \sum_{q \leqslant Q} \sum_{\substack{1 \leqslant a \leqslant q \\ (a,q)=1}} \left|F\left(\frac{a}{q}\right)\right| \leqslant Q^2 \int_0^1 |F(t)|\, dt + \frac{1}{2}\int_0^1 |F'(t)|\, dt.$$

We apply this inequality to the function $F(x) = T^2(x)$, where $T(x) = S(x)\exp(-2\pi i m)$, where for even N we put $m = \frac{1}{2}N + M$, and for odd N we put $m = \frac{1}{2}(N + 1) + M$. Clearly

$$T(x) = \sum_{|k| \leqslant \frac{1}{2}N} a_{m+k}\exp\{2\pi i k x\} \quad \text{and} \quad |S(x)| = |T(x)|.$$

Since

$$\int_0^1 |F(t)|\, dt = \int_0^1 |S^2(t)|\, dt = \sum_{n=1+N}^{N+M} |a_n|^2 = Z$$

and

$$\int\limits_0^1 |F'(t)|\,dt = 2\int\limits_0^1 |T(t)T'(t)|\,dt \leqslant 2\left(\int\limits_0^1 |S^2(t)|\,dt\right)^{\frac{1}{2}}\left(\int\limits_0^1 |T'(t)|^2\,dt\right)^{\frac{1}{2}}$$

$$= 2Z^{\frac{1}{2}}\left(\int\limits_0^1 |T'(t)|^2\,dt\right)^{\frac{1}{2}},$$

but

$$T'(t) = 2\pi i \sum_{|k|\leqslant\frac{1}{2}N} k a_{m+k}\exp\{2\pi ikx\}$$

gives

$$\int\limits_0^1 |T'(t)|^2\,dt = 4\pi^2 \sum_{|k|\leqslant\frac{1}{2}N} k^2 |a_{m+k}|^2 \leqslant \pi^2 N^2 Z,$$

hence the inequality (4.17) leads to

$$\sum_{q\leqslant Q}\sum_{\substack{1\leqslant a\leqslant q\\(a,q)=1}} \left|S\left(\frac{a}{q}\right)\right|^2 \leqslant (Q^2+\pi N)Z. \quad\blacksquare$$

Corollary 1. *If A is a set of M natural numbers lying in an interval of length N, and for a prime p we denote by $Z(a,p)$ the number of elements of A congruent to $a(\bmod\, p)$, then for any $Q \geqslant 2$ we have*

$$\sum_{p\leqslant Q} p \sum_{a=1}^p \left(Z(a,p)-\frac{M}{p}\right)^2 \leqslant (Q^2+\pi N)\,M.$$

Proof. We take in the theorem just proven

$$a_m = \begin{cases} 1 & \text{for} \quad m\in A, \\ 0 & \text{for} \quad m\notin A. \end{cases}$$

Then $M = Z$, and moreover, when p is a prime number,

$$\sum_{a=1}^{p-1}\left|S\left(\frac{a}{p}\right)\right|^2 = \sum_{a=1}^{p-1}\sum_{\substack{n\in A\\m\in A}}\exp\left(2\pi i\,\frac{n-m}{p}\,a\right) = \sum_{m,n\in A}\sum_{a=1}^{p-1}\exp\left(2\pi i\frac{n-m}{p}\,a\right)$$

$$= \sum_{m,n\in A}\left(\sum_{a=1}^{p}\exp\left(2\pi i\frac{n-m}{p}\,a\right)-1\right) = \sum_{\substack{m,n\in A\\m\equiv n(\bmod\, p)}}(p-1)-\sum_{\substack{m,n\in A\\m\not\equiv n(\bmod\, p)}}1$$

by virtue of Lemma 1.5. Furthermore,

$$\sum_{a=1}^{p-1}\left|S\left(\frac{a}{p}\right)\right|^2 = p \sum_{\substack{m,\,n\in A \\ m\,\equiv\,n(\mathrm{mod}\,p)}} 1 - M^2 = p \sum_{a=1}^{p}\left(\sum_{\substack{m\in A \\ m\,\equiv\,a(\mathrm{mod}\,p)}} 1\right)^2 - M^2$$

$$= p \sum_{a=1}^{p} Z^2(a,\,p) - M^2 = p \sum_{a=1}^{p}\left(Z(a,\,p) - \frac{M}{p}\right)^2,$$

and this in turn implies

$$\sum_{p\leqslant Q} p \sum_{a=1}^{p}\left(Z(a,\,p) - \frac{M}{p}\right)^2 = \sum_{p\leqslant Q} \sum_{a=1}^{p-1}\left|S\left(\frac{a}{p}\right)\right|^2$$

$$\leqslant \sum_{q\leqslant Q} \sum_{\substack{1\leqslant a\leqslant q \\ (a,\,q)=1}}\left|S\left(\frac{a}{q}\right)\right|^2 \leqslant (Q^2 + \pi N)\,M. \quad\blacksquare$$

We can now prove Linnik's theorem on the large sieve.

Corollary 2 (Ju. V. Linnik [1]). *Let N be a given natural number and for every prime p not exceeding $N^{\frac{1}{2}}$, $\Phi(p)$ distinct residue classes $(\mathrm{mod}\,p)$ be given, where $0\leqslant\Phi(p)<p$. If now I is any interval of length N, then in I there are at most*

$$\frac{(1+\pi)\,N}{\displaystyle\sum_{p\leqslant\sqrt{N}}\frac{\Phi(p)}{p-\Phi(p)}}$$

integers not lying in any of the given residue classes.

Proof. Let A be the set of those integers $n\in I$ which do not lie in any residue classes chosen. We denote the number of elements of A by M. Then for $p\leqslant N^{\frac{1}{2}}$ we have

$$\sum_{a=1}^{p} Z(a,\,p) = M,$$

$Z(a,\,p)=0$ for $\Phi(p)$ classes of residues $a(\mathrm{mod}\,p)$, and here, by Corollary 1

(4.18) $$\sum_{p\leqslant\sqrt{N}} p \sum_{a=1}^{p}\left(Z(a,\,p) - \frac{M}{p}\right)^2 \leqslant (1+\pi)\,NM.$$

We denote the set of the residue classes chosen for a given prime p by B_p. Then

$$\sum_{a=1}^{p}\left(Z(a,\,p) - \frac{M}{p}\right)^2 = \Phi(p)\frac{M^2}{p^2} + \sum_{a\notin B_p}\left(Z(a,\,p) - \frac{M}{p}\right)^2.$$

Since the minimum of the function

$$\sum_{a \notin B_p} \left(x_a - \frac{M}{p} \right)^2$$

under the condition $\sum_{a \notin B_p} x_a = M$ is equal to

$$\left(\frac{M}{p - \Phi(p)} - \frac{M}{p} \right)^2 (p - \Phi(p)),$$

which can easily be checked e.g. by Legendre's indeterminate multiplier method, we obtain

$$\sum_{a=1}^{p} \left(Z(a, p) - \frac{M}{p} \right)^2 \geq \frac{\Phi(p) M^2}{p^2} + (p - \Phi(p)) \left(\frac{M}{p - \Phi(p)} - \frac{M}{p} \right)^2$$
$$= \frac{M^2 \Phi(p)}{p(p - \Phi(p))},$$

and this, along with (4.18), implies

$$\sum_{p \leq N^{\frac{1}{2}}} \frac{M \Phi(p)}{p - \Phi(p)} \leq (1 + \pi) N,$$

which is equivalent to our assertion. ∎

3. In 1927 Emil Artin conjectured that every integer which is $\neq -1$ and not a square, is a primitive root for infinitely many prime numbers. No proof of this has been found. It is known only that if the extended Riemann hypothesis concerning the zeros of the Dedekind zeta-functions ζ (which are generalizations of the Riemann zeta-function) is true, then Artin's conjecture is also true, and moreover for the number $N_a(x)$ of primes not exceeding x whose primitive root is a, the asymptotic equality

$$N_a(x) = A(a) x \log^{-1} x + O(x \log \log x \log^{-2} x), \qquad A(a) > 0$$

holds (C. Hooley [1]).

Without the use of any unproven hypothesis, P. J. Stephens [1] has shown that for $N > \exp(4 \log^{\frac{1}{2}} x \log \log^{\frac{1}{2}} x)$ we have

$$\frac{1}{N} \sum_{a \leq N} N_a(x) = \prod_p \left(1 - \frac{1}{p(p-1)} \right) \operatorname{li} x + O\left(\frac{x}{\log^D x} \right),$$

where D is any constant > 1. The proof of this result rests on the large sieve.

We shall now prove a result due to P. X. Gallagher [1], implying that almost every natural number is a primitive root for some prime.

Theorem 4.8. *Let I be any interval of length N and let $F(I)$ denote the number of natural numbers in this interval which are not primitive roots for any prime $p \leq N^{\frac{1}{2}}$.*

Then

$$F(I) \ll N^{\frac{1}{2}} \log N.$$

Remark. On the other hand, it is easy to observe that for the interval $I = [1, N]$, we have $F(I) \gg N^{\frac{1}{2}}$, because squares are not primitive roots, and there are $[N^{\frac{1}{2}}]$ squares in I.

Proof. We shall apply Corollary 2 to the preceding theorem. For each prime $p < N^{\frac{1}{2}}$ we choose residues (mod p) consisting of primitive roots (mod p). From the Corollary to Theorem 1.19 it follows that in this case $\Phi(p) = \varphi(p-1)$. Since $F(I)$ is the number of integers not lying in any of these residue classes, we have

(4.19)
$$F(I) \leqslant \frac{(1 + \pi) N}{\displaystyle\sum_{p \leqslant N^{\frac{1}{2}}} \frac{\varphi(p-1)}{p - \varphi(p-1)}}$$

Hence there remains to estimate from below the sum

(4.20)
$$\sum_{p \leqslant N^{\frac{1}{2}}} \frac{\varphi(p-1)}{p - \varphi(p-1)} \geqslant \sum_{p \leqslant N^{\frac{1}{2}}} \frac{\varphi(p-1)}{p-1} = S.$$

Since

$$\frac{N}{\log^2 N} \ll \pi(N^{\frac{1}{2}})^2 = \left(\sum_{p \leqslant N^{\frac{1}{2}}} \left(\frac{\varphi(p-1)}{p-1} \right)^{\frac{1}{2}} \left(\frac{p-1}{\varphi(p-1)} \right)^{\frac{1}{2}} \right)^2 \leqslant S \cdot \sum_{p \leqslant N^{\frac{1}{2}}} \frac{p-1}{\varphi(p-1)},$$

we obtain

(4.21)
$$S \gg \left(\sum_{p \leqslant N^{\frac{1}{2}}} \frac{p-1}{\varphi(p-1)} \right)^{-1} N \log^{-2} N.$$

We shall now need an elementary lemma which is interesting in itself:

Lemma 4.4. *There exists a constant $C > 0$ such that for any natural number n we have*

$$\sigma(n) \varphi(n) \geqslant C n^2.$$

Proof. Since the functions $\sigma(n)$, $\varphi(n)$ and n^2 are multiplicative, we have

$$\frac{\sigma(n) \varphi(n)}{n^2} = \prod_{p^\alpha \| n} \frac{p^{\alpha+1} - 1}{p-1} p^{\alpha-1}(p-1) p^{-2\alpha} = \prod_{p^\alpha \| n} \left(1 - \frac{1}{p^{1+\alpha}} \right)$$

$$\geqslant \prod_{p | n} \left(1 - \frac{1}{p^2} \right) \geqslant \prod_{p} \left(1 - \frac{1}{p^2} \right) = \left(\sum_{n=1}^{\infty} \frac{1}{n^2} \right)^{-1} = \frac{6}{\pi^2}. \quad \blacksquare$$

Corollary. *For any natural number* n, *we have*

$$\sum_{d|n} \frac{1}{d} \gg \frac{n}{\varphi(n)}.$$

Proof. Indeed,

$$\sum_{d|n} \frac{1}{d} = \sum_{d|n} \frac{d}{n} = \frac{\sigma(n)}{n} \gg \frac{n}{\varphi(n)}. \quad \blacksquare$$

Using this corollary, we may write

$$\sum_{p \leqslant N^{\frac{1}{2}}} \frac{p-1}{\varphi(p-1)} \ll \sum_{p \leqslant N^{\frac{1}{2}}} \sum_{d|p-1} \frac{1}{d} = \sum_{d \leqslant N^{\frac{1}{2}}} \frac{1}{d} \sum_{\substack{p \leqslant N^{\frac{1}{2}} \\ p \equiv 1 (\mathrm{mod}\, d)}} 1$$

$$= \sum_{d \leqslant \log N} \frac{\pi(N^{\frac{1}{2}}; d, 1)}{d} + \sum_{\log N < d \leqslant N^{\frac{1}{2}}} \frac{\pi(N^{\frac{1}{2}}; d, 1)}{d}$$

$$\ll \sum_{d \leqslant \log N} \frac{N^{\frac{1}{2}}}{d\varphi(d) \log N} + \sum_{\log N < d \leqslant N^{\frac{1}{2}}} \frac{N^{\frac{1}{2}}}{d^2}.$$

Herein, applying the Brun-Titchmarsh theorem (Corollary to Theorem 4.5) to the first sum, and to the second, the trivial estimate $\pi(N^{\frac{1}{2}}; d, l) \ll N^{\frac{1}{2}} d^{-1}$, we obtain

$$\sum_{p \leqslant N^{\frac{1}{2}}} \frac{p-1}{\varphi(p-1)} \ll \frac{N^{\frac{1}{2}}}{\log N} \sum_{d \leqslant \log N} \frac{1}{d\varphi(d)} + N^{\frac{1}{2}} \sum_{d > \log N} \frac{1}{d^2} \ll N^{\frac{1}{2}} \log^{-1} N,$$

because in view of $\varphi(d) \geqslant \pi(d) + O(1) \gg \dfrac{d}{\log d}$ the series $\displaystyle\sum_{d=1}^{\infty} \frac{1}{d\varphi(d)}$ is convergent.

From the estimate obtained and the formulas (4.19), (4.20) and (4.21) we obtain

$$F(I) \ll N^{\frac{1}{2}} \log N$$

as asserted. $\quad \blacksquare$

Corollary. *Almost every natural number* n *is a primitive root for some prime.* $\quad \blacksquare$

The reader can find many results connected with the large sieve in the book of H. L. Montgomery [1].

Exercises

1. Prove that if $0 \leqslant t < \frac{1}{2}$, then the number of those natural numbers in a given interval of length N whose order in the group $G(p)$ does not exceed N^t for every prime $p \leqslant N^{\frac{1}{2}}$, is equal to $O(N^t \log N)$. Here the constant in the Landau symbol can be chosen independent of t if

$0 \leqslant t < \frac{1}{2} - \epsilon$, where $\epsilon > 0$.

2.   Let $a < b$ and let $f(x)$ be a complex-valued function with a continuous derivative in (a, b). Show that

$$\left| f\left(\frac{a+b}{2}\right) \right| \leqslant \frac{1}{b-a} \int_a^b |f(x)| \, dx + \frac{1}{2} \int_a^b |f'(x)| \, dx.$$

3.   Let $a < b$, $0 < \delta < b - a$ and let X be some finite subset of the interval $[a + \frac{1}{2}$, $b - \frac{1}{2}]$. If $N_c(t)$ denotes the number of elements of X distant from t by less than c, and f satisfies the conditions of the preceding problem, then

$$\sum_{t \in X} \frac{|f(t)|}{N_\delta(t)} \leqslant \frac{1}{\delta} \int_a^b |f(x)| \, dx + \frac{1}{2} \int_a^b |f'(x)| \, dx.$$

4.   Let $S(t)$ be the function appearing in Theorem 4.7, X be a finite subset of the interval $[0, 1)$ and let

$$\delta = \min_{\substack{x_1, x_2 \in X \\ x_1 \neq x_2}} \|x_1 - x_2\|,$$

where $\|t\|$ denotes the distance of t from the nearest integer.

   Show that

$$\sum_{x \in X} |S(x)|^2 \leqslant \left(\frac{1}{\delta} + \pi N \right) \sum_{n = 1 + M}^{N + M} |a_n|^2.$$

(The results of Exercises 1 - 4 are due to P. X. Gallagher [1]. Cf. also H. L. Montgomery [1]).

CHAPTER V

GEOMETRY OF NUMBERS

§1. Convex sets

1. In this chapter we shall be concerned with problems of geometrical nature appearing in number theory. The first section is purely of a geometrical character and in reality has little to do with number theory, but the concepts introduced there will have later purely arithmetic applications, above all in the proof of Waring's theorem on the representation of natural numbers as a sum of k-th powers. The next sections will be concerned with the theory of lattices and the determination of the number of lattice points in various plane regions.

From the theory of convex sets we shall give only basic results which will be needed later. The reader, interested in this theory, should consult Eggleston's book [1].

We define the convex sets in any linear space E of finite dimension over $\mathscr{R}$ in the following manner: a non-empty subset A of E is called *convex* if for any x_1, $x_2 \in A$ and any non-negative real t_1, t_2 whose sum is 1, the point $t_1 x_1 + t_2 x_2$ belongs to A.

Geometrically, this means that together with two points, the line segment joining them is also contained in the set A.

Lemma 5.1. *If A is a convex set, $t_1, \ldots, t_k \geqslant 0$ and $t_1 + \ldots + t_k = 1$, then from $x_1, \ldots, x_k \in A$ follows $\sum_{i=1}^{k} t_i x_i \in A$.*

Proof. Let us apply induction on k. For $k = 2$ the assertion is contained in the definition of a convex set. Suppose that the lemma is true for some k and let

$$x = \sum_{i=1}^{k+1} t_i x_i, \qquad \sum_{i=1}^{k+1} t_i = 1,$$

where $x_i \in A$, $t_i \geqslant 0$ $(i = 1, 2, \ldots, k + 1)$.

Here we may suppose that $t_{k+1} < 1$, because otherwise we would have $x = x_{k+1} \in A$. This gives $t_1 + \ldots + t_k > 0$, therefore we may write

$$x = (t_1 + \ldots + t_k) \left(\frac{t_1}{t_1 + \ldots + t_k} x_1 + \ldots + \frac{t_k}{t_1 + \ldots + t_k} x_k \right) + t_{k+1} x_{k+1}$$

and applying the induction hypothesis we obtain $x \in A$. ∎

In the next chapter, the following simple lemma will be useful:

Lemma 5.2. *If* $T: E \to E$ *is a linear transformation, and* A *is a convex subset of the linear space* E, *then* $T(A)$ *is also a convex set.*

Proof. If $T(x_1)$, $T(x_2) \in T(A)$ and $t_1, t_2 \geqslant 0$, $t_1 + t_2 = 1$, then

$$t_1 T(x_1) + t_2 T(x_2) = T(t_1 x_1 + t_2 x_2) \in T(A) .$$ ∎

If B is an arbitrary subset of E, then the intersection of all convex sets containing B is called the *convex closure* of B and is denoted by $c(B)$.

Lemma 5.3. *The set* $c(B)$ *is a convex set for any non-empty set* $B \subset E$.

Proof. Let $\mathscr{A}$ be the family of all convex sets of E containing B. Then

$$c(B) = \bigcap_{A \in \mathscr{A}} A$$

and if $x_1, x_2 \in c(B)$, $t_1 + t_2 = 1$, $t_1, t_2 \geqslant 0$, then for any $A \in \mathscr{A}$ we have $x_1, x_2 \in A$, and hence $t_1 x_1 + t_2 x_2 \in A$, whence $t_1 x_1 + t_2 x_2 \in c(B)$. ∎

The following lemma describes the structure of the convex closure of an arbitrary subset of the space E:

Lemma 5.4. *If* $B \subset E$, *then the set* $c(B)$ *coincides with the set* $\hat{B}$ *of all the points of* E *expressible in the form*

$$t_1 x_1 + \ldots + t_k x_k ,$$

where k *is a natural number,* $x_1, \ldots, x_k$ *lie in* B, *and the coefficients* t_i *are non-negative and satisfy* $t_1 + \ldots + t_k = 1$.

Proof. From Lemma 5.3 it follows that all the points of the set $\hat{B}$ lie in each convex set containing B, and hence $\hat{B} \subset c(B)$. In order to prove the reverse inclusion let us note that the set $\hat{B}$ is convex. Indeed, if

$$x = \sum_{j=1}^{k} t_j x_j , \qquad y = \sum_{j=1}^{l} u_j y_j$$

$\left(t_j, u_j \geqslant 0, \ \sum_{j=1}^{k} t_j = \sum_{j=1}^{l} u_j = 1; x_1, \ldots, x_k, y_1, \ldots, y_l \in B \right)$ belong to $\hat{B}$ and $s_1 + s_2 = 1$,

$s_1, s_2 \geq 0$, then

$$s_1 x + s_2 y = \sum_{j=1}^{k} s_1 t_j x_j + \sum_{j=1}^{l} s_2 u_j y_j,$$

but

$$\sum_{j=1}^{k} s_1 t_j + \sum_{j=1}^{l} s_2 u_j = \sum_{j=1}^{k} s_1 t_j + \sum_{j=1}^{l} u_j - \sum_{j=1}^{l} s_1 u_j = 1,$$

and so $s_1 x + s_2 y \in \hat{B}$. Since, moreover, $B \subset \hat{B}$, we have $c(B) \subset \hat{B}$. ∎

Corollary (Carathéodory's theorem). *If $dim\ E = n$, $B \subset E$ and $x \in c(B)$, then there exists $s \leq 1 + n$ points $x_1, \ldots, x_s$ of the set B and non-negative numbers $t_1, \ldots, t_s$ with sum 1 such that*

$$x = t_1 x_1 + \ldots + t_s x_s \ .$$

If all the coordinates of the points of B in a certain basis are rational, then we can choose the numbers $t_1, \ldots, t_s \in \mathcal{Q}$.

Proof. If x lies in $c(B)$, then by Lemma 5.4 there exists a natural number k such that

(5.1) $$x = \sum_{j=1}^{k} t_j x_j \ ,$$

where $t_j \geq 0$, $t_1 + \ldots + t_k = 1$ and $x_j \in B$. Thus it suffices to show that if k does not exceed $1 + n$, then there exists an analogous expression for the element x in which there are at most $k - 1$ terms. Hence suppose that $k > n + 1$ and $t_1, \ldots, t_k > 0$. The points $x_2 - x_1, \ldots, x_k - x_1$ are linearly dependent, for there are $k - 1 > n$ of them, and hence there exist real coefficients a_i $(i = 2, \ldots, k)$, not all zero, such that

(5.2) $$\sum_{j=2}^{k} a_j (x_j - x_1) = 0 \ .$$

Therefore we may write x in the form

$$x = \sum_{j=1}^{k} t_j x_j + \lambda \sum_{j=2}^{k} a_j (x_j - x_1) = \left(t_1 - \lambda \sum_{j=2}^{k} a_j \right) x_1 + \sum_{j=2}^{k} (t_j + \lambda a_j) x_j = \sum_{j=1}^{k} u_j x_j,$$

where λ is any real number. Note that for every choice of λ we have $u_1 + \ldots + u_k = 1$. The set of λ's for which

$$u_1 = t_1 - \lambda \sum_{j=2}^{k} a_j \geq 0, \qquad u_j = t_j + \lambda a_j \geq 0 \qquad (j = 2, \ldots, k)$$

is closed and non-empty, because it contains the number 0, and does not coincide with the real line, since at least one of the a_j's does not vanish. If λ_0 is a point belonging to its boundary, then for some j we have $u_j = 0$ and we see that x is expressible in the form (5.1), where in the sum on the right-hand side there are at most $k - 1$ terms. If all the coefficients of elements of B in a certain basis are rational, then the numbers a_j in (5.2) may be chosen from $\mathcal{Q}$ (it is sufficient to regard the points of B as elements of a linear space over $\mathcal{Q}$), and then λ_0 also lies in $\mathcal{Q}$, because we have either

$$t_j + \lambda_0 a_j = 0$$

for some $j \geqslant 2$, or

$$t_1 + \lambda_0 \sum_{j=2}^{k} a_j = 0 \; . \quad \blacksquare$$

2.   If E is an n-dimensional vector space, $e_1, \ldots, e_n$ is its basis, and $a, a_1, \ldots, a_n$ are real numbers, then the set of all elements

$$x = x_1 e_1 + \ldots + x_n e_n \in E$$

satisfying the relation

$$a_1 x_1 + \ldots + a_n x_n = a$$

is called a *hyperplane*. It is not difficult to note that this concept does not depend upon the choice of the basis $e_1, \ldots, e_n$.

We shall also treat the space E as a topological space with product topology, that is the sequence of points

$$\sum_{k=1}^{n} x_k^{(m)} e_k \quad (m = 1, 2, 3, \ldots)$$

will be convergent to a point

$$\sum_{k=1}^{n} x_k e_k$$

iff $\lim_{m \to \infty} x_k^{(m)} = x_k$ holds for $k = 1, 2, \ldots, n$. (This does not depend upon the choice of the basis $\{e_i\}$, either). The space E then becomes a metric space, with metric,

$$d\left(\sum_{k=1}^{n} x_k e_k, \; \sum_{k=1}^{n} y_k e_k \right) = \max_{k=1, \ldots, n} |x_k - y_k|$$

and we can use the notions and theorems of the theory of metric spaces. For convenience let us recall some notions:

We call the intersection of the closure of a set A with the closure of its complement the *boundary* of A.

We call the set of points $x \in A$ for which there exists a disk with center at x, contained in A, the *interior* of the set A (Int A).

If C is a convex set in E, then every hyperplane having a common point with the boundary and disjoint with the interior of C is called an *osculating hyperplane of C.* Note that such a hyperplane divides E into two parts, and the set C has common points with the interior of only one of them.

The fundamental property of this concept is contained in the following theorem:

Theorem 5.1. *If C is a convex set in an n-dimensional vector space E, and P is a point on its boundary, then there exists at least one osculating hyperplane of C containing P.*

Proof. If the smallest subspace $\tilde{C}$ of E containing C is a proper subspace of E, then $\tilde{C}$ is an osculating hyperplane of C, since the interior of C must be empty in this case. Hence we may suppose that C generates the whole plane E.

Lemma 5.5. *If X is an open, convex set and $P \notin X$, then there exists an osculating hyperplane p going through the point P and having no common points with the set X.*

Proof. Fix a basis $e_1, \ldots, e_n$ in the space E and define the scalar product of elements

$$x = \sum_{k=1}^{n} x_k e_k, \quad y = \sum_{k=1}^{n} y_k e_k$$

by

$$x \cdot y = \sum_{k=1}^{n} x_k y_k.$$

Our plane becomes a Euclidean plane and we may speak about orthogonality, orthogonal projection, etc.

Note that it suffices to prove the lemma in the case $P = 0$. Denote by L a subspace of the greatest possible dimension, which is disjoint with X. Its dimension does not exceed $n - 1$. If it is equal to $n - 1$ and $a = \sum_{k=1}^{n} a_k e_k \neq 0$ belongs to the orthogonal complement of L, then the point $x = \sum_{k=1}^{n} x_k e_k$ belongs to L iff $a \cdot x = 0$, that is iff

$$\sum_{k=1}^{n} a_k x_k = 0,$$

and so we may simply take L as p.

Hence suppose now that $\dim L \leqslant n - 2$. If L' denotes the orthogonal complement of L, then we have $\dim L' \geqslant 2$. Project orthogonally E on L'. Then, the projection of 0 is of course 0, and that of X is a certain open, convex subset of

L', which we denote by X_1. Note that 0 does not belong to X_1, since only the points of L give the point 0 under the orthogonal projection on L' and $X \cap L = \emptyset$, and moreover each one-dimensional subspace of L' must contain the points of X_1 by the maximality of L. Select any two-dimensional subspace E_1 of L'. Then the set $X_2 = E_1 \cap X_1$ is an open, convex subset of E_1. If we regard E_1 as a Euclidean plane, then we observe that half-lines starting from the point 0 and having points in common with X_2 form an angle in this plane. In view of the convexity of X_2 this angle cannot exceed π, for, otherwise, the point 0 would have to belong to X_2, which is not the case. Since, however, every line passing through 0 (that is, every one-dimensional subspace of E_1) has common points with X_2, and X_2 is open, this angle must exceed π and we get a contradiction. Hence the case $\dim L \leqslant n - 2$ is impossible. ∎

Lemma 5.6. *If X is any convex set, then its interior* Int X *is either convex or empty.*

Proof. Suppose that the interior of the set X is not empty and let P_1, P_2 be its elements. By S_i $(i = 1, 2)$ we denote an open ball with center P_i, and radius c which is chosen so that it is contained in Int X. We may assume that the balls S_1, S_2 are disjoint, and moreover, that a given point $P = t_1 P_1 + t_2 P_2$ $(t_1 + t_2 = 1, t_1 > 0, t_2 > 0)$ does not lie in either of them. The lemma will be proved if we show that $P \in$ Int X. Suppose the contrary. Then there exists a sequence of points $Q_1, Q_2, \ldots$ not lying in X, converging to P. Denote by p a line going through the points P_1, P_2, and by p_j a line parallel to p and going through Q_j. For any fixed positive ϵ we choose a j so that $d(P, Q_j) < \epsilon$. Since p and p_j are parallel, and $P_i \in p$, we infer that $d(p_j, P_i) < \epsilon$ for $i = 1, 2$ (where for any set A and $P \in E$, $d(A, P) = \inf\limits_{Q \in A} d(Q, P)$). Hence for $i = 1, 2$ there exists a point $R_i^{(j)} \in S_i \cap p_j$. The point Q_j lies on the line segment joining $R_1^{(j)}$ and $R_2^{(j)}$, and in view of $R_i^{(j)} \in X$ we obtain $Q_j \in X$, contrary to the choice of Q_j. ∎

In order to prove the theorem it now suffices to apply Lemma 5.5 to the point P and $X = $ Int C, which is, by virtue of Lemma 5.6, a convex set. ∎

3.   In this subsection we shall consider the concept of the center of mass of a convex body in an n-dimensional space E. Fixing in an arbitrary manner a basis of this space we may identify it with the Cartesian product of n-real lines.

Let p be any hyperplane in E. It divides E into two parts — E_1 and E_2. Let

$$
\delta(p, x) = \begin{cases} d(p, x) & \text{when } x \in E_1 \, , \\[2ex] -d(p, x) & \text{when } x \in E_2 \, . \end{cases}
$$

We call the quantity $m_p(X)$ defined by

$$m_p(X) = \int\limits_X \delta(p, x)\, dx.$$

the *moment* of any convex set X with respect to the hyperplane p.

If P is a point such that for every hyperplane p passing through P the moment $m_p(X) = 0$, then we say that P is the *center of mass* of X. In analysis it is proved that every measurable set of finite positive measure possesses a uniquely determined center of mass.

For an application in Chapter VI we shall require the following simple lemma:

Lemma 5.7. *If X is a non-empty bounded and convex set in an n-dimensional linear space E, not lying in any hyperplane, and P is its center of mass, then P lies in the interior of X.*

Proof. First observe that X must have a non-empty interior. Indeed, from the assumption it follows that X contains a system of $n+1$ points, not lying in any hyperplane, and the simplex determined by them lies in X and has non-empty interior. If P does not belong to Int X, then by Lemmas 5.5 and 5.6 there would exist a hyperplane p passing through P and disjoint with Int X. Then $\delta(p, x)$ would have a fixed sign for $x \in X$, and so we would have $m_p(X) \neq 0$. ∎

Exercises

1. Prove that every convex set is measurable.

2. Show that if $L_1(x_1, \ldots, x_n), \ldots, L_m(x_1, \ldots, x_n)$ are linear forms with real coefficients, and furthermore, $a_1 \leqslant b_1, \ldots, a_m \leqslant b_m$ are given reals, then the set of points $(x_1, \ldots, x_n) \in \mathscr{R}^n$ satisfying

$$a_i \leqslant L_i(x_1, \ldots, x_n) \leqslant b_i \quad , \qquad i = 1, 2, \ldots, m$$

is either convex or empty.

3. Prove that if A, B are disjoint convex sets in a linear space E, then there exists a hyperplane p with the property that for $x \in A$ we have $d(p, x) > 0$, and for $x \in B$, $d(p, x) < 0$.

§2. Minkowski's theorem

1. In this section we shall introduce the concept of a lattice, prove some of its properties and give a proof of one of the fundamental theorems in the geometry of numbers, namely, Minkowski's theorem on convex sets. It has many important applications, mainly in the theory of Diophantine approximations.

We call a subset L of the n-dimensional linear space $\mathscr{R}^n$ a *lattice* if there exist linearly independent points $P_1, \ldots, P_n \in L$ with the property that every point

$P \in L$ can be expressed in the form

(5.3) $$P = m_1 P_1 + \ldots + m_n P_n \qquad (m_i \in \mathscr{Z}) ,$$

and moreover, every point of this form lies in L.

From linear independence of points $P_1, \ldots, P_n$ it follows at once that the representation (5.3) of a given point is unique.

An example of a lattice is $\mathscr{Z}^n$, composed of all those points of $\mathscr{R}^n$ with integer coordinates.

Theorem 5.2. *Every lattice L in $\mathscr{R}^n$ is its subgroup (with addition as its operation), and moreover, there exists a non-zero disc with center at 0, not containing any non-zero point of L.* (In such case we say that L is a *discrete* subgroup of $\mathscr{R}^n$).

Proof. The first part of the assertion follows immediately from the definition, and to prove the second, suppose that there exists a sequence $Q_1, Q_2, \ldots$ of non-zero points of L, converging to 0. Now if integers a_{ij} are determined by

$$Q_i = \sum_{j=1}^{n} a_{ij} P_j \qquad (i = 1, 2, \ldots) ,$$

and furthermore, $a_i = \max_j |a_{ij}|$, then the sequence of points

(5.4) $$Q_i' = \sum_j a_{ij} a_i^{-1} P_j$$

also converges to zero in view of

$$d(Q_i', 0) \leqslant \frac{1}{a_i} \, d(Q_i, 0) \leqslant d(Q_i, 0) \to 0 .$$

Choosing if necessary some subsequence we may suppose without loss of generality that $a_i = |a_{i1}|$ for $i = 1, 2, \ldots$ and also that the sequences $\{a_{ij}/a_i\}$ are convergent for $j = 1, \ldots, n$. Let $b_j = \lim_{i \to \infty} \dfrac{a_{ij}}{a_i}$. We now obtain, letting in formula (5.4) i to pass to infinity, the equality

$$\sum_{j=1}^{n} b_j P_j = 0 .$$

Since $|b_1| = 1$, the points P_j must be linearly dependent, and this contradicts our assumption. ∎

The set of points $P_1, \ldots, P_n$ appearing in the definition of a lattice L is called a *basis*. Of course, it is not uniquely determined — a given lattice may have many bases. Thus, for example, the lattice $\mathscr{Z}^2$ has a basis $(0, 1), (1, 0)$, but has also another basis $(1, 1), (0, -1)$, since its every point (x, y) can be expressed in the form

$$(x, y) = x(1, 1) + (x - y)(0, -1) \; .$$

Note that the mapping $\displaystyle\sum_{k=1}^{n} m_k P_k \to \langle m_1, \ldots, m_k \rangle \in \mathscr{Z}^n$ is an isomorphism

between L and $\mathscr{Z}^n$. Also note that every subgroup of $\mathscr{R}^n$ isomorphic with $\mathscr{Z}^n$ is a lattice.

We shall now prove a theorem describing the behaviour of a lattice under non-singular linear transformations. It allows us, above all things, to determine all the bases of a given lattice, and enables us also to introduce the notion of the discriminant of a lattice which proves to be very useful in the sequel.

Theorem 5.3. *Let L be a lattice in $\mathscr{R}^n$ and let T be a non-singular linear transformation of this space into itself.*

(i)   *The image $T(L)$ of L is again a lattice, and the image of a basis is some basis of $T(L)$.*

(ii)   *If $P_1, \ldots, P_n$ form a basis of L, and hence that of $\mathscr{R}^n$, and in this basis, the transformation T has the matrix with integer entries and with the determinant equal to 1 or -1, then $T(L) = L$. Conversely, every linear transformation T with the property $T(L) = L$ is of this form. (We call such transformations the automorphisms of the lattice). Moreover, to every basis $Q_1, \ldots, Q_n$ of a lattice L there exists precisely one automorphism T such that $T(P_i) = Q_i$ $(i = 1, 2, \ldots, n)$.*

Proof.

(i)   This follows immediately from the definition of a lattice and the observation that non-singular linear transformations preserve linear independence.

(ii)   If $Q_i = T(P_i) = \displaystyle\sum_{j=1}^{n} a_{ij} P_j$, $i = 1, 2, \ldots, n$, where $a_{ij} \in \mathscr{Z}$, $\det[a_{ij}] = \pm 1$,

then the matrix $[a_{ij}]$ has the inverse matrix $[b_{ij}]$ with integer entries and

$$P_i = \sum_{j=1}^{n} b_{ij} Q_j \; .$$

From this it follows that the lattices with bases $\{P_i\}$ and $\{Q_i\}$ coincide, since every linear combination of points $P_1, \ldots, P_n$ with integer coefficients is such a linear combination of points $Q_1, \ldots, Q_n$, and *vice versa*.

Now consider an arbitrary automorphism T of a lattice L. This must be a non-singular transformation, because otherwise, the lattice $L = T(L)$ would lie in some $(n - 1)$-dimensional subspace of $\mathscr{R}^n$.

Since $Q_i = T(P_i) \in L$, we have with some integers a_{ij},

$$Q_i = \sum_{j=1}^{n} a_{ij} P_j \qquad (i = 1, \ldots, n),$$

and so the matrix of T has integer entries with regard to the basis $P_1, \ldots, P_n$. Applying the same reasoning to the transformation T^{-1} we find that the inverse matrix has also integer entries, so that $\det[a_{ij}] = \pm 1$.

The last part of the theorem now follows without difficulty. ∎

Corollary 1. *If L is a lattice with a basis $P_1, \ldots, P_n$, then the volume $V(P_1, \ldots, P_n)$ of the parallelotope built on the line segments $OP_1, \ldots, OP_n$ does not depend upon the choice of the basis.*

Proof. If $Q_1, \ldots, Q_n$ form another basis of L, and T is an automorphism of L, mapping P_i on Q_i for $i = 1, \ldots, n$, then

$$V(Q_1, \ldots, Q_n) = V\big(T(P_1), \ldots, T(P_n)\big) = |\det T|\, V(P_1, \ldots, P_n)$$

$$= V(P_1, \ldots, P_n) \quad . \qquad ∎$$

We call the volume of the parallelotope appearing in this corollary the *discriminant of the lattice L* and denote it by $d(L)$. The argument used in the proof of this corollary leads immediately to the following result:

Corollary 2. *If T is a non-singular linear transformation, L is a lattice in $\mathcal{R}^n$, and $M = T(L)$, then*

$$d(M) = |\det T|\, d(L) \quad . \qquad ∎$$

Theorem 5.4. *If L is a subgroup of a lattice M with finite index k, then L is a lattice and $d(L) = k d(M)$.*

Proof. The first part of the assertion follows from Theorem 1.12 and the observation that if A is a subgroup of $\mathcal{Z}^n$ isomorphic with $\mathcal{Z}^k$ with some $k < n$, then the index of A in $\mathcal{Z}^n$ is infinite. There remains therefore to prove the equality $d(L) = k d(M)$

For this purpose let $P_1, \ldots, P_n$ be a basis of L, $Q_1, \ldots, Q_n$ be a basis of M, and $R_1, \ldots, R_k$ a set of representatives of M/L. Moreover, let

$$X = \left\{ \sum_{i=1}^{n} \lambda_i P_i : 0 \leqslant \lambda_i < 1 \right\},$$

$$Y = \left\{ \sum_{i=1}^{n} \lambda_i Q_i : 0 \leqslant \lambda_i < 1 \right\}$$

and let $f(x)$ be the characteristic function of the set X.

Since the family of disjoint sets $Q + Y$ $(Q \in M)$ covers $\mathcal{R}^n$, we have

$$d(L) = \int_{\mathcal{R}^n} f(x)dx = \sum_{Q \in M} \int_{Q+Y} f(x)dx.$$

Since

$$M = \bigcup_{i=1}^{k} (R_i + L)$$

(and the union is disjoint), we obtain

$$d(L) = \sum_{i=1}^{k} \sum_{P \in L} \int_{R_i + P + Y} f(x)dx = \sum_{i=1}^{k} \sum_{P \in L} \int_{Y} f(y - R_i - P)dy$$

$$= \sum_{i=1}^{k} \int_{Y} \left(\sum_{P \in L} f(y - R_i - P) \right) dy.$$

Note that if $P, P' \in L$ and $f(y - R_i - P) = f(y - R_i - P') = 1$, then $y - R_i - P$, $y - R_i - P'$ lie in X, and hence their difference $P' - P$ must be of the form

$$\sum_{i=1}^{n} \lambda_i P_i$$

with $|\lambda_i| < 1$, which is possible only in the case $P = P'$. Furthermore, it is clear that for any arbitrarily chosen $y \in Y$ and R_i we can find P in L so that $y - R_i - P$ lies in X. We obtain finally the equality

$$\sum_{P \in L} f(y - R_i - P) = 1 \ ,$$

and this leads to $d(L) = k \int_Y dy = kd(M)$. ∎

2.   We can now go on to the announced Minkowski's theorem:

Theorem 5.5. *If X is a convex set in $\mathcal{R}^n$, symmetric with regard to the origin and of measure $V(X)$, and L is a lattice in the same space such that the inequality*

(5.5) $$V(X) > 2^n d(L)$$

holds, then there exists a non-zero point in $X \cap L$.

Moreover, if X is compact, then in place of the inequality (5.5) it suffices to assume the condition

(5.6) $$V(X) \geqslant 2^n d(L) \ .$$

Proof. It suffices to consider the case of a bounded set X, since in the case of an unbounded X, we may consider the intersection of X and a ball with a sufficiently large radius and with center at the origin 0.

In the sequel we confine ourselves to the case $L = \mathscr{Z}^n$, because the general case can be reduced to this owing to the following argument: Let T be a non-singular linear transformation mapping L onto $\mathscr{Z}^n$. Then the absolute value of the determinant of T is equal to $d^{-1}(L)$, and hence

$$V(T(X)) = \int\limits_{T(X)} dx_1 \ldots dx_n = \int\limits_X |\det T| dx_1 \ldots dx_n = \frac{V(X)}{d(L)} .$$

Since $T(X)$ is also a convex set (Lemma 5.2), symmetric with regard to the origin, with $V(T(X)) > 2^n$, and furthermore, for compact X, $T(X)$ is compact and $V(T(X)) \geq 2^n$, the problem is in fact reduced to the case $L = \mathscr{Z}^n$.

Let us begin with a very simple lemma due to R. Remak [1]:

Lemma 5.8. *If $f(x)$ is a non-negative function which is defined and integrable on $\mathscr{R}^n$, then there exists at least one point x_0 such that*

$$\sum_{x \in \mathscr{Z}^n} f(x + x_0) \geq \int\limits_{\mathscr{R}^n} f(x) dx .$$

Proof. Denote by I^n the unit cube in $\mathscr{R}^n$, i.e.

$$I^n = \left\{ (x_1, \ldots, x_n) : 0 \leq x_i < 1 \quad (i = 1, \ldots, n) \right\}$$

and let

$$F(y) = \sum_{x \in \mathscr{Z}^n} f(x + y),$$

Then

$$\int\limits_{I^n} F(y) dy = \int\limits_{I^n} \sum_{x \in \mathscr{Z}^n} f(x + y) dy = \sum_{x \in \mathscr{Z}^n} \int\limits_{I^n} f(x + y) dy$$

$$= \sum_{x \in \mathscr{Z}^n} \int\limits_{I^n + x} f(y) dy = \int\limits_{\mathscr{R}^n} f(y) dy,$$

since $\bigcup\limits_{x \in \mathscr{Z}^n} (I^n + x) = \mathscr{R}^n$, and for $x, y \in \mathscr{Z}^n$, $x \neq y$ we have $(I^n + x) \cap (I^n + y) = \emptyset$.

Therefore,

$$\int\limits_{I^n} F(y) dy = \int\limits_{\mathscr{R}^n} f(y) dy,$$

and hence there must exist a point x_0 such that

$$F(x_0) \geq \int\limits_{\mathscr{R}^n} f(y) dy . \quad \blacksquare$$

Corollary. *Let k be a natural number and X be any bounded, measurable subset of $\mathscr{R}^n$ with measure $V(X) > k$. Then there exists a point x_0 such that the set $X + x_0$ contains at least $k + 1$ points of the lattice $\mathscr{Z}^n$.*

Proof. It is enough to apply the Lemma to the characteristic function of the set X. ∎

(The case $k = 1$ of this corollary is called *Blichfeldt's lemma*).

We can now prove Minkowski's theorem. Consider the set

$$\tfrac{1}{2}X = \left\{ \tfrac{1}{2}P \colon P \in X \right\} \quad .$$

Since $V(\tfrac{1}{2}X) = V(X)2^{-n} > 1$, by the Corollary to the preceding lemma there exists a point x_0 such that the set $\tfrac{1}{2}X + x_0$ contains at least two different points, say P_1 and P_2, of the lattice $\mathscr{L}^n$. Their difference $P_1 - P_2$ is clearly a non-zero point of $\mathscr{L}^n$, and lies in X, since

$$P_1 = \tfrac{1}{2}Q_1 + x_0, \quad P_2 = \tfrac{1}{2}Q_2 + x_0, \quad Q_1, Q_2 \in X \ ,$$

so that

$$P_1 - P_2 = \tfrac{1}{2}(Q_1 - Q_2) = \tfrac{1}{2}Q_1 + \tfrac{1}{2}(-Q_2) \quad .$$

The point $-Q_2$ lies in X, since X is symmetric with regard to the origin, and using the convexity of X we obtain $P_1 - P_2 \in X$. In the case where X is compact and the condition (5.5) is replaced by (5.6), we consider the sets

$$\left(1 + \tfrac{1}{n}\right)X = \left\{ \left(1 + \tfrac{1}{n}\right)x \colon \ x \in X \right\} \quad (n = 1, 2, \ldots) .$$

By

$$V(tX) = t^n V(X) > V(X) \geqslant 2^n \qquad (t > 1)$$

they satisfy the assumption of the first part of the theorem proved above, so that in each of them one can find some non-zero point of the lattice $\mathscr{L}^n$. All of them lie in the compact set $2X$, so that they have an accumulation point, which must lie in $\mathscr{L}^n$. Moreover, it lies in infinitely many sets $(1 + \tfrac{1}{n})X$, therefore it also lies in their intersection which is equal to X. ∎

It is worth noting that Minkowski's theorem cannot be strengthened, that is, in the inequality (5.5) the factor 2^n cannot be replaced by a smaller one. Indeed, for the set

$$X = \left\{ (x_1, \ldots, x_n) \colon \ -1 < x_i < 1 \right\} \quad ,$$

which is obviously symmetric regarding the origin and is convex, we have $V(X) = 2^n$, but it contains no non-zero point of the lattice $\mathscr{L}^n$.

3. As an application of Minkowski's theorem we shall now prove the so-called *theorem on linear forms* (*Linearformensatz*), also due to Minkowski:

Theorem 5.6. *For* $j = 1, 2, \ldots, n$ *let linear forms*

$$F_j(x_1, \ldots, x_n) = \sum_{i=1}^{n} a_{ij} x_i$$

have real coefficients and let L *be a lattice in* $\mathscr{R}^n$ *with discriminant* $d(L)$.

Denote by D *the determinant of the matrix* $[a_{ij}]$. *If* $D \neq 0$, *and* $c_1, \ldots, c_n$ *are given positive numbers satisfying the condition*

(5.7) $$c_1 \ldots c_n \geqslant |D| d(L) \quad .$$

then there exists a non-zero point of L *such that*

$$|F_1(x_1, \ldots, x_n)| \leqslant c_1 \quad ,$$

(5.8) $$|F_j(x_1, \ldots, x_n)| < c_j \qquad (j = 2, \ldots, n) \quad .$$

Proof. First consider the case where the inequality (5.7) is strict. Then the set of points whose coordinates satisfy the inequalities (5.8) is convex, bounded and symmetric with respect to the origin. By computing its volume V, we have obviously

$$V = \int_{-c_1}^{c_1} dx_1 \ldots \int_{-c_n}^{c_n} \det[a_{ij}]^{-1} dx_n = 2^n c_1 \ldots c_n D^{-1} > 2^n d(L)$$

and we may apply Minkowski's theorem. In general, let us consider the set X_ϵ defined by the inequalities

$$|F_1(x_1, \ldots, x_n)| \leqslant c_1 + \epsilon \quad ,$$

$$|F_j(x_1, \ldots, x_n)| < c_j \qquad (j = 2, \ldots, n)$$

(with $0 < \epsilon < 1$) to which we may apply the foregoing argument. Let $P_\epsilon \in X_\epsilon \cap L$. It remains to observe that there are only finitely many possibilities of P_ϵ, and so one of these points satisfies these inequalities for any $\epsilon > 0$. ∎

Corollary 1. *If* a *is any real number, then for each* N *there exists a rational number* $\dfrac{p}{q}$, $(p, q) = 1$ *such that* $0 < q \leqslant N$ *and*

$$\left| a - \frac{p}{q} \right| < \frac{1}{qN} \quad .$$

Proof. Applying the theorem to the form

$$F_1(x_1, x_2) = x_2 \qquad \text{and} \qquad F_2(x_1, x_2) = x_2 a - x_1$$

by taking $c_1 = N$, $c_2 = \dfrac{1}{N}$ and $L = \mathscr{Z}^2$. Here $|D| = 1$, and so there exist

$x_1, x_2 \in \mathscr{Z}$ such that $\left| a - \dfrac{x_1}{x_2} \right| < \dfrac{1}{|x_2| N}$ and $0 < |x_2| \leqslant N$. Taking $q = |x_2|$ and

$p = x_1 \cdot \mathrm{sgn}\, x_2$, we obtain the assertion. ∎

Corollary 2. *If* $D \in \mathscr{N}$ *and* D *is not a square of any natural number, then the equation*

$$X^2 - DY^2 = 1$$

(called the Pell equation) has infinitely many integer solutions X, Y.

Proof. Note that it is enough to construct one solution $\langle \alpha, \beta \rangle$ of this equation with the additional condition $\beta \neq 0$. Indeed, if we define the numbers $\langle \alpha_n, \beta_n \rangle$ by the formula

$$\alpha_n + \beta_n D^{\frac{1}{2}} = \left(\alpha + \beta D^{\frac{1}{2}} \right)^n \quad,$$

then they will be integral and satisfy the equation $\alpha_n^2 - D\beta_n^2 = 1$. Moreover, for $m \neq n$

$$\langle \alpha_m, \beta_m \rangle \neq \langle \alpha_n, \beta_n \rangle \quad .$$

To find a solution we construct a sequence of numbers $1 \leqslant Q_1 < Q_2 < \ldots \in \mathscr{N}$, $x_1, x_2, \ldots \in \mathscr{Z}$, $y_1, y_2, \ldots \in \mathscr{N}$, satisfying the conditions:

$$y_i \leqslant Q_i \qquad (i = 1, 2, \ldots) \quad,$$

$$\frac{x_k}{y_k} \neq \frac{x_l}{y_l} \qquad (k \neq l) \quad,$$

$$\left| y_i D^{\frac{1}{2}} - x_i \right| < \frac{1}{Q_i} \qquad (i = 1, 2, \ldots)$$

and

$$|x_i^2 - Dy_i^2| < 3D^{\frac{1}{2}} \qquad (i = 1, 2, \ldots) \quad .$$

Put $Q_1 = y_1 = 1$, $x_1 = \left[D^{\frac{1}{2}} \right]$, and if the numbers Q_i, y_i, x_i satisfying these conditions are already chosen for $i = 1, 2, \ldots, k - 1$, then take Q_k as any natural number greater than Q_{k-1} for which the inequality

$$\left| y_i D^{\frac{1}{2}} - x_i \right| > \frac{1}{Q_k} \qquad (i = 1, 2, \ldots, k - 1)$$

holds.

Using Corollary 1 we can now find $x_k, y_k \in \mathscr{Z}$ such that $0 < y_k \leqslant Q_k$ and

$$\left| y_k D^{\frac{1}{2}} - x_k \right| < \frac{1}{Q_k} \quad .$$

Then

$$|x_k| < y_k D^{\frac{1}{2}} + 1 \leqslant Q_k D^{\frac{1}{2}} + 1 \quad ,$$

and so

$$|y_k D^{\frac{1}{2}} + x_k| < 2Q_k D^{\frac{1}{2}} + 1 < 3Q_k D^{\frac{1}{2}} \quad ,$$

which leads to

$$|x_k^2 - y_k^2 D| < 3D^{\frac{1}{2}} \quad .$$

By the choice of Q_k we have here $x_k/y_k \neq x_i/y_i$ $(i = 1, 2, \ldots, k-1)$.

Hence there exists an integer $m \neq 0$ such that for infinitely many pairs $\langle x_i, y_i \rangle$ the equality

$$x_i^2 - Dy_i^2 = m$$

holds. Among such pairs one can find two of them, say $\langle x_r, y_r \rangle$ and $\langle x_s, y_s \rangle$ $(r \neq s)$ satisfying

$$x_r \equiv x_s (\mathrm{mod}\ m) , \qquad y_r \equiv y_s (\mathrm{mod}\ m) \quad .$$

If we now put $\xi = x_r x_s - Dy_r y_s$, $\eta = y_r x_s - y_s x_r$, then

$$\left(x_r + y_r D^{\frac{1}{2}} \right) \left(x_s - y_s D^{\frac{1}{2}} \right) = \xi + \eta D^{\frac{1}{2}} \quad .$$

Moreover,

$$\xi^2 - \eta^2 D = (x_r^2 - y_r^2 D)(x_s^2 - y_s^2 D) = m^2$$

and

$$\xi \equiv x_r^2 - Dy_r^2 \equiv 0 (\mathrm{mod}\ m) \quad ,$$

$$\eta \equiv y_r x_s - y_r x_s \equiv 0 (\mathrm{mod}\ m) \quad ,$$

i.e.

$$\xi = m\alpha , \qquad \eta = m\beta \qquad (\alpha, \beta \in \mathscr{Z})$$

and finally

$$\alpha^2 - \beta^2 D = \frac{\xi^2 - \eta^2 D}{m^2} = 1 \quad .$$

If we had $\beta = 0$, then we would have $\eta = 0$, and so $x_r/y_r = x_s/y_s$, which is impossible. ∎

Another application of Minkowski's theorem is the following theorem on the minimum of a quadratic form. We shall formulate it for the lattice $\mathscr{Z}^n$, leaving to the reader the deduction of an analogous result for an arbitrary lattice.

Theorem 5.7. *Let* $F(x_1, \ldots, x_n) = \sum_{i,j} a_{ij} x_i x_j$ *be a positive definite quadratic form with real coefficients and let its discriminant* $D = \det[a_{ij}]$ *be non-zero. If* m_F *denotes the minimum of* $F(x_1, \ldots, x_n)$ *for integral* $x_1, \ldots, x_n$, *not all zero, then*

$$m_F \leqslant \frac{4}{\pi} \left(\Gamma(1 + \tfrac{1}{2}n)^2 D \right)^{1/n} = 4 \left(\frac{D}{V(K_n)^2} \right)^{1/n},$$

where $V(K_n)$ *is the volume of the n-dimensional unit cube.*

Proof. If t is a given positive number, and X_t is the hyperellipsoid defined by the equation

$$F(x_1, \ldots, x_n) \leqslant t \quad,$$

then its volume $V(X_t)$ is equal to $V(K_n) t^{\frac{1}{2}n} D^{-\frac{1}{2}}$.

By Minkowski's theorem, this ellipsoid will contain a non-zero point of the lattice $\mathscr{L}^n$ provided

$$V(K_n) t^{\frac{1}{2}n} D^{-\frac{1}{2}} > 2^n \quad,$$

that is

$$t > 4 \left(V(K_n)^{-2} D \right)^{1/n} = \frac{4}{\pi} \left(\Gamma(1 + \tfrac{1}{2}n)^2 D \right)^{1/n}.$$

Hence for each positive ϵ we have

(5.9) $$m_F \leqslant \frac{4}{\pi} \left(\Gamma(1 + \tfrac{1}{2}n)^2 D \right)^{1/n} + \varepsilon.$$

There are only finitely many points $(x_1, \ldots, x_n)$ for which the inequality

$$F(x_1, \ldots, x_n) \leqslant \frac{4}{\pi} \left(\Gamma(1 + \tfrac{1}{2}n)^2 D \right)^{1/n} + 1,$$

holds, and by arbitrariness of ϵ in the inequality (5.9) we see that

$$m_F \leqslant \frac{4}{\pi} \left(\Gamma(1 + \tfrac{1}{2}n)^2 D \right)^{1/n}. \quad \blacksquare$$

Corollary. *If* $F(x, y)$ *is a positive definite quadratic form with real coefficients and discriminant* D, *different from zero, then there exist integers* X, Y, *not both zero, for which*

$$|F(X, Y)| \leqslant \frac{4}{\pi} D^{\frac{1}{2}} \quad. \quad \blacksquare$$

The estimate given in Theorem 5.6 can be improved, namely for quadratic forms in two variables we have $m_F \leqslant \dfrac{1}{\sqrt{3}} D^{\frac{1}{2}}$ (cf. exercise 5).

If we denote by $m(n)$ the largest possible value of the ratio $\dfrac{m_F}{D^{1/n}}$ for positive quadratic forms in n variables, then there is a conjecture that the ratio $\dfrac{m(n)}{n}$ tends to a certain limit as n tends to infinity. It is known that

$$\limsup \frac{m(n)}{n} \leqslant \frac{1}{\pi e}$$

and

$$\liminf \frac{m(n)}{n} \geqslant \frac{1}{2\pi e}$$

(cf. e.g. Lekkerkerker [1], §38).

Exact values of $m(n)$ are known only for $n \leqslant 8$:

$$m(2) = \frac{2}{\sqrt{3}}, \quad m(3) = 2^{\frac{1}{3}}, \quad m(4) = 2^{\frac{1}{2}}, \quad m(5) = 8^{\frac{1}{5}},$$

$$m(6) = 2 \cdot 3^{-\frac{1}{6}}, \quad m(7) = 64^{-\frac{1}{7}}, \quad m(8) = 2 \quad.$$

4.   We end this section by proving a *theorem of L. Rédei* ([1]), which is a consequence of Minkowski's theorem, and enables us to obtain important arithmetic results.

Theorem 5.8. *Let* $m_1, \ldots, m_k$ *be natural numbers, and* $L_1(P), \ldots, L_k(P)$ *be functions defined for points* P *of the lattice* $\mathscr{Z}^n$, *having integer values. Moreover, suppose that for each* i *the condition* $L_i(x) \equiv L_i(y) \pmod{m_i}$ *implies* $L_i(x - y) \equiv 0 \pmod{m_i}$. *If now* X *is a convex set in* $\mathscr{R}^n$, *symmetric with respect to the origin whose volume* $V(X)$ *exceeds* $m_1 \ldots m_k 2^n$, *then there exists a non-zero point of* $\mathscr{Z}^n$ *contained in* X *and satisfying the system of congruences*

$$(5.10) \qquad\qquad L_i(P) \equiv 0 \pmod{m_i} \qquad (i = 1, 2, \ldots, k) \quad.$$

When X *is compact, then it is enough to suppose* $V(X) \geqslant m_1 \ldots m_k 2^n$.

Proof. Let L be the set of all points P of $\mathscr{Z}^n$ that satisfy the system (5.10). From our assumptions it follows that L is an additive subgroup of $\mathscr{Z}^n$. Moreover, we note that if for $i = 1, \ldots, k$ we have $L_i(x) \equiv L_i(y) \pmod{m_i}$, then x and y lie in the same coset of $\mathscr{Z}^n$ modulo L, and therefore the number of such cosets is at most $m_1 \ldots m_k$. On the basis of Theorem 5.4 we conclude that L is a lattice and that $d(L) \leqslant m_1 \ldots m_k$. Hence $V(X) > 2^n d(L)$ and we can apply Minkowski's theorem. ∎

Corollary 1 (*Aubry-Thue's theorem*). *If* p *is a prime, then for each integer* m *not divisible by* p *one can find integers* x, y *different from zero and satisfying*

$$|x|, \ |y| < p^{\frac{1}{2}} \ \text{and} \ x \equiv my \pmod{p}.$$

Proof. Apply Rédei's theorem with $n = 2$, $k = 1$, $m_1 = p$, $L_1(x_1, x_2) = mx_2 - x_1$ and

$$X = \left\{ (x_1, x_2): \, |x_1|, |x_2| \leqslant p^{\frac{1}{2}} \right\} \; .$$

Then

$$V(X) = 4\left(p^{\frac{1}{2}} \right)^2 = 4p \; . \quad \blacksquare$$

It is also not difficult to give a proof of this theorem without the use of geometric means (cf. e.g. T. Nagell [1]).

Corollary 2 (A. Brauer, T. L. Reynolds [1]). *If* $m_1, \ldots, m_k$ *are natural numbers,* $k < n$, *and* t *is a natural number satisfying* $m_1 \ldots m_k \leqslant t^k$, *then for each system of* k *linear forms* $L_i(x_1, \ldots, x_n)$ *in* n *variables one can find integers* $y_1, \ldots, y_n$, *not all zero and satisfying* $|y_i| < t^{k/n}$ *such that*

$$L_i(y_1, \ldots, y_n) \equiv 0 \pmod{m_i}$$

holds for $i = 1, 2, \ldots, k$.

Proof. Follows immediately from the Theorem. $\quad \blacksquare$

We shall use this last result in the next chapter to prove Lagrange's theorem on the expressibility of natural numbers as a sum of four squares.

Exercises

1. Prove that $L \subset \mathscr{R}^n$ is a lattice iff it is a discrete subgroup of $\mathscr{R}^n$ containing n linearly independent points.

2. Let $X_1, \ldots, X_m$ be bounded and measurable subsets of $\mathscr{R}^n$ with measures $V_1, \ldots, V_m$, respectively. Further, let $c_1, \ldots, c_n$ be any positive numbers. Show that if $N_t(z)$ denotes the number of lattice points contained in the set $X_i - z = \left\{ x_i - z: \, x_i \in X_i \right\}$, then there exists a point z such that

$$\sum_{i=1}^{m} c_i N_i(z) \geqslant \sum_{i=1}^{m} c_i V_i.$$

3. Show that if k is a given natural number and X is a convex set, symmetric with respect to the origin with volume exceeding $k2^n$, then X contains at least $2k$ points of the lattice $\mathscr{Z}^n$.

4. Prove that the theorem on linear forms (Theorem 5.6) remains true for linear forms with complex coefficients provided for each linear form

$$F_i(x_1, \ldots, x_n) = \sum_{j=1}^{n} a_{ij} x_j$$

its conjugate form

$$\bar{F}_i(x_1, \ldots, x_n) = \sum_{j=1}^{n} \bar{a}_{ij} x_j,$$

also belongs to $\left\{ F_i \right\}$, and moreover, if F_i, F_j are conjugates then $c_i = c_j$.

5. Prove that for positive definite quadratic forms in two variables we have $m_F \leqslant \dfrac{2}{\sqrt{3}} D^{\frac{1}{2}}$

and find a form for which the equality holds here.

6.   Prove that if p is a prime of the form $4k + 1$, then in the interval $[1, p^{\frac{1}{2}}]$ lies at least one quadratic non-residue (mod p).

§3. Lattice points in plane regions

1. Many problems in number theory can be reduced to the evaluation of the number of lattice points (mainly, points of the lattice $\mathscr{Z}^2$) in some regions. Thus, for example, if we are interested in the sum of the number $d(n)$ of divisors of n in some interval $[a, b]$, then writing it in the form

$$S(a, b) = \sum_{a \leqslant n \leqslant b} \sum_{d | n} 1 = \sum_{a \leqslant n \leqslant b} \sum_{\substack{x, y \\ xy = n}} 1 = \sum_{a \leqslant xy \leqslant b} 1$$

we see at once that it is equal to the number of lattice points in the region $x > 0$, $a \leqslant xy \leqslant b$, that is, in the region bounded by two hyperbolas.

Similarly, if we denote by $r_2(n)$ the number of representations of a natural number n as a sum of two squares (taking also into account squares of negative integers) then we have clearly

$$\sum_{n \leqslant x} r_2(n) = \sum_{\substack{(a,b) \in \mathscr{Z}^2 \\ a^2 + b^2 \leqslant x}} 1,$$

and hence the problem amounts to the count of lattice points in the circle with center at origin and radius $x^{\frac{1}{2}}$. If we restrict ourselves to the representations $n = a^2 + b^2$ with non-negative a, b then it is necessary to consider a quadrant of a circle instead.

Intuitively it is clear that given a large region D with area $S(D)$ the number $N(D)$ of lattice points lying in the interior of D should be approximately equal to $S(D)$. The precise formulation of this observation is given in the following theorem:

Theorem 5.9. *If A is a region in the plane with area $S(A)$ whose boundary is rectifiable and has the length $L(A)$, then for the number $N(A)$ of lattice points contained in the interior of A we have the estimate*

(5.11) $|N(A) - S(A)| \leqslant \dfrac{8\pi}{\sqrt{2}} L(A) + 8\pi$.

Proof. By γ we denote the boundary of the region A and let $\hat{A}$ denote the set of points whose distance from γ is at most $2^{\frac{1}{2}}$. Further, let $A^+ = A \cup \hat{A}$, $A^- = A \backslash \hat{A}$. Obviously, the inclusions

$$A^- \subset A \subset A^+$$

hold and therefore

$$N(A^-) \leqslant N(A) \leqslant N(A^+)$$.

Moreover, if a lattice point $P = (x, y)$ lies in A, then the square K_p with vertices $(x, y), (x + 1, y), (x, y + 1), (x + 1, y + 1)$ lies in A^+. Hence

$$N(A) = \sum_{P \in A} 1 = \sum_{P \in A} S(K_p) \leqslant S(A^+) \ .$$

Since $A^- \subset \bigcup_{P \in A} K_p$, thus

$$S(A^-) \leqslant \sum_{P \in A} S(K_p) = N(A) \ .$$

From the inequalities obtained above we now get

$$|N(A) - S(A)| \leqslant S(A^+) - S(A^-) = S(A^+ \setminus A^-) = S(\hat{A}) \ .$$

and to complete the proof of the theorem it is sufficient to show that $S(\hat{A})$ does not exceed $\dfrac{8\pi}{\sqrt{2}} L(A) + 8\pi$. For this purpose we divide γ into r arcs $L_1, \ldots, L_r$ of which $r - 1$ have length $2^{\frac{1}{2}}$, and one has length at most $2^{\frac{1}{2}}$. Evidently

$$r = \left[L(A) 2^{-\frac{1}{2}} \right] + 1 < \frac{L(A)}{\sqrt{2}} + 1 \ .$$

Now note that any arc of length a is contained in some disc of radius not exceeding a, because it suffices to describe a disc of radius a and centered at any point of this arc. Applying this remark to our arcs $L_1, \ldots, L_r$ we conclude that the boundary of A is contained in a union of r discs with radius not greater than $2^{\frac{1}{2}}$. Since each point of $\hat{A}$ is distant from this boundary by at most $2^{\frac{1}{2}}$, $\hat{A}$ is contained in a union of r with radius $2 \cdot 2^{\frac{1}{2}}$, whence

$$S(A) \leqslant 8\pi r < \frac{8\pi L(A)}{\sqrt{2}} + 8\pi \ . \qquad \blacksquare$$

Remark. As was proved by V. Járnik [1] (cf. also H. Steinhaus [1]), the right-hand side of the inequality (5.11) can be replaced by $\max\{1, L(A)\}$, and M. Nosarzewska [1] showed that

$$N(A) > S(A) - \tfrac{1}{2} L(A) \ .$$

This result has recently been transferred to the case of higher dimensions by J. Bokowski, H. Hadwiger and J. M. Wills [1]. They have shown that if K is a convex region in the n-dimensional space with volume V and area F, then for the number N of lattice points in the interior of K the inequality

$$N > V - \tfrac{1}{2} F$$

holds where the coefficient $\tfrac{1}{2}$ cannot be replaced by a smaller one.

Corollary 1 (C. F. Gauss). *For* $x \geqslant 2$ *we have*

$$\sum_{n \leqslant x} r_2(n) = \pi x + O\left(x^{\frac{1}{2}}\right).$$

Proof. In fact, here A is the circle with radius $x^{\frac{1}{2}}$, so that

$$S(A) = \pi x \quad , \qquad L(A) = 2\pi x^{\frac{1}{2}} = O(x^{\frac{1}{2}}) \quad . \qquad \blacksquare$$

Corollary 2. *If* A, B *are positive numbers, then*

$$\sum_{\substack{x, y > 0 \\ Ax + By \leqslant T}} 1 = \frac{1}{2AB} T^2 + O(T) \quad . \qquad \blacksquare$$

Corollary 3. *If* A *is a plane region with area* S *and a rectifiable boundary, and we denote by* A_t *the region arising from* A *by a dilatation with coefficient* t, *then as* t *tends to infinity we have for the number* $N(t)$ *of lattice points in* A_t *the estimate:*

$$N(t) = t^2 S(A) + O(t) \quad . \qquad \blacksquare$$

Remark. As proved by H. Weber [1], in the n-dimensional space an analogue of the above result holds. We then have

$$N(t) = ct^n + O(t^{n-1}) \quad ,$$

where c is the n-dimensional volume of the set A.

2. We shall now prove a result of I. M. Vinogradov which in many cases permits to obtain a good estimate of the error term in lattice point problems.

Theorem 5.10. *Let* $f(x)$ *be a function of class* C^2 *in an interval* $[M, M+t-1]$, $(M, t \in \mathscr{Z})$ *(that is, f is a function having the second continuous derivative in this interval) such that the estimate*

$$A^{-1} \leqslant |f''(x)| \leqslant kA^{-1} \quad ,$$

hold there with certain $A > 2$, $k \geqslant 1$.

Then for the sum

$$S = \sum_{x=M}^{M+t-1} f(x) \quad ,$$

one has

$$S = \tfrac{1}{2}t + O\left((k^2 t \log A + kA) A^{-\frac{1}{3}}\right) \quad ,$$

where the constant implied by Landau's symbol $O(\cdot)$ *is absolute.*

Proof. To simplify the notation we put $[A^{\frac{1}{3}}] = \tau$. Let $m_0 = 0$, and define the sequence of natural numbers $m_1, m_2, \ldots$, using Corollary 1 to Theorem 5.6 in such a way that for $i \geqslant 1$ the following conditions are satisfied:

(i) $(a_i, m_i) = 1$,

(ii) $\left| f'(M + m_1 + \ldots + m_{i-1}) - \dfrac{a_i}{m_i} \right| \leqslant \dfrac{1}{\tau m_i}$

and

(iii) $0 < m_i \leqslant \tau$.

Let $M_i = M + m_1 + \ldots + m_{i-1}$ and let s be the smallest natural number satisfying

$$0 \leqslant M + t - 1 - M_{s+1} < \tau \quad .$$

Note that we then have

(5.12) $$0 \leqslant S - \sum_{i=1}^{s} \sum_{x=M_i}^{M_{i+1}-1} \{f(x)\} = \sum_{x=M_{s+1}}^{M+t-1} \{f(x)\} < \tau + 1$$

and there remains to estimate the sum

(5.13) $$S_i = \sum_{x=M_i}^{M_{i+1}-1} \{f(x)\} = \sum_{x=M_i}^{M_i+m_i-1} \{f(x)\}$$

for $i = 1, \ldots, s$.

Since in the interval under consideration the function $f(x)$ has a continuous second derivative, we may write with some ξ_x satisfying $0 \leqslant \xi_x \leqslant m_i - 1$

$$f(x) = f(M_i) + f'(M_i)(x - M_i) + \tfrac{1}{2}(x - M_i)^2 f''(M_i + \xi_x) \quad ,$$

and hence for a c independent of x and satisfying $|c| \leqslant 1$, we have

$$f(x) = f(M_i) + \left(\dfrac{a_i}{m_i} + \dfrac{c}{\tau m_i} \right)(x - M_i) + \tfrac{1}{2}(x - M_i)^2 f''(M_i + \xi_x),$$

that is

(5.14) $$f(x) = \dfrac{a_i + c\tau^{-1}}{m_i}(x - M_i) + \dfrac{R(x)}{m_i},$$

where

$$R(x) = m_i f(M_i) + \tfrac{1}{2} m_i (x - M_i)^2 f''(M_i + \xi_x)$$

and

(5.15) $$m_i f(M_i) - \frac{1}{2}k \leqslant R(x) \leqslant m_i f(M_i) + \frac{1}{2}k \quad .$$

We shall base the estimation of the sums (5.13) on the following lemma which we shall be able to employ owing to (5.14) and (5.15):

Lemma 5.9. *If* m *is a natural number,* $(a, m) = 1$, $h > 0$, c *is a real number, and moreover in the interval* $0 \leqslant x \leqslant m - 1$ *a function* $R_1(x)$ *is defined satisfying the inequality* $c \leqslant R_1(x) \leqslant c + h$, *then for the sum*

$$S_0 = \sum_{x=0}^{m-1} \left\{ \frac{ax + R_1(x)}{m} \right\}$$

the estimate

$$|S_0 - \tfrac{1}{2}m| \leqslant h + \tfrac{1}{2}$$

holds.

Proof. Let $c = [c] + \epsilon$, $0 \leqslant \epsilon < 1$ and let $r = r(x)$ be the least non-negative residue (mod m) of $ax + [c]$. Then we have evidently

$$\left\{ \frac{ax + R_1(x)}{m} \right\} = \left\{ \frac{r - [c] + R_1(x)}{m} \right\}$$

and putting

$$\Phi(r) = R_1(x) - [c] = R_1(x) - c + \epsilon$$

we obtain

$$S_0 = \sum_{r=0}^{m-1} \left\{ \frac{r + \Phi(r)}{m} \right\} \quad ,$$

since if x runs over all residues (mod m), then so does r. We have moreover $\epsilon \leqslant \Phi(r) \leqslant h + \epsilon$.

Note that the assertion of the lemma becomes obvious in the case $m \leqslant 1 + 2h$, since then $0 \leqslant S_0 \leqslant m$, and so

$$|S_0 - \tfrac{1}{2}m| \leqslant \tfrac{1}{2}m \leqslant \tfrac{1}{2}(1 + 2h) = h + \tfrac{1}{2} \quad .$$

We may therefore suppose that $m > 1 + 2h$, that is, $h < \frac{1}{2}(m - 1)$. Now note that if $r < m - h - \epsilon$, then

$$\left\{ \frac{r + \Phi(r)}{m} \right\} = \frac{r + \Phi(r)}{m} \quad ,$$

and hence

$$\frac{\epsilon}{m} \leqslant \left\{\frac{r + \Phi(r)}{m}\right\} - \frac{r}{m} \leqslant \frac{h + \epsilon}{m} \quad,$$

and if $r \geqslant m - h - \epsilon$, then we have on one hand

$$\left\{\frac{r + \Phi(r)}{m}\right\} - \frac{r}{m} \leqslant \frac{r + \Phi(r)}{m} - \frac{r}{m} = \frac{\Phi(r)}{m} \leqslant \frac{h + \epsilon}{m} \quad,$$

and on the other hand

$$\left\{\frac{r + \Phi(r)}{m}\right\} - \frac{r}{m} = \left\{\frac{r + \Phi(r)}{m} - 1\right\} - \frac{r}{m} \geqslant \frac{r + \Phi(r)}{m} - 1 - \frac{r}{m} \geqslant \frac{\epsilon}{m} - 1 \quad,$$

since

$$\frac{r + \Phi(r)}{m} \leqslant \frac{r + h + \epsilon}{m} < \frac{r + \frac{1}{2}(m - 1) + \epsilon}{m} < \frac{1}{2} + \frac{r + 1}{m} < 2 \quad.$$

Hence we see that

$$\sum_{r=0}^{m-1} \left\{\frac{r + \Phi(r)}{m}\right\} - \sum_{r=0}^{m-1} \frac{r}{m} \leqslant m \cdot \frac{h + \epsilon}{m} \leqslant h + \epsilon$$

and, considering two cases $h + \epsilon \in \mathscr{L}$ and $h + \epsilon \notin \mathscr{L}$ separately, that

$$\sum_{r=0}^{m-1} \left\{\frac{r + \Phi(r)}{m}\right\} - \sum_{r=0}^{m-1} \frac{r}{m} \geqslant -h \quad,$$

but

$$\sum_{r=0}^{m-1} \frac{r}{m} = \frac{1}{2}(m - 1) \quad,$$

and hence

$$-h \leqslant S_0 - \frac{1}{2}(m - 1) \leqslant h + \epsilon$$

and

$$\left| S_0 - \frac{1}{2}m \right| \leqslant h + \frac{1}{2} \quad. \quad \blacksquare$$

Now we apply this lemma to the sum (5.13) taking $a = a_i$, $m = m_i$, $R_1(x) = c\tau^{-1}x + R(x + M_i)$. Since we have clearly

$$|c\tau^{-1}x| \leqslant 1 \quad,$$

we get

$$m_i f(M_i) - \tfrac{1}{2}k - 1 \leqslant R_1(x) \leqslant m_i f(M_i) + \tfrac{1}{2}k + 1 \ ,$$

and hence we may take $h = k + 2$, which gives

$$|S - \tfrac{1}{2}m_i| \leqslant k + \tfrac{5}{2}$$

and finally

(5.16) $$|S - \tfrac{1}{2}(m_1 + \ldots + m_s)| \leqslant s(k + \tfrac{5}{2}) + \tau + 1 \ll ks + \tau \ ,$$

where the constant in Vinogradov's symbol is absolute. There remains to estimate the number s. To this end note first that by our assumptions the derivative of the function $f(x)$ is monotone in the interval considered, and hence for a fixed irreducible fraction a/m the inequality

$$\frac{a}{m} - \frac{1}{\tau m} \leqslant f'(x) \leqslant \frac{a}{m} + \frac{1}{\tau m}$$

holds in some interval (α, β), where α, β depends on the fraction a/m. If, moreover, $f''(x)$ is positive, then

$$\frac{2}{\tau m} \geqslant f'(\beta) - f'(\alpha) = \int_\alpha^\beta f''(u)\,du \geqslant \frac{\beta - \alpha}{A} \ ,$$

and if $f''(x)$ is negative, then we have similarly

$$\frac{2}{\tau m} \geqslant f'(\alpha) - f'(\beta) = \int_\beta^\alpha f''(u)\,du = \int_\alpha^\beta |f''(u)|\,du \geqslant \frac{\beta - \alpha}{A}$$

and in both cases we obtain

$$\beta - \alpha \leqslant \frac{2A}{m\tau} \ .$$

This estimate now allows us to estimate s. Indeed, let a/m be an irreducible fraction with a positive denominator which does not exceed τ. Let us estimate the number of indices $i \leqslant s$ for which $a = a_i$, $m = m_i$. For each such index, the number $M + m_1 + \ldots + m_{i-1}$ lies in the interval (α, β). Denote by γ the first, and by δ the last index with this property. Then $M + m_1 + \ldots + m_{\gamma-1}$, $M + m_1 + \ldots + m_\gamma, \ldots, M + m_1 + \ldots + m_{\delta-1}$ lie in (α, β), and so

$$m_\gamma + m_{\gamma+1} + \ldots + m_{\delta-1} \leqslant \beta - \alpha \ .$$

Since $m_\gamma = m_{\gamma+1} = \ldots = m_{\delta-1} = m$, we have

$$(\delta - \gamma)m \leqslant \beta - \alpha \leqslant \frac{2A}{m\tau}$$

and

$$\delta - \gamma \leqslant \frac{2A}{\tau m^2} \quad ,$$

and so the number of indices i for which $a = a_i$, $m = m_i$ holds is at most $1 + 2A/\tau m^2$.

In the following let m be a fixed number and denote by λ, μ respectively the least and the greatest values of a_i corresponding to the indices i for which $m = m_i$. We shall show that

(5.17) $$\frac{\mu - \lambda}{m} - \frac{2}{\tau m} \leqslant \frac{kt}{A} .$$

Indeed, if $f'(x)$ is increasing, then

$$\min_{x \in [M, M+t]} f'(x) = f'(M) , \qquad \max_{x \in [M, M+t]} f'(x) = f'(M+t) ,$$

so that

$$\frac{\mu}{m} \leqslant f'(M+t) + \frac{1}{\tau m} \quad ,$$

$$\frac{\lambda}{m} \geqslant f'(M) - \frac{1}{\tau m}$$

and

$$\frac{\mu - \lambda}{m} \leqslant f'(M+t) - f'(M) + \frac{2}{\tau m} = \int_M^{M+t} f''(u)\,du + \frac{2}{\tau m} \leqslant \frac{kt}{A} + \frac{2}{\tau m} ,$$

and if f' is decreasing, the calculation is identical.

From the inequality (5.17) we now obtain

$$\mu - \lambda + 1 \leqslant \frac{ktm}{A} + \frac{2}{\tau} + 1 \leqslant \frac{ktm}{A} + 3 .$$

Hence the equality $m_i = m$ can hold for at most

$$\left(1 + \frac{2A}{\tau m^2}\right) \cdot \left(\frac{ktm}{A} + 3\right)$$

indices. Finally, summing over $m = 1, \ldots, \tau$, we obtain

$$s \leqslant \sum_{m=1}^{\tau} \left(1 + \frac{2A}{\tau m^2}\right)\left(\frac{ktm}{A} + 3\right) = \sum_{m=1}^{\tau} \left(\frac{ktm}{A} + \frac{2kt}{\tau m} + 3 + \frac{6A}{\tau m^2}\right)$$

$$\ll \frac{kt\tau^2}{A} + \frac{kt}{\tau}\log\tau + \tau + \frac{A}{\tau}$$

$$\ll ktA^{-\frac{1}{3}} + kt\frac{\log A}{A^{\frac{1}{3}}} + A^{\frac{1}{3}} + A^{\frac{2}{3}} \ll \frac{kt\log A}{A^{\frac{1}{3}}} + A^{\frac{2}{3}}.$$

From this estimate of s we obtain, using formula (5.16) and noting that $|m_1 + \ldots + m_s - t| < \tau + 1$, the estimate

$$|S - \tfrac{1}{2}t| \ll \frac{k^2 t \log A}{A^{\frac{1}{3}}} + kA^{\frac{2}{3}} + A^{\frac{1}{3}} \ll (k^2 t \log A + kA)A^{-\frac{1}{3}} . \quad \blacksquare$$

3.   Using the theorem proved in the previous subsection, we now strengthen Theorem 2.12 and prove the following result first obtained by G. Voronoi in 1903 (Voronoi's result was somewhat sharper; $\log x$ appeared there in place of $\log^2 x$):

Theorem 5.11. *If $d(n)$ denotes the number of divisors of a natural number n, then for x tending to infinity we have*

$$\sum_{n \leqslant x} d(n) = x\log x + (2C - 1)x + O\left(x^{\frac{1}{3}}\log^2 x\right).$$

Proof.   In the proof of Theorem 2.12 we have seen that

$$\sum_{n \leqslant x} d(n) = 2 \sum_{m \leqslant x^{\frac{1}{2}}} \left[\frac{x}{m}\right] - [\sqrt{x}\,]^2 .$$

Now let x be a large number. We divide the interval $\left(2x^{\frac{1}{3}}, x^{\frac{1}{2}}\right]$ into subintervals of the shape $(a, a']$, where $a' \leqslant 2a$. We can always do this in such a way, that the number of subintervals obtained is $O(\log x)$. In each such subinterval the function $f(t) = x/t$ satisfies

$$\frac{x}{4a^3} \leqslant f''(t) \leqslant \frac{8x}{4a^3} .$$

Applying the preceding theorem with $A = 4a^3/x$ and $k = 8$, we get

$$\sum_{m \in (a, a']} \left\{\frac{x}{m}\right\} = \tfrac{1}{2}(a' - a) + O\left(x^{\frac{1}{3}}\log x\right),$$

so that

$$\sum_{m \leqslant x^{\frac{1}{2}}} \left\{\frac{x}{m}\right\} = \sum_{2x^{\frac{1}{3}} \leqslant m \leqslant x^{\frac{1}{2}}} \left\{\frac{x}{m}\right\} + O\left(x^{\frac{1}{3}}\right)$$

$$= \sum_a \left(\tfrac{1}{2}(a' - a) + O\left(x^{\frac{1}{3}}\log x\right)\right) + O\left(x^{\frac{1}{3}}\right)$$

$$= \tfrac{1}{2}x^{\frac{1}{2}} + O\left(x^{\frac{1}{3}}\log^2 x\right).$$

But, since by Corollary 2 to Theorem 2.4

$$\sum_{m \leqslant x^{\frac{1}{2}}} \frac{x}{m} = \tfrac{1}{2}x\log x + Cx + (\tfrac{1}{2} - \{\sqrt{x}\})\sqrt{x} + O(1),$$

we have finally

$$\sum_{m \leqslant x^{\frac{1}{2}}} \left[\frac{x}{m}\right] = \sum_{m \leqslant x^{\frac{1}{2}}} \left(\frac{x}{m} - \left\{\frac{x}{m}\right\}\right)$$

$$= \tfrac{1}{2}x\log x + Cx - \{\sqrt{x}\}\sqrt{x} + O(x^{\frac{1}{3}}\log^2 x)$$

and

$$\sum_{n \leqslant x} d(n) = x\log x + 2Cx - 2\{\sqrt{x}\}\sqrt{x} - [\sqrt{x}]^2 + O\left(x^{\frac{1}{3}}\log^2 x\right)$$

$$= x\log x + 2Cx - 2\{\sqrt{x}\}\sqrt{x} - (x - 2\sqrt{x}\{\sqrt{x}\} + \{\sqrt{x}\}^2) + O\left(x^{\frac{1}{3}}\log^2 x\right)$$

$$= x\log x + (2C - 1)x + O\left(x^{\frac{1}{3}}\log^2 x\right). \qquad \blacksquare$$

4. We conclude by proving a stronger form of Corollary 1 to Theorem 5.9.

Theorem 5.12. *If* $r_2(n)$ *denotes the number of representations of a natural number* n *in the form* $a^2 + b^2$, *where* $a, b \in \mathscr{Z}$, *then for* x *tending to infinity we have*

$$\sum_{n \leqslant x} r_2(n) = \pi x + O\left(x^{\frac{1}{3}}\log x\right).$$

Remark. This result is slightly weaker than that of W. Sierpiński [1], who obtained the estimate $O(x^{\frac{1}{3}})$ for the error term in this formula in 1906.

Proof. First note that

$$(5.18) \qquad \sum_{n \leqslant x} r_2(n) = 1 + 4\left[x^{\frac{1}{2}}\right] + 8 \sum_{0 < m \leqslant \frac{1}{\sqrt{2}}x^{\frac{1}{2}}} [\sqrt{x - m^2}] - 4\left[\frac{x^{\frac{1}{2}}}{\sqrt{2}}\right]^2.$$

Indeed, we have

$$\sum_{n \leqslant x} r_2(n) = \sum_{a^2 + b^2 \leqslant x} 1$$

$$= \sum_{|a| \leqslant \frac{1}{\sqrt{2}}x^{\frac{1}{2}}} \sum_{|b| \leqslant \sqrt{x - a^2}} 1 + \sum_{|b| \leqslant \frac{1}{\sqrt{2}}x^{\frac{1}{2}}} \sum_{|a| \leqslant \sqrt{x - b^2}} 1 - \sum_{|a|, |b| \leqslant \frac{1}{\sqrt{2}}x^{\frac{1}{2}}} 1$$

$$= 2 \sum_{|a| \le \frac{1}{\sqrt{2}} x^{\frac{1}{4}}} \sum_{|b| \le \sqrt{x-a^2}} 1 - \sum_{|a|,|b| \le \frac{1}{\sqrt{2}} x^{\frac{1}{4}}} 1$$

$$= 2 \left(2 \sum_{0 < a \le \frac{1}{\sqrt{2}} x^{\frac{1}{4}}} \sum_{|b| \le \sqrt{x-a^2}} 1 + \sum_{|b| \le x^{\frac{1}{4}}} 1 \right) - \left(2 \left[\frac{1}{\sqrt{2}} x^{\frac{1}{2}} \right] + 1 \right)^2$$

$$= 2 \left(2 \sum_{0 < a \le \frac{1}{\sqrt{2}} x^{\frac{1}{4}}} (1 + 2[\sqrt{x-a^2}]) + 2 \left[x^{\frac{1}{2}} \right] + 1 \right)$$

$$- 4 \left[\frac{1}{\sqrt{2}} x^{\frac{1}{2}} \right]^2 - 4 \left[\frac{1}{\sqrt{2}} x^{\frac{1}{2}} \right] - 1$$

$$= 8 \sum_{0 < a \le \frac{1}{\sqrt{2}} x^{\frac{1}{4}}} [\sqrt{x-a^2}] + 4 \left[x^{\frac{1}{2}} \right] - 4 \left[\frac{1}{\sqrt{2}} x^{\frac{1}{2}} \right]^2 + 1 .$$

Using Theorem 2.4 (iii) we get

$$\sum_{0 < a \le \frac{1}{\sqrt{2}} x^{\frac{1}{4}}} \sqrt{x-a^2} = \int_0^{\frac{1}{\sqrt{2}} x^{\frac{1}{4}}} \sqrt{x-t^2} \, dt + \left(\frac{1}{2} - \left\{ \frac{1}{\sqrt{2}} x^{\frac{1}{2}} \right\} \right) (\frac{1}{2}x)^{\frac{1}{4}} - \frac{1}{2} x^{\frac{1}{2}}$$

$$+ \int_0^{\frac{1}{\sqrt{2}} x^{\frac{1}{4}}} (\frac{1}{2} - \{t\}) \, dt - \int_0^{\frac{1}{\sqrt{2}} x^{\frac{1}{4}}} \left(\frac{x}{(x-t^2)^{\frac{3}{2}}} \int_0^t (\frac{1}{2} - \{u\}) \right) \, du \, dt - x^{\frac{1}{2}}$$

$$= \int_0^{\frac{1}{\sqrt{2}} x^{\frac{1}{4}}} \sqrt{x-t^2} \, dt - \frac{\sqrt{2}-1}{2\sqrt{2}} x^{\frac{1}{2}} - \left\{ (\frac{1}{2}x)^{\frac{1}{4}} \right\} (\frac{1}{2}x)^{\frac{1}{4}} + O(1)$$

$$= (\frac{1}{8}\pi + \frac{1}{4})x - \frac{\sqrt{2}-1}{2\sqrt{2}} x^{\frac{1}{2}} - \left\{ (\frac{1}{2}x)^{\frac{1}{4}} \right\} (\frac{1}{2}x)^{\frac{1}{4}} + O(1) ,$$

because in the interval $(0, x^{\frac{1}{2}} 2^{-\frac{1}{2}})$ we have for the function $f(t) = \sqrt{x - t^2}$ the equalities:

$$f'(t) = \frac{-t}{\sqrt{x - t^2}} , \qquad f''(t) = \frac{-x}{(x - t^2)^{\frac{3}{2}}}$$

and

$$\frac{1}{x^{\frac{1}{2}}} \leqslant |f''(t)| \leqslant \frac{2\sqrt{2}}{x^{\frac{1}{2}}} \ .$$

By Theorem 5.10 we have

$$\sum_{0 < a \leqslant \frac{1}{\sqrt{2}} x^{\frac{1}{2}}} \{\sqrt{x-a^2}\} = \tfrac{1}{2}(\tfrac{1}{2}x)^{\frac{1}{2}} + O\left(\frac{x^{\frac{1}{2}} \log x + x^{\frac{1}{2}}}{x^{\frac{1}{6}}}\right)$$

$$= \frac{1}{2\sqrt{2}} x^{\frac{1}{2}} + O\left(x^{\frac{1}{3}} \log x\right),$$

and so

$$\sum_{0 < a \leqslant \frac{1}{\sqrt{2}} x^{\frac{1}{2}}} [\sqrt{x-a^2}] = (\tfrac{1}{8}\pi + \tfrac{1}{4})x - \tfrac{1}{2}x^{\frac{1}{2}} - \{(\tfrac{1}{2}x)^{\frac{1}{2}}\} \sqrt{\tfrac{1}{2}x} + O\left(x^{\frac{1}{3}} \log x\right)$$

and finally

$$\sum_{n \leqslant x} r_2(n) = (\pi + 2)x - 4x^{\frac{1}{2}} - 8(\tfrac{1}{2}x)^{\frac{1}{2}}\{(\tfrac{1}{2}x)^{\frac{1}{2}}\} + 4x^{\frac{1}{2}} - 4\left((\tfrac{1}{2}x)^{\frac{1}{2}} - \{(\tfrac{1}{2}x)^{\frac{1}{2}}\}\right)^2$$

$$+ O\left(x^{\frac{1}{3}} \log x\right) \ = \ \pi x + O\left(x^{\frac{1}{3}} \log x\right) \ . \quad ■$$

Let us denote by $r_k(n)$ the number of representations of a given natural number n as a sum of k squares. A. Walfisz's book ([1]) is devoted to the mean value of this function for $k \geqslant 4$. It is not difficult to prove that

$$\sum_{n \leqslant x} r_k(n) = \frac{\pi^{\frac{1}{2}k}}{\Gamma(1+\frac{1}{2}k)} x^{\frac{1}{2}k} + R_k(x),$$

where $R_k(x) = o(x^{\frac{1}{2}k})$ and the problem consists of finding the best estimate of the error. It can be shown that for $k \geqslant 5$ we have $R_k(x) = O(x^{\frac{1}{2}k-1})$, but the estimate $R_k(x) = o(x^{\frac{1}{2}k-1})$ is not true for any k. For $k = 4$ we have $R_4(x) = O(x \log x (\log \log x)^{-1})$, and here the estimate $R_4(x) = o(x \log \log x)$ is false.

Exercises

1. Prove that if $r_3(n) = \displaystyle\sum_{n = x_1^2 + x_2^2 + x_3^2} 1$, where $x_i \in \mathscr{Z}$, then

$$\sum_{n \leqslant x} r_3(n) = \tfrac{4}{3}\pi x^{\frac{3}{2}} + O(x) \ .$$

2.   Examine the asymptotic behaviour of the sum

$$\sum_{\substack{x_1,\ldots,x_n > 0 \\ A_1 x_1 + \ldots + A_n x_n \leqslant x}} 1,$$

where $A_1, \ldots, A_n$ are given positive numbers.

3.   Examine the asymptotic behaviour of the number of points of an arbitrary lattice L in

$$\frac{x^2}{a^2} + \frac{y^2}{b^2} = t, \text{ as } t \to \infty.$$

4.   Prove that in formula (5.11) the right-hand side can be replaced by max $\{1, L(A)\}$.

CHAPTER VI

ADDITIVE NUMBER THEORY

§1. Schnirelman's density

1. If A, B are subsets of $\mathcal{N}_0$, then by their sum we understand the set of all non-negative integers of the form $a + b$, where $a \in A$ and $b \in B$. More generally, if $A_1, \ldots, A_m$ are subsets of $\mathcal{N}_0$, then we call the set of all non-negative integers of the form $a_1 + \ldots + a_m$, where $a_i \in A_i$, their sum. Here we do not assume that the A_i's are distinct. In the case $A_1 = \ldots = A_n$, we write nA in place of $A_1 + \ldots + A_n$. It is clear that the operation defined in this way in the set of all non-empty subsets of $\mathcal{N}_0$ is associative, commutative and $\{0\}$ is its neutral element.

The determination of the sum is usually very difficult even for simple sets. Already in the eighteenth century there was a certain interest in the sets nA_k, where A_k denotes the set of the k-th powers, i.e. $A_k = \{0, 1^k, 2^k, 3^k, \ldots\}$. In 1770 J. L. Lagrange showed that $4A_2 = \mathcal{N}_0$, i.e., every natural number is a sum of four squares. We shall prove this result in §2. Its generalization to arbitrary k-th powers is called *Waring's problem*, after the name of an English mathematician who was the first to formulate it. A proof in the general case was first given by D. Hilbert ([1]) in 1909. His proof was later simplified by W. J. Ellison and we shall present it in the following section.

Among famous mathematical problems an important place takes the *Goldbach conjecture* which asserts that every natural number $\neq 1$ is a sum of at most three primes. In complete generality it is not settled yet, but in 1937 I. M. Vinogradov proved that every sufficiently large odd integer is a sum of three odd primes. (A proof may be found e.g. in Pracher's book [1]). Concerning even integers, we know only that for the number $N(x)$ of even integers $\leq x$ that are not sums of two primes, the estimate $N(x) = O(x^{1-c})$ is valid with some positive c (see, H. L. Montgomery and R. C. Vaughan [2]). Moreover, J. J. Chen [3] has shown that every sufficiently large even integer $2N$ is expressible in the form

$$2N = p + a \quad,$$

where p is a prime and $\Omega(a) \leqslant 2$ (such an a is called an *almost prime*). G. H. Hardy and J. E. Littlewood [1] have shown that if all the zeros of the Dirichlet L-functions lying in the strip $0 < \mathrm{Re}\, s < 1$ have the real part equal to $\frac{1}{2}$, then for any positive ϵ the estimate $N(x) = O\left(x^{\frac{1}{2}+\epsilon}\right)$ holds.

These results are obtained by difficult analytic methods with which we shall not be concerned in this book. In 1930 the Russian mathematician, L. G. Schnirelman gave a new method for solving additive problems, using which he showed very simply that there exists a constant C with the property that every natural number greater than 1 is a sum of at most C primes. The present section is devoted to the exposition of Schnirelman's method.

In Chapter II we have encountered the concept of natural density of a given subset of the set of natural numbers. In 1930 Schnirelman introduced some modifications of this notion, which proved to be very useful in additive problems. If $A \subset \mathcal{N}_0$, and we denote the number of non-zero elements of A not exceeding n, by $A(n)$ then we call the quantity

$$\delta(A) = \inf_{n=1,2,\ldots} A(n)n^{-1}$$

the *Schnirelman density* of the set A.

Unlike the natural density, every set clearly has Schnirelman density.

The fundamental properties of the Schnirelman density are contained in the following theorem:

Theorem 6.1.

(i)   In order that $\delta(A) \neq 0$, it is necessary and sufficient that $d_*(A) \neq 0$ and $1 \in A$.

(ii) In order that $\delta(A) = 1$ it is necessary and sufficient that the set A contains all natural numbers.

(iii) If $1 \in A$ and $0 \in B$, then

$$\delta(A + B) \geqslant \delta(A) + \delta(B) - \delta(A) \cdot \delta(B) .$$

Proof.

(i)   From the definition we have $\dfrac{A(n)}{n} \geqslant \delta(A)$ for $n = 1, 2, \ldots$, and hence

$d_*(A) \geqslant \delta(A)$ and $A(1) \geqslant \delta(A)$, which proves the necessity of the conditions. To prove sufficiency, we note that from $d_*(A) \neq 0$ follows the existence of a positive constant c_1 with the property that for sufficiently large n, say $n \geqslant N$, we have $A(n) \geqslant c_1 n$. Since $A(1) \neq 0$, the number $c_2 = \inf_{1 \leqslant n \leqslant N} A(n)n^{-1}$ is positive, and putting $c = \min(c_1, c_2)$, we obtain $A(n)/n \geqslant c$ for $n = 1, 2, \ldots$, which shows that $\delta(A) \geqslant c > 0$.

(ii)   The condition $\delta(A) = 1$ is equivalent to the equality $A(n) = n$ for $n = 1, 2, \ldots$, and this means that A contains all the natural numbers.

(iii) This is the most interesting part of the theorem and, as we shall see below, it has many important consequences.

Let $B(n) = \sum\limits_{1 \leqslant b_i \leqslant n} 1$. Select a natural number n, put $A(n) = k$ and let

$$1 = a_1 < a_2 < \ldots < a_k \leqslant n$$

be all the elements of A in the interval $[1, n]$. Between the numbers a_i and a_{i+1} there is a gap containing $g_i = a_{i+1} - a_i - 1$ numbers, and if $n \in A$, then at the end of the interval there appears a gap containing $g_k = n - a_k$ numbers. Observe that at least $B(g_i)$ numbers lying in the i-th gap belong to $A + B$. Indeed, this gap consists of the numbers $a_i + 1, a_i + 2, \ldots, a_i + g_i$ and if $0 < b_j \leqslant g_i$, $b_j \in B$, then $a_i + b_j$ lies in it, and here we have distinct $a_i + b_j$ for distinct b_j, since a_i is fixed. Moreover, since $0 \in B$, every number of A belongs to $A + B$. Thus we obtain

$$(A+B)(n) \geqslant A(n) + \sum_{i=1}^{k} B(g_i) \geqslant A(n) + \delta(B) \sum_{i=1}^{k} g_i$$

$$= A(n) + (n - A(n))\delta(B) = A(n)(1 - \delta(B)) + n\delta(B)$$

$$\geqslant n[\delta(A) + \delta(B) - \delta(A)\delta(B)]. \quad \blacksquare$$

Corollary 1. *If* $0, 1 \in A$, *then for any natural number* k *we have*

$$\delta(kA) \geqslant 1 - (1 - \delta(A))^k .$$

Proof. For $k = 1$ the assertion is trivial, and if we assume its truth for $k = n - 1$, then by the theorem proved above, we have

$$\delta(nA) = \delta((n-1)A + A) \geqslant \delta((n-1)A) + \delta(A) - \delta(A)\delta((n-1)A)$$

$$= \delta((n-1)A)(1 - \delta(A)) + \delta(A)$$

$$\geqslant (1 - (1 - \delta(A))^{n-1})(1 - \delta(A)) + \delta(A)$$

$$= 1 - (1 - \delta(A))^n. \quad \blacksquare$$

Corollary 2. *If* $0 \in A$ *and* $\delta(A) \neq 0$, *then there exists a natural number* k *such that* $kA = \mathcal{N}_0$, *and hence every natural number is a sum of* k *elements of the set* A.

Proof. By Corollary 1 there exists a natural number m such that

$$\delta(mA) > \tfrac{1}{2} ,$$

since $(1 - \delta(A))^k$ tends to zero as k increases. Now fix an n. The number of b_j's of the set mA not exceeding n is greater than $\tfrac{1}{2}n$, and therefore, the number of natural numbers of the form $n - b_i$ $(b_i \in mA)$ is also greater than $\tfrac{1}{2}n$. The sets

$$\{b_j : b_j \in mA, \ b_j \leqslant n\} \quad \text{and} \quad \{n - b_i : b_i \in mA, \ b_i \leqslant n\}$$

cannot be disjoint and we find i, j such that $b_j = n - b_i$, that is $n = b_i + b_j \in 2mA$. Since n is arbitrary, we obtain $\mathcal{N}_0 \subset 2mA$.   ∎

Every set such that for some natural number k we have $kA \supset \mathcal{N}$ is called a *basis of the set of natural numbers*, and the least integer k with this property is called the *order* of this basis. The conclusion may therefore be expressed as follows:

If a set A contains 0 and has positive Schnirelman density, then it is a basis of the set of natural numbers.

It is worth noting that in the case $A \neq \mathcal{N}$ the condition $0 \in A$ is necessary, since if $0 \notin A$, then we do not have $1 \in kA$ for any $k > 1$. However, the condition $\delta(A) \neq 0$ is not necessary. Indeed, as we shall see below, the set of squares of integers is a basis (see Theorem 6.4), but its natural density as well as its Schnirelman density is equal to 0.

2.   As an application of the concepts introduced in the preceding subsection we shall now prove *Schnirelman's theorem on prime numbers:*

Theorem 6.2. *Every natural number distinct from 1 can be expressed as a sum of at most C primes, where C is a certain constant.*

Proof. Denote by P_2 the set of all natural numbers that are expressible as a sum of two primes. Let us show that this set has positive natural density. To this end let $P_2(x)$ denote the number of elements of this set which do not exceed x, and let $p(m)$ denote the number of representations of m as a sum of two primes. Then from the Cauchy-Schwarz inequality we obtain

$$\left(\sum_{n \leqslant x} p(n) \right)^2 = \left(\sum_{\substack{n \leqslant x \\ n \in P_2}} p(n) \right)^2 \leqslant \sum_{n \leqslant x} p^2(n) \cdot P_2(x),$$

and so

$$P_2(x) \geqslant \frac{\left(\sum_{n \leqslant x} p(n) \right)^2}{\sum_{n \leqslant x} p^2(n)} .$$

Now observe that we have on the one hand for sufficiently large x

$$\sum_{n \leqslant x} p(n) = \sum_{p_1 + p_2 \leqslant x} 1 \geqslant \sum_{p_1, p_2 \leqslant \frac{1}{2} x} 1 = \pi^2(\tfrac{1}{2} x) \geqslant B_1 \frac{x^2}{\log^2 x} ,$$

with some positive constant B_1 in view of the prime number theorem, and on the other hand

$$\sum_{n \leqslant x} p^2(n) \leqslant \frac{B_2 x^3}{\log^4 x}$$

holds with some constant B_2 by the Corollary to Theorem 4.6 and we see that for sufficiently large x the inequality

$$P_2(x) \geqslant B_1^2 B_2^{-1} x ,$$

holds, that is, the set P_2 has positive natural density. Now put $A = P_2 \cup \{0, 1\}$ and apply Theorem 6.1. From (i) we get $\delta(A) \neq 0$, and by Corollary 2 we obtain the existence of a number k such that $kA \supset \mathcal{N}$. Hence every natural number n can be expressed in the form

$$n = a_1 + \ldots + a_k \; ,$$

where the numbers a_i are either the sums of two primes or equal to 1, that is

$$n = p_1 + \ldots + p_r + s \; ,$$

where $r + s \leqslant 2k$, and p_i are primes. If now s is an even integer, say, $s = 2s_1$, then we obtain

$$n = p_1 + \ldots + p_r + \underbrace{2 + \ldots + 2}_{s_1 \text{ times}}$$

and n is a sum of $r + s_1 \leqslant 2k$ primes, and if s is an odd integer different from 1, that is, $s = 2s_1 + 1$, $s_1 \neq 0$, then

$$n = p_1 + \ldots + p_r + \underbrace{2 + \ldots + 2}_{s_1 - 1 \text{ times}} + 3 \; ,$$

and n is a sum of $r + s_1 \leqslant 2k$ primes.

Finally, if s is equal to 1 and $n \geqslant 4$, then

$$n - 2 = q_1 + \ldots + q_t + u \; ,$$

with primes $q_1, \ldots, q_t$ and $t + u \leqslant 2k$. If $u \neq 1$, then as we have just seen, $n - 2$ is a sum of at most $2k$ primes, and hence $n = (n - 2) + 2$ is a sum of at most $(2k + 1)$ primes, and if $u = 1$, then

$$n = q_1 + \ldots + q_t + 3$$

is a sum of at most $2k$ primes.

The cases $n = 2, 3$ are trivial. ∎

3.   Making use of the Corollary to Theorem 6.1, we may determine a number k for which $kA \supset \mathcal{N}$ for a set A with positive Schnirelman density and containing 0. Indeed, a simple calculation shows that is suffices to take

$$k \geqslant \frac{\log 4}{\log(1 - \delta(A))^{-1}} \; .$$

This estimate can be improved by using the following strengthening of Theorem 6.1 (iii) proved by H. B. Mann.

Theorem 6.3 (H. B. Mann [1]). *If the sets A and B contain 0, then*

$$\delta(A + B) \geqslant \min(1, \delta(A) + \delta(B)) \; .$$

Proof. Let A, B be sets containing 0. As before, for any set S we write

$$S(n) = \sum_{\substack{k \leqslant n \\ k \in S}} 1 \, .$$

The assertion of the theorem follows from the following result concerning finite sets:

(*) *If n is a given natural number, $0 < c \leqslant 1$ and $A, B \subset [0, n]$, and moreover, for $m = 1, 2, \ldots, n$ the inequality $A(m) + B(m) \geqslant cm$ holds, then $(A + B)(m) \geqslant cm$ holds for $m = 1, 2, \ldots, n$.*

Indeed, it suffices to apply (*) to $A \cap [0, n]$, $B \cap [0, n]$, taking $c = \delta(A) + \delta(B)$.

Supposing that (*) is false for suitably chosen n, c, A and B, choose the smallest n for which (*) fails. Note that for $n = 1$ (*) is clearly true, since by assumption, $1 \in A \cup B$ and so $1 \in A + B$ and $(A + B)(1) = 1 \geqslant c$. After fixing n, we choose A, B so that (*) is false and the quantity $B(n)$ is minimal. We may suppose that $B(n)$ is positive, since otherwise we would have $B = \{0\}$, and in that case (*) is trivially valid.

If we now construct the sets A', B' satisfying the following conditions:

(i)   for $m = 1, \ldots, n$ we have $A'(m) + B'(m) \geqslant cm$,

(ii) $A' + B' \subset A + B$,

(iii) $B'(n) < B(n)$,

then we see that for them (*) is false and this gives a contradiction owing to (iii) and the minimal choice of B.

To do this, note that there exist elements $a \in A$ such that $(a + B)\backslash A$ is non-empty. Indeed, otherwise, we would have $A + B \subset A$, but this is impossible, since the greatest element of $(A + B)$ surely does not lie in A, because B contains non-zero elements. We denote the smallest such element $a \in A$ by a_+.

If $a_+ > 0$ and $0 \leqslant r < a_+$, then for every b in B we have

(6.1) $b + (A \cap [0, r]) \subset A$,

and hence we also have

(6.2) $(A + B) \cap [0, r] \subset A$.

We may now define the sets A' and B' which satisfy the conditions (i) - (iii) Put

$$B'' = \left\{ b'' \in B \colon a_+ + b'' \notin A \right\} \, .$$

This set is non-empty in view of the choice of a_+. Finally put

$$B' = B \setminus B''$$

and

$$A' = A \cup \left\{ x \colon x \leqslant n, \ x \in a_+ + B'' \right\} \, .$$

Now we show that these sets satisfy (i), (ii), (iii), and moreover, both contain 0. This last condition is satisfied, since $0 \in A \subset A'$, and if we had $0 \notin B'$, then $0 \in B''$, that is, $a_+ \notin A$, which is apparently absurd. The condition (iii) is satisfied by the fact that the set B'' is non-empty, and in order to check (ii), we consider an arbitrary element $a_1 + b_1$, where $a_1 \in A'$, $b_1 \in B'$. If a_1 lies in A, then $a_1 + b_1 \in A + B$, and if a_1 does not belong to A, then $a_1 = a_+ + b_2$, with $b_2 \in B''$, and hence $a_1 + b_1 = (a_+ + b_1) + b_2$. From the definition of B' it follows that $a_+ + b_1$ lies in A, and so we have $a_1 + b_1 \in A + B$ in this case too. Hence the condition (ii) is satisfied and there remains to check (i).

In the case $a_+ = 0$, we have

$$B'' = B \setminus A, \quad B' = B \setminus B'', \quad A' = A \cup B'',$$

and hence

$$A'(m) + B'(m) = A(m) + B(m)$$

and the condition (i) coincides with the assumption of (*). Therefore we may suppose that a_+ is a positive integer. In this case the set A' arises by adding to the set A the elements of B'' shifted by a_+. If such an element $a_+ + b''$ (with $b'' \in B''$) does not belong to the interval $[1, m]$, then obviously $b'' \in [m - a_+ + 1, n]$. Therefore it suffices to show that

$$A(m) + C(m) \geqslant cm ,$$

where the set C arises by removing from the set B all positive elements contained in the interval $[m - a_+ + 1, m]$.

We denote by b_1 the least positive element of B contained in $[m - a_+ + 1, m]$. (If there are no such elements, then $B''(m) = C(m) = B(m)$ and the assertion follows from the assumption). Write $m = b_1 + r$. Then $0 \leqslant r < a_+ \leqslant n$, and so by the choice of n we must have

$$(A + B)(r) \geqslant cr .$$

Since, furthermore, $0 \in A + B$ and $c \leqslant 1$,

$$|(A + B) \cap [0, r]| = 1 + (A + B)(r) \geqslant 1 + cr \geqslant c(1 + r)$$

and formula (6.2) leads to $|A \cap [0, r]| \geqslant c(1 + r)$.

From formula (6.1) it follows that the interval $[b_1, b_1 + r]$ contains at least as many elements of the set A as the interval $[0, r]$ does, that is, at least $c(1 + r)$. Hence

$$A(m) - A(b_1 - 1) = A(b_1 + r) - A(b_1 + 1) \geqslant c(1 + r)$$

and we obtain eventually

$$A\,(m) + C\,(m) = A\,(b_1 + r) + C\,(b_1 + r) \geqslant A\,(b_1 + r) + B\,(b_1 - 1)$$

$$= A\,(b_1 - 1) + B\,(b_1 - 1) + A\,(b_1 + r) - A\,(b_1 - 1)$$

$$\geqslant c\,(b_1 - 1) + c\,(1 + r) = c\,(b_1 + r) = cm,$$

and so the pair A', B' satisfies the conditions (i) - (iii), which contradicts the choice of A, B. ∎

Exercises

1.   Calculate the Schnirelman density of an arbitrary arithmetic progression formed of natural numbers.

2.   Prove that if A is a basis of the set of natural numbers, then there exists a positive number c such that for sufficiently large n the inequality $A(n) \geqslant n^c$ holds. Moreover, if A is a basis of order k, then there exists a positive constant B such that for sufficiently large n we have $A(n) \geqslant Bn^{1/k}$.

3.   Give an example of a set A containing 1 and satisfying the condition $A(n) \geqslant n^{\frac{1}{2}}$ for sufficiently large n, which is not a basis of the set of natural numbers.

4.   Let $g \geqslant 2$ be a given natural number. For any set A of natural numbers we denote by $\hat{A}$ the set of all natural numbers of the form

$$\sum_{i \in A} a_i g^i \qquad (0 \leqslant a_i \leqslant g - 1),$$

where only finitely many a_i's are different from zero.

Show that if $N_1, \ldots, N_k$ are disjoint subsets of $\mathcal{N}$ and their union is $\mathcal{N}$, then the set

$$B = \bigcup_{j=1}^{k} \hat{N}_j$$

is a basis of the set of natural numbers of order k.

5.   Give an example of a basis B of the set of natural numbers which is of order k and satisfies $B(n) = O(n^{1/k})$.

(Hint. Apply the preceding problem to the case when N_j are suitably chosen arithmetic progressions.)

§2. The Waring-Hilbert theorem

1.   In 1770 an English mathematician, Edward Waring, gave in his book *Meditationes algebraicae* the following theorem without proof:

> *Every natural number is a sum of four squares, nine cubes, nineteen fourth powers, etc.*

In the case of squares, this theorem had been formulated earlier by Fermat, *Descartes et al.*, and Euler had several times attempted a proof of it. The first proof was finally given by J. L. Lagrange in 1770.

Presently, under *Waring's theorem* (or more precisley, *the theorem of Waring-Hilbert*, because D. Hilbert proved it in 1909) we understand the following:

For each natural number k, there exists a natural number n_k such that every natural number is a sum of at most n_k k-th powers of natural numbers.

The smallest of such numbers n_k is denoted traditionally by $g(k)$ and is called the *Waring's constant* for the exponent k. Hence we have two problems here — prove the existence of $g(k)$ and determine its value.

Lagrange's result just quoted gives $g(2) = 4$ (since 7 is not a sum of three squares). The problem on fourth powers was attacked in the middle of the 19th Century by Liouville, who obtained the inequality $g(4) < 53$. The first result concerning the cubes is due to E. Maillet (1878): $g(3) < 21$. In 1896 this author obtained the estimate $g(5) < 192$. The existence of $g(8)$ was proven again by Maillet in 1907, $g(10)$ was estimated by I. Schur (1909) and finally, $g(7)$ by Wieferich (1909). In the same year there appeared the fundamental work of D. Hilbert containing a complete proof of the existence of the number $g(k)$ for arbitrary k. Hilbert's proof was based on the evaluation of complicated multi-dimensional integrals and was rather complicated. Later many authors simplified Hilbert's argument and we shall present such a modified proof of the Waring-Hilbert theorem in the following section.

2.   First we shall prove the equality $g(2) = 4$, since in the first place it does not require such complicated arguments as in the full theorem of Waring-Hilbert, and in the second, it is used in the proof of the general theorem. Thus we shall now prove:

Theorem 6.4. *Every natural number is a sum of four squares of integers.*

Lemma 6.1. *If a, b are the sum of four squares of integers, then their product is also such a sum.*

Proof. If $a = x_1^2 + x_2^2 + x_3^2 + x_4^2$ and $b = y_1^2 + y_2^2 + y_3^2 + y_4^2$, then

$$(6.3) \quad ab = (x_1^2 + \ldots + x_4^2)(y_1^2 + \ldots + y_4^2)$$

$$= (x_1 y_1 + x_2 y_2 + x_3 y_3 + x_4 y_4)^2 + (x_1 y_2 - x_2 y_1 + x_3 y_4 - x_4 y_3)^2$$

$$+ (x_1 y_3 - x_3 y_1 + x_4 y_2 - x_2 y_4)^2 + (x_1 y_4 - x_4 y_1 + x_2 y_3 - x_3 y_2)^2 . \quad \blacksquare$$

The identity (6.3) is called *Euler's identity*.

By Theorem 1.6, it suffices thus to prove that every prime number p can be expressed as a sum of four squares. For $p = 2$ this is clear, hence we suppose that $p > 2$. First let us show that some multiple mp of this number is a sum of four squares, and then we may take $m = 1$ here.

Lemma 6.2. *If p is an odd prime, then there exists an integer $0 < m < p$ such that $pm = 1 + X^2 + Y^2$ with some integral X, Y.*

Proof. The integers $0^2, 1^2, \ldots, k^2$, where $k = \frac{1}{2}(p-1)$, are incongruent $(\bmod\, p)$ (since from $r^2 \equiv s^2 \pmod{p}$ with $0 < r < s < k$, it would follow that either $r \equiv s \pmod{p}$ or $r + s \equiv 0 \pmod{p}$, and both these cases are excluded), and there are $k + 1$ of them. Similarly, the integers $-1, -1-1^2, -1-2^2, \ldots, -1-k^2$ are incongruent $(\bmod\, p)$ and there are again $k + 1$ of them. Since $2(k+1)$ exceeds p, there exist integers x, y not exceeding k such that

$$x^2 \equiv -1 - y^2 \,(\text{mod } p) \ ,$$

that is, $1 + x^2 + y^2 = mp$ with some natural m. Moreover, $mp \leqslant 1 + 2k^2$ $< 1 + \frac{1}{2}p^2 \leqslant p^2$ and $m < p$. ∎

Now we may complete the proof. Suppose that the integers X, Y satisfy the assertion of Lemma 6.2. By Corollary 2 to Theorem 5.8 there exist integers $x_1, \ldots, x_4$, not all zero and not exceeding $p^{\frac{1}{2}}$ in absolute value which satisfy the system of congruences

$$x_1 \equiv Xx_2 + Yy_3 \ (\text{mod } p) \ ,$$

$$x_4 \equiv Yx_2 - Xx_3 \ (\text{mod } p) \ .$$

We have then $x_1^2 + x_4^2 \equiv (X^2 + Y^2)x_2^2 + (X^2 + Y^2)x_3^2 \equiv -x_2^2 - x_3^2 \ (\text{mod } p)$, and hence $\sum\limits_{i=1}^{4} x_i^2 = Cp$, where $C = 1, 2$ or 3. If $C = 1$ we are ready. If $C = 2$, then we have e.g. $x_1 \equiv x_2 \,(\text{mod } 2)$ and $x_3 \equiv x_4 \,(\text{mod } 2)$, whence

$$p = \left(\frac{x_1 + x_2}{2}\right)^2 + \left(\frac{x_1 - x_2}{2}\right)^2 + \left(\frac{x_3 - x_4}{2}\right)^2 + \left(\frac{x_3 + x_4}{2}\right)^2 \ .$$

If, finally, $C = 3$, then one of the integers x_i is divisible by 3 and we may suppose that it is x_1. Changing if necessary the signs of x_2, x_3, x_4, we may assume that they are congruent to one another (mod 3), and then we have

$$p = \left(\frac{x_2 + x_3 + x_4}{3}\right)^2 + \left(\frac{x_1 + x_3 - x_4}{3}\right)^2 + \left(\frac{x_1 - x_2 + x_4}{3}\right)^2 + \left(\frac{x_1 + x_2 - x_3}{3}\right)^2 \ .$$

This proof is due to A. Brauer and T. L. Reynolds [1].

One can also prove (see e.g. W. Sierpiński [3], Vol. I, Chap. XVI) that if $n = 2^k n_1$, where $k \geqslant 0$ and n_1 is an odd integer, then for the number $r_4(n)$ of representations of n as a sum of four squares, where representations differing in order of factors are counted as one, the formula

$$r_4(n) = \begin{cases} 8\sigma(n_1), & \text{if } k = 0 \ , \\ \\ 24\sigma(n_1), & \text{if } k \neq 0 \ . \end{cases}$$

holds.

3.   We shall now consider the problem of representing natural numbers as sums of two squares. It is settled completely by the following theorem:

Theorem 6.5. *For a natural number n to be expressed as a sum of two squares it is necessary and sufficient that*

(6.4) $$n = 2^k \cdot p_1^{2a_1} \ldots p_i^{2a_i} q_1^{b_1} \ldots q_s^{b_s},$$

where $s, t \geq 0$, $a_1, \ldots, a_t, b_1, \ldots, b_s$ are any natural numbers, $k \geq 0$, p_i are prime numbers $\equiv 3 \pmod 4$ and q_i are prime numbers $\equiv 1 \pmod 4$.

In particular, a prime p can be expressed as a sum of two squares iff $p = 2$ or $p \equiv 1 \pmod 4$.

Proof. Let us first show that the condition formulated in the theorem is necessary. If an integer n is not of the form (6.4) but can be expressed as a sum of two squares, then there exist a prime $p \equiv 3 \pmod 4$ and an odd index $2r + 1$ such that $p^{2r+1} \| n$. Let $n = x^2 + y^2$ and $d = (x, y)$. Writing $x = dx_1$, $y = dy_1$, we have $(x_1, y_1) = 1$ and $n = d^2(x_1^2 + y_1^2)$. If $N = x_1^2 + y_1^2$, then one can see that $p^{2s+1} \| N$, where $s \geq 0$. Hence N is divisible by p, and the integers x_1, y_1 must both be indivisible by p. By Theorem 1.22 the congruence $y_1 \equiv X x_1 \pmod p$ is solvable and we obtain

$$x_1^2(1 + X^2) \equiv x_1^2 + y_1^2 \equiv 0 \pmod p,$$

which implies that $1 + X^2 \equiv 0 \pmod p$, i.e. -1 is a quadratic residue $\pmod p$, which is impossible, for $p \equiv 3 \pmod 4$ and $\left(\dfrac{-1}{p}\right) = -1$.

To prove the sufficiency, we shall establish an analogue of Lemma 6.1:

Lemma 6.3. *If the integers a, b are sums of two squares, then so is their product ab.*

Proof. If $a = x_1^2 + x_2^2$, $b = y_1^2 + y_2^2$, then

$$ab = (x_1^2 + x_2^2)(y_1^2 + y_2^2) = (x_1 y_1 + x_2 y_2)^2 + (x_1 y_2 - x_2 y_1)^2. \quad \blacksquare$$

Since $2 = 1^2 + 1^2$ and for any prime number p we have $p^2 = 0^2 + p^2$, it now suffices to prove that every prime $p \equiv 1 \pmod 4$ can be expressed as a sum of two squares. We denote the least positive integer by M with the property that Mp can be expressed as a sum of two squares, $Mp = x^2 + y^2$, where x, y are not divisible by p. Since $p \equiv 1 \pmod 4$, -1 is a quadratic residue $\pmod p$ and there exists an integer a such that $a^2 \equiv -1 \pmod p$, so that $a^2 + 1 = mp$ with some m, which implies that the integer M is well-defined and does not exceed $m < p$. If $M = 1$, then p is a sum of two squares. Hence suppose that $M > 1$ and let $Mp = X^2 + Y^2$. If X, Y were divisible by M, then M^2 would divide Mp, and hence M would be a divisor of p, which is not possible.

Now we choose in the interval $[-M/2, M/2]$ the integers x_0, y_0 congruent respectively to $x, y \pmod M$. Since $x_0^2 + y_0^2 \equiv x^2 + y^2 \equiv 0 \pmod M$ and $x_0^2 + y_0^2 > 0$, we have with some integer k

$$0 < kM = x_0^2 + y_0^2 < 2(M/2)^2 < M^2 \; ,$$

which leads to $0 < k < M$.

Since

$$(xx_0 + yy_0)^2 + (xy_0 - x_0 y)^2 = (x^2 + y^2)(x_0^2 + y_0^2) = M^2 kp$$

and furthermore

$$xx_0 + yy_0 \equiv xy_0 - x_0 y \equiv 0 (\text{mod } M) \; ,$$

we may write with suitable integers A, B

$$A^2 M^2 + B^2 M^2 = kM^2 p$$

and we obtain $A^2 + B^2 = kp$, which contradicts the choice of M.   ∎

4.   We now set out to prove the Waring-Hilbert theorem and we begin with its formulation.

Theorem 6.6 *If* $k \in \mathcal{N}$, *then there exists an integer* $s = s(k)$ *with the property that every natural number is expressible as a sum of at most* s *k-th powers of natural numbers.*

We follow the method given by W. J. Ellison [1].

We shall prove the theorem in the following somewhat different form:

Theorem 6.7. *If* $k \in \mathcal{N}$, *then there exist a natural number* M *and positive rational numbers* $c_1, \ldots, c_M$ *depending solely on* k *with the property that every sufficiently large natural number* N, *say* $N \geqslant A$, *is expressible in the form*

$$(6.5) \qquad\qquad N = \sum_{i=1}^{M} c_i n_i^k \qquad (n_i \in \mathcal{N}) \; .$$

First let us show that Theorem 6.6 follows from Theorem 6.7: Denote by m the least common multiple of denominators of c_i $(i = 1, 2, \ldots, M)$. Then $mc_i = d_i$ are all natural numbers.

Now if $N \geqslant mA$, then

$$N = mq + r, \qquad 0 \leqslant r < m \; ,$$

and $q \geqslant A$, hence by Theorem 6.7 we have

$$q = \sum_{i=1}^{M} c_i n_i^k \qquad (n_i \in \mathcal{N}) \; ,$$

so that

$$mq = \sum_{i=1}^{M} d_i n_i^k$$

is a sum of $d_1 + \ldots + d_M$ k-th powers of natural numbers. Since r is a sum of at most $r < m$ such powers in view of $r = 1^k + 1^k + \ldots + 1^k$, N is a sum of at most $d_1 + \ldots + d_M + m$ such powers. It remains to observe that every integer not exceeding mA is a sum of at most mA k-th powers, and this gives for $s(k)$ the estimation

$$s(k) \leqslant d_1 + \ldots + d_M + m + mA \quad .$$

Hence we shall prove Theorem 6.7. We begin with a lemma found by Hilbert and proved by him using rather tedious methods. The idea of using the notion of the center of mass is due to E. Schmidt [1].

Lemma 6.4. *For every natural number* k *there exist positive rationals* $c_0, \ldots, c_N$ *and integers* a_{ij} $(i = 1, \ldots, N, \ j = 1, 2, \ldots, 5)$ *such that*

(6.6)
$$(x_1^2 + \ldots + x_5^2)^k = \sum_{i=0}^{N} c_i \left(\sum_{j=1}^{5} a_{ij} x_j \right)^{2k}.$$

Proof. Consider the linear space E over the field of real numbers whose elements are homogeneous forms of degree $2k$ in 5 variables and with real coefficients. The forms of the shape

$$x_1^{\alpha_1} x_2^{\alpha_2} \ldots x_5^{\alpha_5} \qquad (\alpha_1 + \ldots + \alpha_5 = 2k, \ \alpha_i \geqslant 0)$$

make up a basis of this space, and so the dimension of E is equal to the number of representations of the integer $2k$ as a sum of at most 5 summands. Denote it by N. (It is easy to see that $N = \frac{1}{24} \prod_{j=1}^{4} (2k + j)$, but this information will not be of any use to us).

Let A be the set of all forms of E of the form $\left(\sum_{i=1}^{5} \lambda_i x_i \right)^{2k}$ with rational λ_i's and let $c(A)$ be the convex closure of this set. By the Corollary to Lemma 5.4 it now suffices to prove that for some rational number $r > 0$ we have

(6.7)
$$r(x_1^2 + \ldots + x_5^2)^k \in c(A) \quad ,$$

because, in that case, we obtain immediately

$$r(x_1^2 + \ldots + x_5^2)^k = \sum_{i=0}^{N} \mu_i \left(\sum_{j=1}^{5} a_{ij} x_j \right)^{2k} \qquad \text{with} \qquad \mu_i \geqslant 0, \ a_{ij} \in Q,$$

and hence also

$$(x_1^2 + \ldots + x_5^2)^k = \sum_{i=0}^{N} v_i \left(\sum_{j=1}^{5} a_{ij} x_j \right)^{2k},$$

where $v_i = \mu_i/r$. If M denotes the least common multiple of the denominators of α_{ij}, then $\beta_{ij} = M\alpha_{ij} \in \mathcal{Z}$ and

$$(x_1^2 + \ldots + x_5^2)^k = \sum_{i=0}^{N} \frac{v_i}{M^{2k}} \left(\sum_{j=1}^{5} \beta_{ij} x_j \right)^{2k}.$$

Hence there remains to prove (6.7). Denote by $g(x_1, \ldots, x_5)$ the center of mass of the set

$$T = \{(\alpha_1 x_1 + \ldots + \alpha_5 x_5)^{2k} : \alpha_1^2 + \ldots + \alpha_5^2 \leqslant 1, \ \alpha_i \in \mathcal{R}\} \ .$$

Then g belongs to $c(A)$ and

$$g(x_1, \ldots, x_5) = \frac{\displaystyle\int_{a_1^2 + \ldots + a_5^2 \leqslant 1} (a_1 x_1 + \ldots + a_5 x_5)^{2k} da_1 \ldots da_5}{\displaystyle\int_{a_1^2 + \ldots + a_5^2 \leqslant 1} da_1 \ldots da_5}.$$

Now let $[b_{ij}]$ be a 5×5 orthogonal matrix for which

$$b_{1j} = x_j (x_1^2 + \ldots + x_5^2)^{-\frac{1}{2}} \qquad (j = 1, 2, \ldots, 5).$$

Making the substitution

$$t_i = \sum_{j=1}^{5} b_{ij} a_j \qquad (i = 1, \ldots, 5)$$

we obtain

$$g(x_1, \ldots, x_5) = \frac{\displaystyle\int_{t_1^2 + \ldots + t_5^2 \leqslant 1} t_1^{2k} dt_1 \ldots dt_5}{\displaystyle\int_{t_1^2 + \ldots + t_5^2 \leqslant 1} dt_1 \ldots dt_5} (x_1^2 + \ldots + x_5^2)^k = c(x_1^2 + \ldots + x_5^2)^k,$$

since our transformation preserves the unit cube. Now if $0 < r < c$ is a rational number, then using $0 < r/c < 1$ we obtain

$$r(x_1^2 + \ldots + x_5^2)^k = \frac{r}{c} g(x_1, \ldots, x_5) \in c(A) \ . \qquad \blacksquare$$

Corollary 1. *If k and y are given natural numbers, then one can find integers $a_0, \ldots, a_N, \ b_0, \ldots, b_N$ and positive rational numbers $c_0, \ldots, c_N$, where N and $c_0, \ldots, c_N$ depend exclusively on k, such that the identity*

$$(x^2+y)^k = \sum_{i=0}^{N} c_i (a_i x + b_i)^{2k}$$

holds.

Proof. By Theorem 6.4, we may write $y = x_2^2 + \ldots + x_5^2$. Putting $x_1 = x$ in the last in Lemma 6.4 and writing $\sum_{j=2}^{5} a_{ij} x_j = b_i$ we obtain the assertion. ∎

Corollary 2. *If Theorem 6.7 is true for* $k = m$, *then it is also true for* $k = 2m$.

Proof. If in the last Corollary we put $k = m$ and $x = 0$, then we see that every k-th power can be expressed in the form $\sum_{i=0}^{N} c_i b_i^{2k}$, where the non-negative rationals c_i and the natural number N depend solely on k, and hence it suffices to make use of the assumed truth of Theorem 6.7 for $k = m$, which we suppose, to obtain its truth for $k = 2m$. ∎

Corollary 3. *If* m *and* $r < m$ *are natural numbers, then there exist natural numbers* $B_i^{(r)}$ $(i = 1, 2, \ldots, r-1)$ *such that for all natural* x *and* y *we have*

$$\sum_{i=0}^{r} B_i^{(r)} x^{2i} (x^2+y)^{m-i} = \sum_{j=1}^{M} c_j n_j^m,$$

where $n_i = n_i(x, y) \in \mathcal{N}$ *and the numbers* $c_j > 0$ *are rational and depend only on* m.

For the simplicity of notation we shall denote by the symbol $\Sigma(m)$ every natural number that can be expressed in the form

$$\sum_{j=1}^{M} c_j n_j^m \qquad (n_j \in \mathcal{N}),$$

where c_j are positive rational numbers depending only on m, and $M \in \mathcal{N}$ also depends only on m. In this way we may write the right-hand side of the formula in Corollary 3 in the form $\Sigma(m)$.

Proof. We put $k = m + r$ in Corollary 1 and differentiate $2r$-times the equality obtained. ∎

Lemma 6.5. *Let* A *be a natural number. For each* k *there exists a natural number* n_0 *such that if* $T > n_0$, *then every natural number* n *in the interval*

$$[AT^k, A(T+1)^k]$$

can be expressed in the form

(6.8)
$$n = \sum_{j=0}^{k-1} b_j T^j + A T^k,$$

where $0 \leq b_i \leq T-1$, $b_i \in \mathcal{Z}$ $(i = 0, \ldots, k-1)$.

Proof. Observe that for sufficiently large T we have

$$AT^k + T^k - T > A(T+1)^k \quad,$$

so that

$$A(T+1)^k < AT^k + (T-1)(T + T^2 + \ldots + T^{k-1})$$

and it suffices to write n in the T-adic system to obtain the assertion. ∎

Now we can prove Theorem 6.7. We apply induction on k. For $k = 2$ the assertion follows from Theorem 6.4. Suppose that the theorem is true for natural numbers $k < m$. By Corollary 2 to Lemma 6.4 it will (and hence also Theorem 6.3) be true for all even integers $< 2m$.

Let $n_1, \ldots, n_{m-1}, T$ be natural numbers and let $T > \max_{1 \leqslant i \leqslant m-1} n_i$. By induction hypothesis there exists an integer h such that

$$n_r = \sum_{i=1}^{h} x_{ir}^{2r} \quad (r = 1, 2, \ldots, m-1, \ x_{ir} \in \mathscr{Z}, \ x_{ir} \geqslant 0).$$

Now fix r and write for brevity x_i for x_{ir}. If now $y_i = T - x_i^2$, then Corollary 3 to Lemma 6.4 gives us

$$\sum_{k=0}^{r} B_k^{(r)} x_i^{2k} T^{m-k} = \Sigma(m)$$

and adding the equalities obtained, we arrive at

$$\sum_{i=1}^{h} \sum_{k=0}^{r} B_k^{(r)} x_i^{2k} T^{m-k} = \Sigma(m),$$

and so

(6.9) $$h B_0^{(r)} T^m + \sum_{k=1}^{r-1} B_k^{(r)} \left(\sum_{i=1}^{h} x_i^{2k} \right) T^{m-k} + B_r^{(r)} n_r T^{m-r} = \Sigma(m)$$

(since $\sum_{i=1}^{h} x_i^{2r} = n_r$).

We now prove a simple lemma:

Lemma 6.6. *If for some $j \leqslant m$ we have $x_1^{2j} + \ldots + x_h^{2j} < T$, $a > 1$ and*

$$T \geqslant (h+1) \frac{a^{2j}}{a-1}, \text{ then}$$

$$a \left(1 + \sum_{i=1}^{h} x_i^{2(j-1)} \right) < T.$$

Proof. We have

$$a\left(1+\sum_{i=1}^{h}x_i^{2(j-1)}\right) = a\left(1+\sum_{x_i\leqslant a}x_i^{2(j-1)}\right)+a\sum_{x_i>a}x_i^{2(j-1)}$$

$$\leqslant (1+h)a^{2j-1}+\frac{1}{a}\sum_{x_i>a}x_i^{2j}\left(\frac{a}{x_i}\right)^2 \leqslant (1+h)a^{2j-1}+\frac{1}{a}\sum_{i=1}^{h}x_i^{2j}$$

$$< (1+h)a^{2j-1}+\frac{T}{a} \leqslant \frac{a-1}{a}T+\frac{T}{a} = T. \qquad \blacksquare$$

Corollary. *If*

$$C_k^{(r)} = B_k^{(r)}\sum_{i=1}^{h}x_i^{2k},$$

then for $T > T_0(m)$ *we have*

$$C_k^{(r)} < T \qquad \text{for } k = 0, 1, \ldots, r-1 .$$

Proof. It suffices to apply the lemma with $a = B_k^{(r)}$, $j = r$ and to note that

$$\max_{k,r}\left\{(h+1)\frac{(B_k^{(r)})^{2r}}{B_k^{(r)}-1}\right\}$$

depends only on m. $\qquad \blacksquare$

Putting $n_r = T - N_r$ in formula (6.9), we obtain that if $N_1, \ldots, N_{m-1}$ are any natural numbers in the interval $(0, T)$, where T is any given natural number, then for $r \leqslant m - 1$ we have

$$(6.10) \qquad \sum_{k=0}^{r-2}C_k^{(r)}T^{m-k}+(C_{r-1}^{(r)}+B_r^{(r)})T^{m-r+1}-B_r^{(r)}N_rT^{m-r} = \Sigma(m).$$

It is worth noting here that for $T > T_1(m)$ we have

$$(6.11) \qquad C_{r-1}^{(r)} + B_r^{(r)} < T .$$

Indeed,

$$C_{r-1}^{(r)}+B_r^{(r)} = B_r^{(r)}+B_{r-1}^{(r)}\sum_{i=1}^{h}x_i^{2r-2} \leqslant \max\,(B_{r-1}^{(r)}, B_r^{(r)})\left(1+\sum_{i=1}^{h}x_i^{2r-2}\right) < T$$

by Lemma 6.6.

We now use the estimate obtained to show that for some A and a sufficiently large T every integer n of the form (6.8) for which $b_0 = 0$, satisfies

$$n = \Sigma(m) .$$

This is the case $q = m - 1$ of the next lemma.

Lemma 6.7. *Let* $0 \leqslant q \leqslant m - 1$. *There exists a number* T_0 *depending only on* m *and a number* A_q *such that if* $T \geqslant T_0$, *then for any integers* $b_{m-q}, \ldots, b_{m-1} \in \mathscr{Z}$, $-T \leqslant b_i < T$ $(i = m - q, \ldots, m - 1)$ *we have*

$$A_q T^m + \sum_{i=m-q}^{m-1} b_i T^i = \Sigma(m).$$

Proof. For $q = 0$ we may take $A_0 = 1$. Suppose that the lemma is true for all natural numbers $< q$. First let us show that for sufficiently large T and all integral R in the interval $[-T, T)$ there exist integers $a_1, \ldots, a_{q-1}$ satisfying the condition

$$0 < a_i < T \quad (i = 1, \ldots, q - 1)$$

and natural numbers a_0, b depending only on q and m such that

$$(6.12) \qquad a_0 T^m + \sum_{i=1}^{q-1} a_i T^{m-i} + bR T^{m-q} = \Sigma(m).$$

In order to get this, we apply formula (6.9) for $R > 0$, and formula (6.10) for $R < 0$, and to prove the estimation $0 < a_i < T$, we make use of the Corollary to Lemma 6.6 and formula (6.11). Finally for $R = 0$ we make use of the induction hypothesis for $q - 1$. Let λ be the smallest natural number satisfying $\lambda b \geqslant A_{q-1} + a_0$. Then $\lambda b - A_{q-1} - a_0 \leqslant b$, where b depends solely on m, and so $(\lambda b - A_{q-1} - a_0) T^m = \Sigma(m)$.

From the induction hypothesis we obtain

$$A_{q-1} T^m - \sum_{i=0}^{q-1} a_i T^{m-i} = \Sigma(m)$$

and

$$(6.13) \quad \lambda b T^m - \sum_{i=0}^{q-1} a_i T^{m-i} = (\lambda b - A_{q-1} - a_0) T^m + A_{q-1} T^m - \sum_{i=1}^{q-1} a_i T^{m-i} = \Sigma(m).$$

Adding side by side (6.12) to (6.13), we obtain

$$b(\lambda T^m + R T^{m-q}) = \Sigma(m) \quad,$$

and so we have also

$$(6.14) \qquad \lambda T^m + R T^{m-q} = \Sigma(m) \ .$$

By induction we have for any $b_i \in [-T, T)$

$$A_{q-1} T^m + b_{m-q+1} T^{m-q+1} + \ldots + b_{m-1} T^{m-1} = \Sigma(m)$$

and adding this to (6.14), we obtain the assertion of the Lemma, on taking $A_q = A_{q-1} + \lambda$, $b_{m-q} = R$. ∎

Corollary. *There exist natural numbers* A, T_0 *depending only on* m *with the property that if* $T \geqslant T_0$, *then for every integer* n *satisfying* $|n| < T^{m-1}$ *we have*

$$A T^m + n T = \Sigma(m) \quad .$$

Proof. It suffices to apply Lemma 6.5. ∎

Now we may complete the proof of the Theorem. Let

$$T_0' = \max\left(T_0, \frac{2}{\left(1 + \frac{1}{A}\right)^{1/m} - 1}\right),$$

$$N \geqslant A\{(T_0'+1)^m + T_0'^m\}$$

and let T be the largest natural number satisfying the condition

$$N \geqslant A\{(T+1)^m + T^m\}.$$

Clearly, $T \geqslant T_0$, and a simple calculation shows that

$$0 \leqslant r = N - A\{(T+1)^m + T^m\} \leqslant T^m.$$

Since $(T, T+1) = 1$, we may find integers x, y such that

$$r = x(1+T) - yT \quad.$$

Here we may suppose that $0 \leqslant x \leqslant T-1$ (on taking as x the least non-negative solution of the congruence $r \equiv x(1+T) \pmod{T}$), and then, since

$$yT = x(1+T) - r \quad,$$

it follows that if $x(1+T) \geqslant r$, then $0 \leqslant yT \leqslant x(1+T) \leqslant T^2 - 1 < T^2$ and $0 \leqslant y < T$, and if $x(1+T) > r$, then $|yT| \leqslant r < T^m$ and $|y| < T^{m-1}$. Thus in either case $|x|, |y| < T^{m-1}$.

Since

$$N = A(T+1)^m + x(1+T) + AT^m - yT \quad,$$

applying the Corollary to the last lemma successively to the numbers

$$A(T+1)^m + x(1+T) \quad \text{and} \quad AT^m - yT$$

and adding the obtained equalities, we obtain $N = \Sigma(m)$. ∎

5. The value of the Waring constant $g(k)$ is known in principle for all $k \neq 4$. If we put

$$A(k) = [(\tfrac{3}{2})^k] + 2^k - 2$$

and

$$B(k) = [(\tfrac{3}{2})^k] + [(\tfrac{4}{3})^k] + 2^k - 2 \quad,$$

then the following theorem holds:

If $k \geqslant 6$ and $3^k - 2^k + 2 < (2^k - 1)[(\tfrac{3}{2})^k]$, then $g(k) = A(k)$. If, on the contrary, this inequality does not hold, then $g(k) = B(k)$ or $g(k) = B(k) + 1$ according *as the equality*

$$[(\tfrac{3}{2})^k][(\tfrac{4}{3})^k] + [(\tfrac{3}{2})^k] + [(\tfrac{4}{3})^k] = 2^k$$

is true or not.

(L. E. Dickson [1], S. S. Pillai [1], R. K. Rubugunday [1], I. Niven [1]).

It is conjectured that for all k we have $g(k) = A(k)$, but it is known only that this equality holds for sufficiently large k (K. Mahler [1]) and for $6 \leqslant k \leqslant 200\,000$ (R. M. Stemmler [1]).

For $k = 3$ we have the equality $g(3) = A(3)$ (A. Wieferich [1]), and for $k = 5$ the equality $g(5) = A(5)$ (J. J. Chen [2]). In the case of $k = 4$, we know only that $A(4) = 19 \leqslant g(4) \leqslant 22$. The inequality $g(4) \geqslant 19$ follows from Theorem 6.8, and the inequality $g(4) \leqslant 22$ has been proved by H. E. Thomas, Jr. [1].

Theorem 6.8. *For* $k \geqslant 2$ *we have* $g(k) \geqslant A(k)$.

Proof. Consider the integer $N = 2^k q - 1$, where $q = [(\tfrac{3}{2})^k]$. Since $N < 3^k$ and $N < 2^k q$,

$$N = (q - 1)2^k + (N - (q - 1)2^k)1^k$$

is a representation of N as a sum of k-th powers with the least number of summands. Hence

$$g(k) \geqslant q - 1 + N - (q - 1)2^k = 2^k + q - 2 = A(k) \ . \qquad \blacksquare$$

For $k \geqslant 3$ all the sufficiently large natural numbers require fewer number of k-th powers than $g(k)$. We denote by $G(k)$ the smallest integer s with the property that every sufficiently large natural number is a sum of at most s k-th powers. Exact values of $G(k)$ are known only in two cases: $G(2) = 4$, $G(4) = 16$. Moreover,

$$4 \leqslant G(3) \leqslant 7, \quad G(5) \leqslant 23, \quad G(6) \leqslant 36, \quad G(7) \leqslant 52,$$
$$G(8) \leqslant 73, \quad G(9) \leqslant 96, \quad G(10) \leqslant 121 \ .$$

For $k \geqslant 170\,000$ the inequality

$$G(k) < 2k \log k + 4k \log \log k + 2k \log \log \log k + 13k$$

holds (I. M. Vinogradov [1]), and J. J. Chen [1] has given the best estimate valid for all k:

$$G(k) < 3k \log k + \tfrac{26}{5} k \ .$$

The reader may find an extensive survey of the results concerning the Waring problem and its generalization in the article by W. J. Ellison [1].

Exercises

1.   Prove that none of the integers of the form $4^a(8b + 7)$ $(a, b \geqslant 0)$ can be represented as a sum of three squares.

Remark.   The converse theorem is also true.

2.   Give a necessary and sufficient condition for a given rational number to be a sum of two squares of rationals.

3.   Prove that infinitely many natural numbers are not sums of four squares of natural numbers.

4.   Show that the set $\{x^2 + x + 1 : x \in \mathcal{N}_0\}$ is a basis of the set of natural numbers.

5.   Prove the inequalities $G(3) \geqslant 4$, $G(4) \geqslant 16$.

6.   Let k, l be given natural numbers and let $A_{k,l}(x)$ denote the number of natural numbers $\leqslant x$, which are the sums of l k-th powers. Show that

$$A_{k,l}(x) \geqslant C(k, l)x^{1-\alpha},$$

where $C(k, l)$ is positive and does not depend on x, and $\alpha = (1 - \frac{1}{k})^l$.

§ 3. Other additive problems

1. In this section we shall be concerned mainly with an application of the method of generating functions, i.e. the method of using the function

$$F(z) = \sum_{n=1}^{\infty} z^{a_n} \qquad (|z| < 1).$$

in the investigation of additive properties of a sequence a_n.

In certain cases the investigation of analytic properties of the function $f(z)$ is necessary but it is sometimes sufficient to treat $f(z)$ as an element of the ring of formal power series. The method of generating functions relies on the following simple lemma:

Lemma 6.8. *If $\{a_n^{(1)}\}, \ldots, \{a_n^{(k)}\}$ are sequences of non-negative integers and $f(m)$ is the number of representations of a natural number m in the form*

$$m = a_{n_1}^{(1)} + \ldots + a_{n_k}^{(k)},$$

and

$$f_i(z) = \sum_{n=1}^{\infty} z^{a_n^{(i)}} \qquad (i = 1, \ldots, k),$$

then for $|z| < 1$ the equality holds:

$$\sum_{m=0}^{\infty} f(m)z^m = \prod_{i=1}^{k} f_i(z).$$

Proof. For $|z| < 1$ we have

$$\prod_{i=1}^{k} f_i(z) = \prod_{i=1}^{k} \sum_{n_i=1}^{\infty} z^{a_{n_i}^{(i)}} = \sum_{n_1,\ldots,n_k=1}^{\infty} z^{a_{n_1}^{(1)}+\ldots+a_{n_k}^{(k)}}$$

$$= \sum_{m=0}^{\infty} \left(\sum_{a_{n_1}^{(1)}+\ldots+a_{n_k}^{(k)} = m} 1 \right) z^m = \sum_{m=0}^{\infty} f(m) z^m. \quad \blacksquare$$

Applying here Cauchy's integral formula, we may determine in principle $f(m)$. This idea is due to G. H. Hardy and J. E. Littlewood, who used it in their famous series *Partitio numerorum* to give a new proof of Waring's theorem, and exploited it to treat other additive problems. The Hardy-Littlewood method was simplified afterwards in the work of I. M. Vinogradov (see, e.g. I. M. Vinogradov [2]), who reduced it to the estimation of certain trigonometrical sums.

The exposition of this method is beyond the scope of this book. (See, e.g. K. Prachar [1], Chap. VI).

2.   In this subsection we apply the method of generating functions to prove the following theorem of P. Erdös and W. H. J. Fuchs [1]:

Theorem 6.9. *If* $a_1 < a_2 < \ldots$ *is a sequence of non-negative integers and*

$$f(n) = \sum_{\substack{i,j \\ n=a_i+a_j}} 1,$$

then no asymptotic identity of the form

$$\sum_{k \leqslant n} f(k) = cn + r_n$$

with a positive constant c *and* $r_n = o(n^{\frac{1}{4}} \log^{-\frac{1}{2}} n)$ *can hold.*

Proof. Let

$$F(z) = \sum_{n=1}^{\infty} z^{a_n} \quad (|z| < 1).$$

From Lemma 6.8 we obtain the equality

$$F^2(z) = \sum_{n=0}^{\infty} f(n) z^n \quad (|z| < 1) ,$$

and multiplying it by the equality

$$\frac{1}{1-z} = \sum_{n=0}^{\infty} z^n$$

we arrive at

$$\frac{F^2(z)}{1-z} = \sum_{n=0}^{\infty} \left(\sum_{k=0}^{n} f(k) \right) z^n.$$

If now we suppose that for the sequence $\{a_n\}$ we have

$$\sum_{k=0}^{n} f(k) = cn + r_n, \qquad (c > 0), \qquad r_n = o(n^{\frac{1}{4}} \log^{-\frac{1}{2}} n),$$

then for $|z| < 1$ we obtain the equality

$$\frac{F^2(z)}{1-z} = \frac{cz}{(1-z)^2} + h(z),$$

where $h(z) = \sum_{n=0}^{\infty} r_n z^n$ and

(6.15) $$r_n = o(n^{\frac{1}{4}} \log^{-\frac{1}{2}} n) \quad .$$

Let us show that this leads to a contradiction. For this purpose we shall need the following lemma from complex function theory:

Lemma 6.9. *Let the series* $\varphi(z) = \sum_{n=0}^{\infty} b_n z^n$ *be convergent within the unit disc and suppose that its coefficients are non-negative real numbers. Then for every value of* t *in the interval* $(0, \pi]$ *and* $0 < r < 1$ *we have*

$$\frac{1}{2t} \int_{-t}^{t} |\varphi(re^{i\vartheta})|^2 d\vartheta \geq \frac{1}{6\pi} \int_{-\pi}^{\pi} |\varphi(re^{i\vartheta})|^2 d\vartheta.$$

Proof. Consider the auxiliary function

$$h(\vartheta) = \begin{cases} 1 - \left| \dfrac{\vartheta}{t} \right| & \text{for} \quad |\vartheta| < t, \\[2ex] 0 & \text{for} \quad t \leq |\vartheta| \leq \pi. \end{cases}$$

Its Fourier series is, as can be easily checked, the series

$$\sum_{k=-\infty}^{\infty} c_k e^{ik\vartheta}$$

(where $c_k = \dfrac{(1 - \cos tk)}{\pi t k^2} \geq 0$), which is absolutely and uniformly convergent to $h(\vartheta)$.
Applying the Parseval identity to the product $h(\vartheta) \varphi(re^{i\vartheta})$ we obtain

$$\int_{-\pi}^{\pi} |h(\vartheta)\varphi(re^{i\vartheta})|^2 d\vartheta = 2\pi \sum_{j=-\infty}^{\infty} \left(\sum_{k+l=j} c_k b_l r^l \right)^2 \geq 2\pi \sum_{j=-\infty}^{\infty} \sum_{k+l=j} c_k^2 b_l^2 r^{2l}$$

$$= 2\pi \sum_{k=-\infty}^{\infty} c_k^2 \sum_{l=0}^{\infty} b_l^2 r^{2l} = \frac{1}{2\pi} \int_{-\pi}^{\pi} h^2(\vartheta) d\vartheta \int_{-\pi}^{\pi} |\varphi(re^{i\vartheta})|^2 d\vartheta$$

$$= \frac{t}{3\pi} \int_{-\pi}^{\pi} |\varphi(re^{i\vartheta})|^2 d\vartheta,$$

since

$$\int_{-\pi}^{\pi} h^2(\vartheta) d\vartheta = \int_{-t}^{t} \left(1 - \left|\frac{\vartheta}{t}\right|\right)^2 d\vartheta = \tfrac{2}{3} t.$$

Making use of this inequality and the remark that for $|\vartheta| < t$ we have $|h(\vartheta)| \leq 1$, and for $t \leq |\vartheta| \leq \pi$ we have $h(\vartheta) = 0$, we may write

$$\int_{-t}^{t} |\varphi(re^{i\vartheta})|^2 d\vartheta \geq \int_{-\pi}^{\pi} |h(\vartheta)\varphi(re^{i\vartheta})|^2 d\vartheta \geq \frac{t}{3\pi} \int_{-\pi}^{\pi} |\varphi(re^{i\vartheta})|^2 d\vartheta. \quad \blacksquare$$

In the following we shall try to find mutually exclusive estimations from above and from below of the integral

$$I(r, t) = \int_{-t}^{t} |F(re^{i\vartheta})|^2 d\vartheta$$

for suitably chosen r and t.

Let us begin with the estimation of $I(r, t)$ from above:

Lemma 6.10. *If $\epsilon > 0$ is given and $1 > r \geq r(\epsilon) > \frac{1}{2}$, $1 - r < t < \pi$, then*

$$I(r, t) < C_1 \log \frac{1}{1-r} + C_2 \,\epsilon t^{\frac{3}{2}} \cdot (1-r)^{-\frac{3}{4}} \log^{-\frac{1}{2}} \frac{1}{1-r},$$

where C_1, C_2 are constants independent of ϵ, r and t.

Proof. Writing $z = re^{i\vartheta}$, we obtain

(6.16)
$$I(r, t) = \int_{-t}^{t} \left| \frac{cz}{1-z} + (1-z)h(z) \right| d\vartheta$$

$$\leq c \int_{-\pi}^{\pi} \frac{d\vartheta}{|1-z|} + \int_{-t}^{t} |h(z)| \, |1-z| d\vartheta.$$

Since

$$(1-z)^{-\frac{1}{2}} = \sum_{k=0}^{\infty} (-1)^k \binom{-\frac{1}{2}}{k} z^k,$$

and moreover, for $k \geq 1$

$$\left| \binom{-\frac{1}{2}}{k} \right| = \left(1 - \frac{1}{2}\right)\left(1 - \frac{1}{4}\right) \cdots \left(1 - \frac{1}{2k}\right) = \exp \sum_{j \leq k} \log\left(1 - \frac{1}{2j}\right)$$

$$\ll \exp\left\{-\sum_{j \leq k} \frac{1}{2j}\right\} \leq \exp\left(-\tfrac{1}{2}\log k + O(1)\right) \ll k^{-\frac{1}{2}},$$

we have

(6.17)
$$\int_{-\pi}^{\pi} |1 - z|^{-1} d\vartheta = \int_{-\pi}^{\pi} \left| \sum_{k=0}^{\infty} (-1)^k \binom{-\frac{1}{2}}{k} z^k \right|^2 d\vartheta$$

$$= 2\pi \sum_{k=0}^{\infty} \left| \binom{-\frac{1}{2}}{k} \right|^2 r^{2k} \ll \sum_{k=1}^{\infty} \frac{r^{2k}}{k} \ll \log\frac{1}{1-r},$$

which gives the estimate for the first term in (6.16).

To estimate the second term, we note that if ϑ lies in the interval $[-t, t]$, then $|1 - z| \leq Ct$, with some constant C. Hence

(6.18)
$$\int_{-t}^{t} |1 - z| |h(z)| d\vartheta \leq Ct \int_{-t}^{t} |h(z)| d\vartheta \leq Ct \left(\int_{-t}^{t} d\vartheta\right)^{\frac{1}{2}} \left(\int_{-\pi}^{\pi} |h^2(z)| d\vartheta\right)^{\frac{1}{2}}$$

$$\ll t^{\frac{3}{2}} \left(\sum_{n=0}^{\infty} |r_n|^2 r^{2n}\right)^{\frac{1}{2}}.$$

We estimate the sum $\sum_{n=0}^{\infty} |r_n|^2 r^{2n}$ using (6.15). If $\epsilon > 0$ is given, then for sufficiently large n we have

$$|r_n| \leq \epsilon n^{\frac{1}{4}} \log^{-\frac{1}{2}} n,$$

and so this inequality holds for $n > (1-r)^{-\frac{1}{2}}$, if r is sufficiently near to 1. Hence for such r we have

$$\sum_{n=0}^{\infty} |r_n|^2 r^{2n} \leq \sum_{n \leq (1-r)^{-\frac{1}{2}}} |r_n|^2 r^{2n} + \epsilon^2 \sum_{n > (1-r)^{\frac{1}{2}}} \frac{n^{\frac{1}{2}} r^{2n}}{\log n}.$$

In the first sum we have at most $(1-r)^{-\frac{1}{2}}$ terms not exceeding

$$\frac{B n^{\frac{1}{2}} r^{2n}}{\log n} < B n^{\frac{1}{2}} r^{2n} \leq B n^{\frac{1}{2}} \leq B(1-r)^{-\frac{1}{4}},$$

and so this sum does not exceed $B(1-r)^{-\frac{3}{4}}$ with a suitable constant B.

Next, the second sum does not exceed

$$\varepsilon^2\log\frac{1}{1-r}\sum_{n=1}^{\infty}n^{\frac{1}{2}}r^{2n}\ll\varepsilon^2(1-r^2)^{-\frac{3}{2}}\log\frac{1}{1-r}\ll\varepsilon^2(1-r)^{-\frac{3}{2}}\log\frac{1}{1-r},$$

and so

$$\sum_{n=0}^{\infty}|r_n|^2r^{2n}\ll(1-r)^{-\frac{3}{4}}+\varepsilon^2(1-r)^{-\frac{3}{2}}\log\frac{1}{1-r}\ll\varepsilon^2(1-r)^{-\frac{3}{2}}\log\frac{1}{1-r}.$$

This inequality together with (6.18) now lead to

(6.19) $$\int_{-t}^{t}|1-z|\,|h(z)|\,d\vartheta\ll\varepsilon t^{\frac{3}{2}}(1-r)^{-\frac{3}{4}}\log^{\frac{1}{2}}\frac{1}{1-r}.$$

The assertion of the lemma is a consequence of formulas (6.17) and (6.19). ∎

The estimate of $I(r,t)$ from below is given in the following lemma:

Lemma 6.11. *There exists a positive constant C such that for $t\le\pi$ and $r<1$ we have*

$$\int_{-t}^{t}|F(re^{i\vartheta})|^2d\vartheta\ge Ct(1-r)^{-\frac{1}{2}}.$$

Proof. We apply Lemma 6.9 to the function $F(z)$ to obtain the estimation

$$I(r,t)=\int_{-t}^{t}|F(re^{i\vartheta})|^2d\vartheta\ge t\int_{-\pi}^{\pi}|F(re^{i\vartheta})|^2d\vartheta=t\sum_{j=1}^{\infty}r^{2a_j}=tF(r^2),$$

but by formula (6.15)

$$F^2(r^2)=\frac{Cr^2}{1-r^2}+(1-r^2)h(r^2)=\frac{Cr^2}{1-r^2}+O\left((1-r^2)\sum_{n=1}^{\infty}n^{-\frac{1}{4}}r^{2n}\right)$$

$$\ge\frac{1}{1-r}+O\left((1-r)(1-r)^{-\frac{5}{4}}\right)\ge(1-r)^{-1},$$

so that

$$\int_{-t}^{t}|F(re^{i\vartheta})|^2d\vartheta\ge t(1-r)^{-\frac{1}{2}}. \quad\blacksquare$$

In order to complete the proof of the theorem we take

$$t=\varepsilon^{-\frac{2}{3}}(1-r)^{\frac{1}{2}}\log\frac{1}{1-r}.$$

and apply the last two lemmas. This leads us to

$$\varepsilon^{-\frac{2}{3}}\log\frac{1}{1-r} \ll \log\frac{1}{1-r} + \varepsilon\log^{-\frac{1}{2}}\frac{1}{1-r} \cdot (1-r)^{-\frac{3}{4}} \cdot \varepsilon^{-1} \cdot (1-r)^{\frac{3}{4}}\log^{\frac{3}{2}}\frac{1}{1-r},$$

i.e.

$$\varepsilon^{-\frac{2}{3}}\log\frac{1}{1-r} \ll \log\frac{1}{1-r}$$

or

$$\varepsilon^{-\frac{2}{3}} \ll 1,$$

which is impossible in view of the arbitrariness of ϵ and the independence of the constant implied by the symbol " $\ll$ " from ϵ.

Corollary. *If* $r_2(n) = \sum\limits_{n=a^2+b^2} 1$, *then in the equality*

$$\sum_{n \leqslant x} r_2(n) = \pi x + R(x)$$

the remainder term $R(x)$ *cannot be equal to* $o\left(x^{\frac{1}{4}}\log^{-\frac{1}{2}}x\right)$. ∎

Remark. G. H. Hardy [1] and A. E. Ingham [1] showed that

$$R(x) \neq o\left(x^{\frac{1}{4}}\log x\right) \quad \text{and} \quad \limsup\frac{R(x)}{x^{\frac{1}{4}}} = +\infty ,$$

respectively, where the first equality means that there exists a positive constant K such that the inequality

$$|R(x)| > -Kx^{\frac{1}{4}}\log x$$

holds for an unbounded sequence of values of x.

3. We shall now be concerned with partitions. We call any representation of a natural number n in the form

$$n = \sum_{i=1}^{k} n_i \qquad (n_i \in \mathcal{N}) ,$$

its *partition*, where the case $k = 1$ is also admitted. We identify those partitions which differ only by the order of terms. The number of all the partitions of an integer n is denoted by $p(n)$. Thus, e.g. we have $p(1) = 1$, $p(2) = 2$, $p(3) = 3$, $p(4) = 5$.

Lemma 6.12.

(i) *The number of solutions of the equation*

(6.20) $$x_1 + 2x_2 + \ldots + (n-1)x_{n-1} + nx_n = n$$

in non-negative integers $x_1, \ldots, x_n$ is equal to $p(n)$.

(ii)   If $m < n$, then $p(m) < p(n)$.

Proof.

(i)   If $n_1 + n_2 + \ldots + n_k = n$ is a partition of an integer n and $x_j = \displaystyle\sum_{n_i = j} 1$

($j = 1, \ldots, n$), then $x_1, \ldots, x_n$ satisfy the equation (6.20), and moreover a solution of (6.20) also gives a partition of n.

(ii)   It suffices to show that for $n = 2, 3, \ldots$ we have $p(n-1) < p(n)$. If $n - 1 = a_1 + \ldots + a_k$ is a partition of $n - 1$, then $n = 1 + a_1 + \ldots + a_k$ is a partition of n, and moreover we do not obtain in this way all the partitions of n. (For instance, $n = n$ does not appear).   ∎

It is worth noting that the function $p(n)$ increases rather rapidly, for e.g. $p(200) = 3\ 972\ 999\ 029\ 388$ (P. Mc Mahon) and the number $p(14\ 031)$ is a 127-place number in decimal notation (D. H. Lehmer [1]).

We now introduce the generating function for $p(n)$, namely,

$$(6.21) \qquad F(z) = 1 + \sum_{n=1}^{\infty} p(n) z^n \ .$$

Lemma 6.13. *The series* (6.13) *is almost uniformly convergent in the disc* $|z| < 1$, *and the following equality holds there:*

$$F(z) = \prod_{k=1}^{\infty} (1 - z^k)^{-1} \ ,$$

the product on the right-hand side being absolutely and almost uniformly convergent.

Proof. Absolute and almost uniform convergence of the product follows from the corresponding properties of the series $\displaystyle\sum_{k=1}^{\infty} z^k$. Moreover we have for natural N the equality

$$\prod_{k=1}^{N} (1 - z^k)^{-1} = \prod_{k=1}^{N} \sum_{j=0}^{\infty} z^{jk} = \sum_{j_1, \ldots, j_N \geq 0} z^{j_1 + 2j_2 + \ldots + Nj_N} = 1 + \sum_{n=1}^{\infty} p_N(n) z^n,$$

where $p_N(n)$ denotes the number of solutions of the equation (6.20) under the additional condition $x_j = 0$ ($j > N$). From Lemma 6.12 (i) it follows that for $n \leq N$ we have $p_N(n) = p(n)$, and for $n > N$ we have $0 \leq p_N(n) \leq p(n)$, and so for real z satisfying $0 \leq z < 1$ we have

$$1 + \sum_{n=1}^{N} p(n) z^n \leq \prod_{k=1}^{\infty} (1 - z^k)^{-1} \leq 1 + \sum_{n=1}^{\infty} p(n) z^n.$$

The radius of convergence of the series (6.21) is therefore equal to 1. Passing now from

N to infinity, we obtain the assertion of the lemma for $z \in [0, 1)$, and by the holomorphy of the functions appearing here we have the assertion also for all z belonging to the interior of the unit disc. ∎

4. In 1918 G. H. Hardy and S. S. Ramanujan [2] proved that

$$p(n) = \left(\frac{1}{4\sqrt{3}} + o(1)\right) \frac{\exp\left(\pi\sqrt{\frac{2}{3}}\, n^{\frac{1}{2}}\right)}{n} ,$$

which implies

$$\log p(n) = \pi\sqrt{\frac{2}{3}}\, n^{\frac{1}{2}} + o\left(n^{\frac{1}{2}}\right).$$

We now prove this estimation of $\log p(n)$:

Theorem 6.10. *For* $n \geqslant 1$ *we have*

$$\pi\sqrt{\frac{2}{3}}\, n^{\frac{1}{2}} + o\left(n^{\frac{1}{2}}\right) \leqslant \log p(n) < \pi\sqrt{\frac{2}{3}}\, n^{\frac{1}{2}}.$$

Proof. We begin with the estimation from above and for this purpose we first prove some simple identities:

Lemma 6.14. *For* $n \geqslant 1$ *we have the identity*

$$np(n) = \sum_{r=1}^{n} r \sum_{1 \leqslant k \leqslant n/r} p(n - rk).$$

Proof. If $\Pi(n) = \{\langle a_1, \ldots, a_k \rangle : a_1 + \ldots + a_k = n\}$ is the set of all partitions of n and

$$S = \sum_{\langle a_1, \ldots, a_k \rangle \in \Pi(n)} (a_1 + \ldots + a_k),$$

then clearly

$$S = \sum_{\langle a_1, \ldots, a_k \rangle \in \Pi(n)} n = np(n).$$

Denoting by $N(n, k, r)$ the number of partitions of n containing exactly k terms equal to r, we obtain

$$S = \sum_{r=1}^{n} \sum_{1 \leqslant k \leqslant n/r} krN(n, k, r).$$

Since we have $p(n - kr)$ partitions of n containing at least k terms equal to r,

$$N(n, k, r) = p(n - kr) - p(n - (k - 1)r) ,$$

so that

$$S = \sum_{r=1}^{n} r \sum_{1 \leqslant k \leqslant n/r} k\big(p(n - kr) - p(n - (k - 1)r)\big) = \sum_{r=1}^{n} r \sum_{1 \leqslant k \leqslant n/r} p(n - kr). \quad ∎$$

Now we shall prove by induction the inequality

(6.22) $$p(n) < \exp(\alpha n^{\frac{1}{2}})$$

with $\alpha = \pi \sqrt{\frac{2}{3}}$.

For $n = 1$ this inequality is clearly true. We assume its validity now for $n < N$. Using Lemma 6.14, we get

$$p(N) = \frac{1}{N} \sum_{r \leqslant N} r \sum_{1 \leqslant k \leqslant N/r} p(N - kr) < \frac{1}{N} \sum_{r \leqslant N} \sum_{1 \leqslant k \leqslant N/r} r \exp\left\{\alpha(N - kr)^{\frac{1}{2}}\right\}$$

$$\leqslant \frac{1}{N} \sum_{r \leqslant N} \sum_{k \leqslant N} r \exp\left\{\alpha N^{\frac{1}{2}}\left(1 - \frac{kr}{N}\right)^{\frac{1}{2}}\right\} \leqslant \frac{1}{N} \sum_{r \leqslant N} \sum_{k \leqslant N} r \exp\left\{\alpha N^{\frac{1}{2}}\left(1 - \frac{kr}{2N}\right)\right\}$$

$$= \frac{1}{N} \exp\left(\alpha N^{\frac{1}{2}}\right) \sum_{r \leqslant N} \sum_{k \leqslant N} r \exp\left(-\frac{\alpha kr}{2N^{\frac{1}{2}}}\right),$$

since we obtain from Bernoulli's inequality

$$\left(1 - \frac{kr}{N}\right)^{\frac{1}{2}} \leqslant 1 - \frac{kr}{2N} .$$

It now remains to show that

(6.23) $$S_0 = \sum_{r \leqslant N} \sum_{k \leqslant N} r \exp\left\{-\frac{\alpha kr}{2N^{\frac{1}{2}}}\right\} < N.$$

To this end we consider for $x > 0$ the function

$$f(x) = \sum_{m=1}^{\infty} m e^{-mx}.$$

It is the derivative of the function $-\sum_{m=1}^{\infty} e^{-mx} = (1 - e^x)^{-1}$, so that

$$f(x) = \frac{1}{e^x + e^{-x} - 2} = \frac{1}{4 \sinh^2 \frac{x}{2}} \leqslant \frac{1}{x^2} .$$

We have therefore

$$S_0 < \sum_{k \leqslant N} f\left(\frac{\alpha k}{2N^{\frac{1}{2}}}\right) \leqslant \sum_{k \leqslant N} \frac{4N}{\alpha^2 k^2} < \frac{4N}{\alpha^2} \cdot \frac{\pi^2}{6} = N,$$

which proves (6.23), and also (6.22).

The proof of the second part of the assertion follows from the following lemma.

Lemma 6.15. *For any positive ϵ there exists a constant $A = A(\epsilon)$ such that* *for* $n \geqslant 1$ *we have*

(6.24) $$p(n) > A \exp\left(\left(\pi\sqrt{\tfrac{2}{3}} - \epsilon\right)\right)n^{\frac{1}{2}} \quad .$$

Proof. Let N be a natural number whose exact value shall be fixed later. Let also $\alpha = \pi\sqrt{\tfrac{2}{3}}$ and

$$A_n = p(n)\exp\left(-(\alpha - \epsilon)n^{\frac{1}{2}}\right) \quad .$$

We take $A = A(\epsilon, N) = \min_{n \leqslant N} A_n$.

For $n \leqslant N$ the estimation (6.24) is clearly true. Now let $n > N$ and suppose that the inequality (6.24) is true for all natural numbers less than n. Making use of this assumption and Lemma 6.14, we obtain that for any positive δ, the following estimation is true:

$$np(n) > A \sum_{k \leqslant n} \sum_{r \leqslant n/k} r \exp\left\{(\alpha - \epsilon)(n - kr)^{\frac{1}{2}}\right\}$$

$$> A \sum_{k \leqslant n^{\frac{1}{3}}} \sum_{r \leqslant n^{\frac{2}{3} - \delta}} r \exp\left\{(\alpha - \epsilon)(n - kr)^{\frac{1}{2}}\right\}.$$

Now observe that for any positive ϑ the inequality

(6.25) $$(n - kr)^{\frac{1}{2}} = n^{\frac{1}{2}}\left(1 - \frac{kr}{n}\right)^{\frac{1}{2}} > n^{\frac{1}{2}}\left(1 - \frac{kr}{2n}(1 + \vartheta)\right)$$

holds if only the quotient kr/n is sufficiently near to zero. Hence if $k \leqslant n^{\frac{1}{3}}$, $r \leqslant n^{\frac{2}{3} - \delta}$, then this inequality will be true provided that $n \geqslant N_1 = N_1(\delta, \vartheta)$.

Therefore if the number N is chosen so that $N \geqslant N_1$, then from (6.25) follows

$$np(n) > A\exp\left\{(\alpha - \epsilon)n^{\frac{1}{2}}\right\} \sum_{k \leqslant n^{\frac{1}{3}}} \sum_{r \leqslant n^{\frac{2}{3} - \delta}} r \exp\left\{-\frac{(\alpha - \epsilon)(1 + \vartheta)kr}{2n^{\frac{1}{2}}}\right\}.$$

In view of the identity $\sum_{r=1}^{\infty} r e^{-rx} = \dfrac{e^{-x}}{(1 - e^{-x})^2}$ valid for $x > 0$, we have

$$\sum_{r \leqslant n^{\frac{2}{3} - \delta}} r\exp(-rx) = e^{-x}(1 - e^{-x})^{-2} - \sum_{r > n^{\frac{2}{3} - \delta}} r e^{-rx}.$$

$$= e^{-x}(1-e^{-x})^{-2} - \int_{n^{\frac{2}{3}-\delta}}^{\infty} r\exp(-rx)\,dr + O\left(n^{\frac{2}{3}-\delta}\exp\left(-n^{\frac{2}{3}-\delta}x\right)\right)$$

$$= e^{-x}(1-e^{-x})^{-2} + O\left(n^{\frac{2}{3}-\delta}\exp\left(-n^{\frac{2}{3}-\delta}x\right)\right),$$

and so for $x = \dfrac{(\alpha-\epsilon)(1+\vartheta)}{2n^{\frac{1}{2}}}k$ we have

$$\sum_{r\leqslant n^{\frac{2}{3}-\delta}} r\exp\left(-\frac{(\alpha-\epsilon)(1+\vartheta)}{2n^{\frac{1}{2}}}kr\right) = \frac{\exp\left(-\dfrac{(\alpha-\epsilon)(1+\vartheta)}{2n^{\frac{1}{2}}}k\right)}{\left(1-\exp\left(-\dfrac{(\alpha-\epsilon)(1+\vartheta)}{2n^{\frac{1}{2}}}k\right)\right)^2}$$

$$+ O\left(n^{\frac{2}{3}-\delta}\exp\left(-kn^{\frac{1}{6}-\delta}\right)\right)$$

and we arrive at

$$(6.26) \quad np(n) > A\exp\left\{(\alpha-\epsilon)n^{\frac{1}{2}}\right\}\sum_{k\leqslant n^{\frac{1}{3}}} \frac{\exp\left\{-\dfrac{(\alpha-\epsilon)(1+\vartheta)k}{2n^{\frac{1}{2}}}\right\}}{\left(1-\exp\left\{-\dfrac{(\alpha-\epsilon)(1+\vartheta)k}{2n^{\frac{1}{2}}}\right\}\right)^2}$$

$$+ O\left(A\exp\left\{(\alpha-\epsilon)n^{\frac{1}{2}}\right\}n^{\frac{2}{3}-\delta}\sum_{k\leqslant n^{\frac{1}{3}}}\exp\left(-kn^{\frac{1}{6}-\delta}\right)\right).$$

Now observe that if $0 < u < 1$ and $0 < x < \sqrt{6u}$, then $x^2 e^{-x}(1-e^{-x})^{-2} \geqslant 1-u$. Indeed, expanding into series, we see that

$$\frac{x^2 e^{-x}}{(1-e^{-x})^2} = \left(\sum_{k=0}^{\infty} \frac{x^{2k}}{(2k+1)!2^{2k}}\right)^{-1},$$

but for $x < \sqrt{6u} < \sqrt{6}$ we have

$$\sum_{k=1}^{\infty} \frac{x^{2k}}{(2k+1)!2^{2k}} < \frac{x^2}{12(1-\frac{1}{80}x^2)} < \frac{x^2}{6},$$

hence

$$\frac{x^2 e^{-x}}{(1-e^{-x})^2} \geqslant \frac{1}{1+\frac{1}{6}x^2} \geqslant \frac{1}{1+u} = \frac{1-u}{1-u^2} \geqslant 1-u.$$

For $\beta = \dfrac{(\alpha - \epsilon)(1 + \vartheta)}{2}$ and $k \leqslant n^{\frac{1}{3}}$ we have

$$\frac{\beta k}{n^{\frac{1}{2}}} < \beta n^{-\frac{1}{6}} < \sqrt{6u}$$

if only $n \geqslant N_2 = N_2(\epsilon, \vartheta, u)$. Therefore if $N \geqslant N_2$, then we have

$$\sum_{k \leqslant n^{\frac{1}{3}}} \exp\left(-\frac{\beta k}{n^{\frac{1}{2}}}\right)\left(1 - \exp\left(-\frac{\beta k}{n^{\frac{1}{2}}}\right)\right)^{-2} \geqslant \frac{(1-u)n}{\beta^2} \sum_{k \leqslant n^{\frac{1}{3}}} \frac{1}{k^2} \geqslant \frac{\pi^2(1-u)n}{6\beta^2} - \frac{C}{n^{\frac{2}{3}}}$$

with some positive constant C.

Since the remainder term in (6.26) does not exceed $AC_1 n^{\frac{2}{3}}\exp\{(\alpha - \epsilon)n^{\frac{1}{2}}\}$ for some constant C_1, the last inequality together with the inequality (6.26) implies

$$np(n) > A\exp\{(\alpha - \epsilon)n^{\frac{1}{2}}\}\frac{\pi^2}{6\beta^2}(1-u)n$$

$$- ACn^{-\frac{2}{3}}\exp\{(\alpha - \epsilon)n^{\frac{1}{2}}\} - AC_1 n^{\frac{2}{3}}\exp\{(\alpha - \epsilon)n^{\frac{1}{2}}\} \,,$$

and so we also have

$$p(n) > A\exp\{(\alpha - \epsilon)n^{\frac{1}{2}}\}\frac{\pi^2}{6\beta^2}(1-u) - R(n) \,,$$

where

$$R(n) = A\exp\{(\alpha - \epsilon)n^{\frac{1}{2}}\}(Cn^{-\frac{5}{3}} + C_1 n^{-\frac{1}{3}}) \,.$$

Finally, let η be any given small constant. Choose ϑ and u so that

$$\frac{\pi^2(1-u)}{6\beta^2} \geqslant 1 + \eta \,.$$

Furthermore, let $N_3 = N_3(\epsilon, \eta)$ be chosen so that for $n \geqslant N_3$ the inequality

$$R(n) < \frac{1}{2}A\eta\exp\{(\alpha - \epsilon)n^{\frac{1}{2}}\}$$

holds. Now, if $N \geqslant N_1, N_2, N_3$, then we obtain

$$p(n) > A(1 + \eta)\exp\{(\alpha - \epsilon)n^{\frac{1}{2}}\} - A\frac{\eta}{2}\exp\{(\alpha - \epsilon)n^{\frac{1}{2}}\}$$

$$= A(1 + \tfrac{1}{2}\eta)\exp\{(\alpha - \epsilon)n^{\frac{1}{2}}\} > A\exp\{(\alpha - \epsilon)n^{\frac{1}{2}}\} \,. \qquad \blacksquare$$

From the lemma just proved the assertion of the theorem follows immediately. Indeed, the inequality (6.24) implies

$$\log p(n) > \left(\pi\sqrt{\frac{2}{3}} - \epsilon \right) n^{\frac{1}{2}} + \log A(\epsilon) \quad,$$

and so we have

$$\liminf_{n \to \infty} \frac{\log p(n)}{n^{\frac{1}{2}}} \geq \pi\sqrt{\frac{2}{3}} - \epsilon \quad,$$

which leads by arbitrariness of ϵ to

$$\liminf_{n \to \infty} \frac{\log p(n)}{n^{\frac{1}{2}}} \geq \pi\sqrt{\frac{2}{3}} \quad. \quad \blacksquare$$

Exercises

1. Prove that

$$p_3(n) = \frac{(n+3)^2}{12} - \frac{7}{72} + \frac{(-1)^n}{8} + \frac{2}{9} \cos\left(\frac{2n\pi}{3}\right) \quad.$$

2. Show that the integer nearest to $\dfrac{(n+3)^2}{12}$ is $p_3(n)$.

3. Find the generating functions for

    a) the number of partitions with terms in a given arithmetic progression,

    b) the number of partitions with all terms different.

4. Prove that the number of partitions of an integer n with odd terms is equal to that of partitions of this number with different terms.

5. Show that for $n = 1, 2, 3, \ldots$ we have

$$n p(n) = \sum_{k=0}^{n} \sigma(k) p(n-k) \quad.$$

CHAPTER VII

PROBABILISTIC NUMBER THEORY

§1. The Turán-Kubilius inequality

1. The present chapter is devoted to the distribution of values of arithmetic functions. Both the methods and problems have a strong probabilistic flavour and this explains the title of this chapter. This field has been developed relatively recently though some results of Ćebyśev are connected with it. The first conscious appeal to probabilistical arguments was done by P. Turán ([1], [2]). He used the inequality of Ćebyśev to prove the theorem of Hardy and Ramanujan concerning the values of functions ω and Ω (see Corollary 3 to Theorem 3.2). Principal results of this theory were later obtained by P. Erdös, M. Kac and A. Wintner. The monograph by I. P. Kubilius [2] treats many aspects of modern probabilistic number theory.

The following argument demonstrates, why it is easiest to apply probabilistical methods to additive functions.

Let $f(n)$ be an additive function which we shall assume for the sake of simplicity to be *strongly additive*, i.e. for each prime number p and any natural number k the equality $f(p^k) = f(p)$ holds, and so for an arbitrary n we have

$$(7.1) \qquad f(n) = \sum_{p|n} f(p) .$$

If $e_p(n)$ is the characteristic function of the set of integers divisible by a prime number p, then from (7.1) we obtain

$$(7.2) \qquad f(n) = \sum_p e_p(n) f(p) .$$

Now let N be a fixed natural number. We shall treat the set $A_N = \{1, 2, \ldots, N\}$ as a probability space with a probability measure P defined for any set $X \subset A_N$ by the formula

$$P_N(X) = \frac{1}{N} \sum_{n \in X} 1 \ .$$

Then we may regard the functions $f(n)$ and $e_p(n)$ as random variables on A_N. Let us note that for expectations E_N the following formulas hold:

$$E_N(e_p) = \frac{1}{N}\left[\frac{N}{p}\right] = \frac{1}{p} + O\left(\frac{1}{N}\right) \ ,$$

(7.3) $$E_N(e_p e_q) = \frac{1}{N}\left[\frac{N}{pq}\right] = \frac{1}{pq} + O\left(\frac{1}{N}\right) \qquad (p \neq q) \ ,$$

and so

$$E_N(e_p e_q) = E_N(e_p) E_N(e_q) + O(1/N) \ .$$

If the term $O(1/N)$ were absent in the last equality then the random variables e_p would be independent, but this term tends to zero as N increases and it may be expected that for large N, these random variables will behave almost independently. Therefore it can be expected that an analogue of classical limit theorems of probability theory will be applicable in this situation. In fact, thanks to the work of P. Erdös, M. Kac, I. P. Kubilius and other mathematicians, it was possible to utilize this approach to the study of additive functions.

Let us return to formula (7.3). Applying it to the variance

$$D_N^2(X) = E_N(X^2) - E_N^2(X)$$

we obtain

$$D_N^2(e_p) = E_N(e_p^2) - E_N^2(e_p) = E_N(e_p) - E_N^2(e_p) = \frac{1}{p}\left(1 - \frac{1}{p}\right) + O\left(\frac{1}{N}\right) \ .$$

If f is a strongly additive function, then by formulas (7.2) and (7.3) we obtain

(7.4) $$E_N(f) = \sum_{p \leq N} f(p) E_N(e_p) = \sum_{p \leq N} f(p) p^{-1} + O\left(\frac{1}{N} \sum_{p \leq N} |f(p)|\right) \ .$$

In the case when we only assume the additivity of the function f, we obtain in place of (7.4) the formula

(7.5) $$f(n) = \sum_{p^k \| n} f(p^k) = \sum_{p^k \leq n} \hat{e}_{p^k}(n) f(p^k) \ ,$$

where $\hat{e}_{p^k}(n)$ is the characteristic function of the set of integers divisible by p^k but not divisible by p^{k+1}. Evidently, we have

$$E_N(f) = \sum_{p^k \leqslant n} f(p^k) E_N(\hat{e}_{p^k}) \quad ,$$

but

$$E_N(\hat{e}_{p^k}) = \frac{1}{N}\left(\left[\frac{N}{p^k}\right] - \left[\frac{N}{p^{k+1}}\right]\right) = \frac{1}{p^k}\left(1 - \frac{1}{p}\right) + O\left(\frac{1}{N}\right) \quad ,$$

and hence

$$(7.6) \quad E_N(f) = \sum_{p^k \leqslant N} \frac{f(p^k)}{p^k}\left(1 - \frac{1}{p}\right) + O\left(\frac{1}{N}\sum_{p^k \leqslant N}|f(p^k)|\right)$$

$$= \sum_{p \leqslant N} f(p)p^{-1} - \sum_{p \leqslant N} f(p)p^{-2} + \sum_{\substack{p^k \leqslant N \\ k \geqslant 2}} \frac{f(p^k)}{p^k}\left(1 - \frac{1}{p}\right) + O\left(\frac{1}{N}\sum_{p^k \leqslant N}|f(p^k)|\right).$$

In particular, if the values of $f(p^k)$ are uniformly bounded, the last formula takes the form

$$(7.7) \qquad E_N(f) = \sum_{p \leqslant n} f(p)p^{-1} + O(1) \quad .$$

2. The above calculations show that the function $f(n)$ should behave like $\sum_{p \leqslant N} f(p)p^{-1}$ at least in the case when the values of $f(p^k)$ are uniformly bounded. This is actually the case and constitutes the contents of the inequality of Turán-Kubilius which is an analogue of the inequality of Čebyšev in probability theory and will be proved later. We first introduce some notations. Let $f(n)$ be an arbitrary additive function and let

$$A_n = \sum_{p \leqslant n} f(p)p^{-1} \quad ,$$

$$B_n^2 = \sum_{p \leqslant n} |f(p)|^2 p^{-1} \quad , \qquad B_n \geqslant 0 \quad ,$$

$$D_n^2 = \sum_{p^k \leqslant n} |f(p^k)|^2 p^{-k} \quad , \qquad D_n \geqslant 0 \quad .$$

(Recall that the index p and later, q run through prime numbers). With these notations the following theorem holds:

Theorem 7.1 (The inequality of Turán-Kubilius). *There exist an absolute constant C such that for all additive functions and all natural numbers n the inequality*

$$\sum_{m \leqslant n} |f(m) - A_n|^2 \leqslant CnD_n^2$$

holds.

Proof. In spite of the simple formulation, the proof of this inequality is not simple at all. We therefore divide it into steps. First, by a simple change of the summation we obtain the formula

$$(7.8) \qquad \sum_{m \leqslant n} f(m) = nK_n + O\left(\frac{nD_n}{(\log n)^{\frac{1}{2}}}\right) \quad ,$$

where

$$K_n = \sum_{p^k \leqslant n} \frac{f(p^k)}{p^k}\left(1 - \frac{1}{p}\right) = \sum_{p^k \leqslant n} \frac{f(p^k)\varphi(p^k)}{p^{2k}} \quad .$$

This formula is of interest in itself because it enables us to express the average value of an additive function by its values at the points p^k.

In an analogous way we deduce next the formula

$$(7.9) \qquad \sum_{m \leqslant n} |f(m)|^2 = nL_n + O(nD_n^2) \quad ,$$

where

$$L_n = \sum_{p^k q^l \leqslant n} f(p^k)\overline{f(q^l)}p^{-k}q^{-l}\left(1 - \frac{1}{p}\right)\left(1 - \frac{1}{q}\right) \quad .$$

Next we use the inequality $|a + b|^2 \leqslant 2(|a|^2 + |b|^2)$ to write

$$\sum_{m \leqslant n} |f(m) - A_n|^2 \leqslant 2 \sum_{m \leqslant n} |f(m) - K_n|^2 + 2n|K_n - A_n|^2 \quad ,$$

and estimate the first term in the sum obtained using (7.8) and (7.9), while the second can be easily estimated by the inequality $|K_n - A_n| < MD_n$ which is not difficult to prove, where M is a constant independent of n and f. In the course of proving Theorem 7.1, all the constants implied by the symbol $O(\cdot)$ and $\leqslant$ will be independent of n and function f.

We begin with the deduction of formula (7.8). Using (7.5) we can write

$$\sum_{m \leqslant n} f(m) = \sum_{p^k \leqslant n} f(p^k) \sum_{\substack{m \leqslant n \\ p^k \| n}} 1 = \sum_{p^k \leqslant n} f(p^k)\left(\left[\frac{n}{p^k}\right] - \left[\frac{n}{p^{k+1}}\right]\right)$$

$$= nK_n + O\left(\sum_{p^k \leqslant n} |f(p^k)|\right)$$

and it remains to show that the obtained error is $O(nD_n/\sqrt{\log n})$. In order to get this we write

$$\sum_{p^k \leqslant n} |f(p^k)| = \sum_{p^k \leqslant n} \sqrt{p^k} \, |f(p^k)| \sqrt{p^{-k}} \quad ,$$

and use Cauchy's inequality which leads us to

$$\left(\sum_{p^k \leqslant n} |f(p^k)| \right)^2 \leqslant \sum_{p^k \leqslant n} p^k \sum_{p^k \leqslant n} |f^2(p^k)| p^{-k} \leqslant n D_n^2 \sum_{p^k \leqslant n} 1 \ll \frac{n^2 D_n^2}{\log n},$$

because

$$\sum_{p^k \leqslant x} 1 = \pi(x) + \pi(x^{1/2}) + \ldots + \pi(x^{1/m})$$

(where m is the greatest natural number for which $x^{1/m} \geqslant 2$ holds), and moreover

$$\pi(x) \ll \frac{x}{\log x} \text{ and } \sum_{k=2}^{m} \pi(x^{1/k}) \leqslant m x^{1/2} \ll x^{1/2} \log x.$$

In this way we obtain (7.8). The proof of (7.9) is similar, with some slight complications:

$$
\begin{aligned}
(7.10) \qquad \sum_{m \leqslant n} |f(m)|^2 &= \sum_{\substack{m \leqslant n \\ q^l \| m}} \sum_{p^k \| m} f(p^k) \overline{f(q^l)} \\
&= \sum_{\substack{p^k, q^l \leqslant n \\ p \neq q}} f(p^k) \overline{f(q^l)} \sum_{\substack{m \leqslant n \\ p^k, q^l \| m}} 1 + \sum_{p^k \leqslant n} |f(p^k)|^2 \sum_{\substack{m \leqslant n \\ p^k \| m}} 1.
\end{aligned}
$$

Note that in the first sum the condition p^k, $q^l \leqslant n$ can be replaced by $p^k q^l \leqslant n$, because for $p^k q^l > n$ the inner sum is empty. It is not difficult to estimate the second term of the sum obtained because it is $\ll R_n$, where

$$(7.11) \qquad R_n = \sum_{p^k \leqslant n} |f(p^k)|^2 \sum_{\substack{m \leqslant n \\ p^k | m}} 1 \ll n D_n^2.$$

For the estimation of the first term we must determine the behavior of the expression

$$\sum_{\substack{m \leqslant n \\ p^k, q^l \| m}} 1$$

as n tends to infinity, where p^k, q^l are fixed and $p \neq q$. Furthermore we must take care of the uniformity of the error term with respect to p, q, k, l. The required result is a consequence of the following slightly general lemma:

Lemma 7.1. *If d is a given natural number, then as n tends to infinity we have*

$$\sum_{\substack{m \leqslant n \\ d \mid m \\ (d, m/d)=1}} 1 = \frac{\varphi(d)}{d^2} n + O(2^{\omega(d)}),$$

where the constant in the symbol $O(2^{\omega(d)})$ *does not depend on* d.

Proof. Using Corollary 3 to Theorem 2.7 and Theorem 2.8, we can write

$$\sum_{\substack{m \leqslant n \\ d \mid m \\ (d, m/d)=1}} 1 = \sum_{\substack{r \leqslant n/d \\ (r,d)=1}} 1 = \sum_{r \leqslant n/d} \sum_{\delta \mid (d,r)} \mu(\delta) = \sum_{\delta \mid d} \mu(\delta) \sum_{\substack{r \leqslant n/d \\ \delta \mid r}} 1 = \sum_{\delta \mid d} \mu(\delta) \left[\frac{n}{d\delta} \right]$$

$$= \frac{n}{d} \sum_{\delta \mid d} \frac{\mu(\delta)}{\delta} + O\left(\sum_{\delta \mid d} |\mu(\delta)| \right) = \frac{\varphi(d)}{d^2} n + O(2^{\omega(d)}),$$

where the constant implied by the symbol $O(\cdot)$ does not exceed 1. ∎

Corollary. *If* p^k, q^l *are given powers of distinct primes* p, q, *then as* n *tends to infinity, we have*

$$\sum_{\substack{m \leqslant n \\ p^k \| m, q^l \| m}} 1 = \frac{n}{p^k q^l} \left(1 - \frac{1}{p}\right)\left(1 - \frac{1}{q}\right) + O(1),$$

where the constant implied by $O(1)$ *does not depend on* p, q, k, l.

Proof. It suffices to note that the sum which we have to estimate is equal to

$$\sum_{\substack{m \leqslant n \\ d \mid m, \ (d, m/d) = 1}} 1$$

with $d = p^k q^l$ and to apply the proved lemma. In our case we have

$$\frac{\varphi(d)}{d^2} = \frac{1}{p^k q^l}\left(1 - \frac{1}{p}\right)\left(1 - \frac{1}{q}\right),$$

and $\omega(d) = 2$, so that $2^{\omega(d)} = O(1)$. ∎

Applying the obtained estimate to (7.10) and using (7.11), we may write

$$(7.12) \qquad \sum_{m \leqslant n} |f(m)|^2 = n \sum_{\substack{p^k q^l \leqslant n \\ p \neq q}} \frac{f(p^k)\overline{f(q^l)}}{p^k q^l} \left(1 - \frac{1}{p}\right)\left(1 - \frac{1}{q}\right)$$

$$+ O\left(\sum_{\substack{p^k q^l \leqslant n \\ p \neq q}} |f(p^k)\overline{f(q^l)}| \right) + O(n D_n^2).$$

If in the first term of (7.12) we omit the condition $p \neq q$, then this term alters at most by

$$n \sum_{\substack{p^k, p^l \\ p^{k+l} \leqslant n}} \frac{|f(p^k)f(p^l)|}{p^{k+l}} \left(1 - \frac{1}{p}\right)^2 \leqslant n \sum_{\substack{p^k, p^l \\ p^{k+l} \leqslant n}} \frac{|f(p^k)|}{p^{\frac{k+l}{2}}} \cdot \frac{|f(p^l)|}{p^{\frac{k+l}{2}}}$$

$$\leqslant n \left(\sum_{\substack{p^k, p^l \\ p^{k+l} \leqslant n}} \frac{|f(p^k)|^2}{p^{k+l}} \sum_{\substack{p^k, p^l \\ p^{k+l} \leqslant n}} \frac{|f(p^l)|^2}{p^{k+l}} \right)^{\frac{1}{2}}$$

$$= n \sum_{\substack{p^k, p^l \\ p^{k+l} \leqslant n}} \frac{|f(p^k)|^2}{p^{k+l}} = n \sum_{p^k \leqslant n} \frac{|f(p^k)|^2}{p^k} \sum_{\substack{l \\ p^l \leqslant np^{-k}}} \frac{1}{p^l}$$

$$\leqslant n \sum_{p^k \leqslant n} \frac{|f(p^k)|^2}{p^k} \left(1 + \frac{1}{p} + \frac{1}{p^2} + \ldots\right) \ll nD_n^2.$$

The middle term is in turn easily estimated due to

$$\sum_{p^k q^l \leqslant n} |f(p^k)f(q^l)| = \sum_{p^k q^l \leqslant n} \frac{|f(p^k)f(q^l)|}{p^{\frac{1}{2}k} q^{\frac{1}{2}l}} p^{\frac{1}{2}k} q^{\frac{1}{2}l}$$

$$\leqslant \left(\sum_{p^k q^l \leqslant n} p^k q^l \sum_{p^k q^l \leqslant n} \frac{|f(p^k)|^2|f(q^l)|^2}{p^k q^l} \right)^{\frac{1}{2}}$$

$$\leqslant \left(n \sum_{p^k q^l \leqslant n} 1 \sum_{p^k, q^l \leqslant n} \frac{|f(p^k)|^2|f(q^l)|^2}{p^k q^l} \right)^{\frac{1}{2}}$$

$$\ll (n^2 D_n^4)^{\frac{1}{2}} = nD_n^2.$$

The obtained estimates give us the formula (7.9). Now we utilize the inequality $|a + b|^2 \leqslant 2|a|^2 + 2|b|^2$ to write

(7.13) $$\sum_{m \leqslant n} |f(m) - A_n|^2 \leqslant 2 \sum_{m \leqslant n} |f(m) - K_n|^2 + 2n|K_n - A_n|^2$$

We now estimate the second term of this sum. We have

$$K_n - A_n = -\sum_{p \leqslant n} f(p)p^{-2} + \sum_{\substack{p^k \leqslant n \\ k \geqslant 2}} f(p^k) p^{-k} \left(1 - \frac{1}{p}\right),$$

but

$$\sum_{p \leqslant n} |f(p)|p^{-2} = \sum_{p \leqslant n} |f(p)|p^{-\frac{1}{2}} \cdot p^{-\frac{3}{2}}$$

$$\leqslant \left(\sum_{p \leqslant n} |f(p)|^2 p^{-1} \sum_{p \leqslant n} p^{-3} \right)^{\frac{1}{2}} = O(D_n),$$

and also

$$\sum_{\substack{p^k \leqslant n \\ k \geqslant 2}} \frac{|f(p^k)|}{p^k}\left(1 - \frac{1}{p}\right) \leqslant \sum_{\substack{p^k \leqslant n \\ k \geqslant 2}} \frac{|f(p^k)|}{p^{\frac{1}{2}k}} \cdot \frac{1}{p^{\frac{1}{2}k}}$$

$$\leqslant \left(\sum_{\substack{p^k \leqslant n \\ k \geqslant 2}} \frac{|f(p^k)|^2}{p^k} \cdot \sum_{\substack{p^k \leqslant n \\ k \geqslant 2}} \frac{1}{p^k}\right)^{\frac{1}{2}} = O(D_n) \quad .$$

To obtain the assertion of the theorem it remains to prove the inequality

(7.14)
$$\sum_{m \leqslant n} |f(m) - K_n|^2 = O(nD_n^2) \quad .$$

For this purpose we transform the sum appearing in this inequality:

$$\sum_{m \leqslant n} |f(m) - K_n|^2 = \sum_{m \leqslant n} \left(f(m) - K_n\right)\left(\overline{f(m)} - \overline{K_n}\right)$$

$$= \sum_{m \leqslant n} |f(m)|^2 - K_n \sum_{m \leqslant n} \overline{f(m)} - \overline{K_n} \sum_{m \leqslant n} f(m) + n|K_n|^2,$$

and by formulas (7.8) and (7.9), it is equal to

$$nL_n - 2n|K_n|^2 + n|K_n|^2 + O(nD_n^2) + O(nD_n|K_n| \log^{-\frac{1}{2}} n) \quad .$$

Since by Theorem 3.2

$$|K_n|^2 \ll \sum_{p^k \leqslant n} \frac{1}{p^k} \sum_{p^k \leqslant n} \frac{|f(p^k)|^2}{p^k} \ll D_n^2 \log\log n$$

we have

$$\sum_{m \leqslant n} |f(m) - K_n|^2 = n(L_n - |K_n|^2) + O(nD_n^2) \quad .$$

But

$$|L_n - |K_n|^2| \leqslant \sum_{\substack{p^k q^l > n \\ p^k, q^l \leqslant n}} |f(p^k)f(q^l)| p^{-k} q^{-l}$$

$$\leqslant \left(\sum_{\substack{p^k q^l > n \\ p^k, q^l \leqslant n}} p^{-k} q^{-l} \sum_{\substack{p^k \leqslant n \\ q^l \leqslant n}} \frac{|f(p^k)|^2}{p^k} \cdot \frac{|f(q^l)|^2}{q^l}\right)^{\frac{1}{2}}$$

$$= D_n^2 \sum_{\substack{p^k q^l > n \\ p^k, q^l \leqslant n}} p^{-k} q^{-l}$$

and it remains to show that

$$S_n = \sum_{\substack{p^k q^l > n \\ p^k, q^l \leqslant n}} p^{-k} q^{-l} = O(1).$$

Now

$$S_n = 2 \sum_{p^k \leqslant \sqrt{n}} p^{-k} \sum_{np^{-k} < q^l \leqslant n} q^{-l} + \left(\sum_{\sqrt{n} < p^k \leqslant n} p^{-k} \right)^2$$

$$= 2 \sum_{p^k \leqslant \sqrt{n}} p^{-k} \left(\log \frac{\log n}{\log n - \log(p^k)} + O\left(\frac{1}{\log n} \right) \right) + O(1),$$

but

$$\left| \log \frac{\log n}{\log n - \log p^k} \right| \ll \frac{\log(p^k)}{\log n} \quad,$$

and hence

$$S_n \ll \sum_{p^k \leqslant \sqrt{n}} \frac{\log p^k}{p^k} \cdot \frac{1}{\log n} + O(1) = \frac{1}{\log n} \sum_{p \leqslant \sqrt{n}} \frac{\log p}{p} + O(1) = O(1),$$

since

$$\sum_{p \leqslant n} \frac{\log p}{p} \ll \log n \quad,$$

and the series

$$\sum_{p^k, \, k \geqslant 2} \frac{\log(p^k)}{p^k}$$

is convergent in view of the fact that

$$\sum_{\substack{p^k \leqslant n \\ k \geqslant 2}} \frac{\log(p^k)}{p^k} \ll \sum_{\substack{p^k \leqslant n \\ k \geqslant 2}} \frac{1}{p^{k-\frac{1}{4}}} = \sum_{p \leqslant n} p^{\frac{1}{4}} \sum_{k=2}^{\infty} p^{-k}$$

$$= \sum_{p \leqslant n} p^{\frac{1}{4}} \cdot p^{-2} \cdot \left(1 - \frac{1}{p} \right)^{-1} \ll \sum_{p \leqslant n} p^{-1-\frac{3}{4}} \ll 1. \quad \blacksquare$$

3.   From the theorem proved above, we shall deduce some consequences.

Corollary 1.   *If a function $f(n)$ is strongly additive, then*

$$\sum_{m \leqslant n} |f(m) - A_n|^2 \leqslant C n B_n^2 \quad,$$

where C is a constant independent of n and f.

Proof. In fact, for strongly additive functions we have

$$D_n^2 = B_n^2 + \sum_{\substack{p^k \leqslant n \\ k \geqslant 2}} |f(p^k)|^2 p^{-k} = B_n^2 + O\left(\sum_{p \leqslant n} |f(p)|^2 \sum_{k \geqslant 2} p^{-k}\right)$$

$$= B_n^2 + O\left(\sum_{p \leqslant n} |f(p)|^2 \frac{1}{p(p-1)}\right) \ll B_n^2. \quad \blacksquare$$

Corollary 2 (P. Turán [2]). *If f is an additive function and there exists a constant B such that for all prime numbers p we have $0 \leqslant f(p) \leqslant B$, and moreover $D_n = o(A_n)$, then the function A_n is a non-decreasing normal order for f.*

Proof. First of all note that by the non-negativity of $f(p)$, the function A_n is actually non-decreasing. If we take an arbitrary $\epsilon > 0$ and denote by $N_\epsilon(n)$ the number of integers $m \leqslant n$ for which

$$|f(m) - A_n| \geqslant \epsilon A_n \quad,$$

then we obtain

$$N_\epsilon(n) \epsilon^2 A_n^2 \leqslant \sum_{m \leqslant n} |f(m) - A_n|^2 \leqslant C n D_n^2 \quad,$$

and therefore

(7.15) $$N_\epsilon(n) \leqslant \frac{C n D_n^2}{\epsilon^2 A_n^2} = o(n) \quad.$$

Now let us note that for integers m in the interval $[n^{\frac{1}{2}}, n]$ we have

$$|A_m - A_n| = \left| \sum_{m < p \leqslant n} \frac{f(p)}{p} \right| \leqslant B \sum_{\sqrt{n} \leqslant p \leqslant n} \frac{1}{p} = B\left(\log\log n - \log\log \sqrt{n} + O(1)\right) \leqslant C_1,$$

with some constant C_1, and so for these m we have

$$|f(m) - A_m| \leqslant |f(m) - A_n| + |A_n - A_m| \leqslant |f(m) - A_n| + C_1 \quad.$$

Now if $m \leqslant n$, then from the inequality

(7.16) $$|f(m) - A_m| \geqslant \epsilon A_m$$

it follows that either $m \leqslant n^{\frac{1}{2}}$ holds or

$$|f(m) - A_n| \geqslant \epsilon A_m - C_1 \geqslant \epsilon(A_n - C_1) - C_1 \geqslant \frac{1}{2}\epsilon A_n$$

holds if n is sufficiently large. Finally we see that the number of those $m \leqslant n$ for which the inequality (7.16) holds does not exceed

$$n^{\frac{1}{2}} + N_{\frac{1}{2}\epsilon}(n) = o(n)$$

for sufficiently large n by virtue of formula (7.15). ∎

Exercises

1. Deduce the theorem of Hardy-Ramanujan from Corollary 2 to Theorem 7.1.
2. Find the explicit value of the constant C appearing in Corollary 1 to Theorem 7.1.
3. Find a non-decreasing normal order for the function

$$\omega_E(n) = \sum_{p \mid n,\ p \in E} 1 \ ,$$

where E is a given arithmetic progression.

§2. The Erdös-Kac theorem

1. In Chapter III (see Corollary 3 to Theorem 3.2) we have seen that the function $\omega(n)$ has a non-decreasing normal order, viz. the function $\log \log n$. From formula (7.7) it follows that $\log \log n$ differs little from the expectation of the function ω treated as a random variable on the set $\{1, 2, \ldots, N\}$ for large N. In view of formula (7.2), we can consider ω as a sum of random variables which behave according to (7.3) like independent random variables. Therefore it can be expected that there exists an analogue of the central limit theorem for the function $\omega(n)$. Such an analogue was discovered in 1940 by P. Erdös and M. Kac ([1]) and it turned out later that this result is a particular case of a general theorem. We shall now prove the theorem of Erdös-Kac in the version of P. Billingsley ([1]).

Theorem 7.2. *We denote by* $N(x; a, b)$ *the number of natural numbers* m *in the interval* $[3, x]$ *for which the inequality*

$$(7.17) \qquad a \leqslant \frac{\omega(m) - \log \log m}{\sqrt{\log \log m}} \leqslant b$$

holds, where $a < b$ *are real numbers. We also allow the possibility* $a = -\infty$ *or* $b = \infty$. *Then, as* x *tends to infinity, we have*

$$N(x; a, b) = \left(x + o(x)\right) \frac{1}{\sqrt{2\pi}} \int_a^b \exp\left(-\tfrac{1}{2}t^2\right) dt.$$

Proof. The proof of this theorem requires the knowledge of elements of probability theory, and also some results of this theory which are not commonly taught in elementary courses, but can be found in any good book on probability theory, e.g.

W. Feller ([1]). We shall formulate these results below by referring to the corresponding pages in Feller's book. Let us begin with the notations. We shall always denote the integral

$$\int_{-\infty}^{x} \exp(-\tfrac{1}{2}t^2)dt$$

by $\Phi(x)$. We denote the probability in the probability space under consideration by P, $E(X)$ will denote the expectation of the random variable X, $V(X)$ its variation, and $\mu_r(X)$ its r-th moment. Hence if $\psi(x)$ is the distribution function of the random variable X, then

$$\mu_r(X) = \int_{-\infty}^{\infty} x^r d\psi(x) \ .$$

We shall also write

$$\mu_r^* = \int_{-\infty}^{\infty} x^r d\Phi(x) \ .$$

Here are the fundamental results which we rely on for the proof of the theorem:

(i) If X_n is a sequence of random variables with distribution functions ψ_n and for every r, the r-th moment of X_n is finite and tends to μ_r^* as n tends to infinity, then $\psi_n \to \Phi$. (W. Feller [1], Vol. II, p. 247).

(ii) If F_n is a sequence of distribution functions, for every x we have $\lim_{n \to \infty} F_n(x) = \Phi(x)$, and there exist constants C_r independent of n such that for all r, n we have

$$\int_{-\infty}^{\infty} |x|^r dF_n(x) \leqslant C_r \ ,$$

then

$$\lim_{n \to \infty} \mu_r(F_n) = \mu_r^* \ .$$

(W. Feller [1], Vol. II, p. 231).

(iii) (A special case of the central limit theorem.) If X_n is a sequence of independent and uniformly bounded random variables with expectation 0 and with the divergent sum of squares of variations

$$\sum_{n=1}^{\infty} V_n^2 \ ,$$

where $V_n = V(X_n)$, then for any real x we have

$$\lim_{n \to \infty} P\left\{ \frac{X_1 + \ldots + X_n}{\sqrt{V_1^2 + \ldots + V_n^2}} \leqslant x \right\} = \Phi(x) \ .$$

(W. Feller [1], Vol. II, p. 264).

Now we shall give a sketch of the proof of Theorem 7.2. The first step will be the reduction of it to the proof of the following asymptotic formula:

$$(7.18) \qquad \lim_{n \to \infty} \frac{1}{x} M(x; a, b) = \Phi(b) - \Phi(a) \quad ,$$

where by $M(x; a, b)$ we denote the number of those $m \leqslant x$ for which

$$a \leqslant \frac{\omega(m) - \log \log x}{\sqrt{\log \log x}} \leqslant b \ .$$

The difference between $M(x; a, b)$ and $N(x; a, b)$ lies in the replacement of $\log \log m$ by $\log \log x$.

In the second step we shall reduce our problem to a similar one in which the function $\omega(m)$ is replaced by $\omega_x(m)$, defined to be the number of prime divisors of m, not exceeding $x^{1/\log \log x}$.

The next reasoning will have probabilistic character — we shall introduce a certain sequence of independent random variables which is chosen so that (iii) is applicable, and next we use it to check the validity of our assertion for the function ω_x. Due to the preceding reduction this will complete the proof of the theorem.

The first step of the proof is contained in the following lemma:

Lemma 7.2. *If for arbitrary real $a < b$ the formula (7.18) holds, then the assertion of the theorem is true.*

Proof. By A_x we denote the set

$$\left\{ x^{\frac{1}{2}} \leqslant m \leqslant x : \ \omega(m) \leqslant 2 \log \log x \right\} \quad .$$

By Corollary 3 to Theorem 3.2, A_x has $x + o(x)$ elements. Moreover for $m \in A_x$ we have

$$\log \log m = \log \log x + O(1)$$

and

$$(\log \log m)^{\frac{1}{2}} = (\log \log x)^{\frac{1}{2}} + o(1) \quad ,$$

and hence

$$\left| \frac{\omega(m) - \log \log m}{(\log \log m)^{\frac{1}{2}}} - \frac{\omega(m) - \log \log x}{(\log \log x)^{\frac{1}{2}}} \right| = o(1)$$

and we see that for each positive ϵ and sufficiently large x we have

$$M(x; a + \epsilon, b - \epsilon) \leqslant N(x; a, b) + o(x) \leqslant M(x; a - \epsilon, b + \epsilon) \quad ,$$

which leads to

$$\Phi(b - \epsilon) - \Phi(a + \epsilon) \leqslant \liminf_{x \to \infty} \frac{1}{x} N(x; a, b)$$

$$\leqslant \limsup_{x \to \infty} \frac{1}{x} N(x; a, b) \leqslant \Phi(b + \epsilon) - \Phi(a - \epsilon) \quad .$$

Since ϵ is arbitrary and the function Φ is continuous, we obtain the assertion of the lemma. ∎

The second step is in the following lemma:

Lemma 7.3. *If $\{a_n\}$ is an increasing sequence of real numbers having the following properties:*

(i) *For any $\epsilon > 0$ we have $a_n = o(n^\epsilon)$,*

(ii) $\displaystyle \sum_{an < p \leqslant n} \frac{1}{p} = o\left((\log \log n)^{\frac{1}{2}} \right)$,

and furthermore

$$\omega_x(m) = \sum_{\substack{p \mid m \\ p \leqslant a_{[x]}}} 1, \quad R(x; a, b) = \sum_{\substack{m \leqslant x \\ m \in B_x}} 1,$$

$$B_x = \left\{ m : a \leqslant \frac{\omega_x(m) - \log \log x}{(\log \log x)^{\frac{1}{2}}} \leqslant b \right\}$$

and

$$\lim_{n \to \infty} \frac{R(x; a, b)}{x} = \Phi(b) - \Phi(a) \quad ,$$

then Theorem 7.2 is true.

Proof. It suffices to show that the conditions of the preceding lemma are satisfied. For this purpose we notice that

$$\omega(m) - \omega_x(m) = \sum_{p > a_N} e_p(m) \qquad (N = [x])$$

and

$$E_N(\omega - \omega_x) = \sum_{p > a_N} \frac{1}{N} \left[\frac{N}{p} \right] \leqslant \sum_{a_N < p \leqslant N} \frac{1}{p} = o\left((\log \log N)^{\frac{1}{2}} \right) = o\left((\log \log x)^{\frac{1}{2}} \right),$$

which implies

$$\lim_{N\to\infty} P_N \left\{ \frac{|\omega(m) - \omega_x(m)|}{(\log\log x)^{\frac{1}{2}}} > \epsilon \right\} = 0 \quad,$$

and so

$$R(x; a, b) = M(x; a, b) + o(x) \quad. \quad \blacksquare$$

We observe that the set of sequences satisfying the conditions of Lemma 7.3 is not empty; for example the sequence $a_n = n^{1/\log\log n}$ belongs to it.

Now let us choose in some probability space a suitably selected sequence of independent random variables $\{X_p\}$, indexed by the successive prime numbers. About this sequence we shall assume that each of its terms takes only two values, 0 and 1, and moreover for every p we have

$$P(X_p = 0) = 1 - \frac{1}{p}, \quad P(X_p = 1) = \frac{1}{p} \quad.$$

The behavior of this sequence is similar to that of the random variable e_p, moreover its terms are independent, which allows us to apply the central limit theorem in its simplest form (iii). Now we shall compare the behavior of the sums

$$w_n = \sum_{p \leqslant a_n} e_p \quad, \quad S_n = \sum_{p \leqslant a_n} X_p \quad,$$

where a_n is any sequence satisfying the conditions of Lemma 7.3.

Lemma 7.4. *Let c_n denote the expectation of the random variable S_n, and s_n^2 its variance. If for $r = 1, 2, \ldots$ we have*

$$(7.19) \qquad \lim_{N\to\infty} E_N \left\{ \frac{(w_N - c_N)^r}{s_N^r} \right\} = \mu_r^* = \mu_r(\Phi) \quad,$$

then Theorem 7.2 is true.

Proof. By the independence of X_p we have

$$c_n = E(S_n) = \sum_{p \leqslant a_n} E(X_p) = \sum_{p \leqslant a_n} \frac{1}{p} = \log\log n + o\left((\log\log n)^{\frac{1}{2}}\right),$$

and also

$$E(S_n^2) = E\left(\left(\sum_{p \leqslant a_n} X_p\right)^2\right) = \sum_{p \leqslant a_n} E(X_p^2) + \sum_{p \leqslant a_n} \sum_{\substack{q \leqslant a_n \\ p \neq q}} E(X_p)E(X_q)$$

$$= \sum_{p \leqslant a_n} \frac{1}{p} + \sum_{\substack{p, q \leqslant a_n \\ p \neq q}} \frac{1}{pq},$$

hence

$$s_n^2 = E(S_n^2) - E^2(S_n) = \sum_{p \le a_n} \frac{1}{p}\left(1 + \sum_{\substack{q \le a_n \\ q \ne p}} \frac{1}{q}\right) - \sum_{p \le a_n} \frac{1}{p} \sum_{q \le a_n} \frac{1}{q}$$

$$= \sum_{p \le a_n} \frac{1}{p}\left(1 - \frac{1}{p}\right) = \log\log n + o\left((\log\log n)^{\frac{1}{2}}\right).$$

Therefore, in order to prove the theorem it is necessary to show that

$$\lim_{N \to \infty} P_N\left\{\frac{w_N - c_N}{s_N} \le x\right\} = \Phi(x) \quad ;$$

but considering (i), we see that it suffices to check that the sequence of r-th moments of the random variable $\dfrac{w_N - c_N}{s_N}$ tends to μ_r^* for $r = 1, 2, \dots$. This coincides, however, with the assumption of our lemma. ∎

The direct evaluation of the limit (7.19) causes some difficulty, therefore we shall use an alternative method. In the lemma below we shall prove that for $r = 1, 2, \dots$ the sequence

$$E_N\left(\frac{(w_N - c_N)^r}{s_N^r}\right) - E\left(\frac{(S_N - c_N)^r}{s_N^r}\right)$$

tends to zero, and next we shall show that for $r = 1, 2, \dots$ we have

$$\lim_{N \to \infty} E\left(\frac{(S_N - c_N)^r}{s_N^r}\right) = \mu_r^*.$$

This result in conjunction with the preceding lemma and Lemma 7.4, gives us the assertion of the theorem.

Lemma 7.5. *For $r = 1, 2, \dots$ we have*

$$\lim_{N \to \infty} \left\{E_N\left(\frac{(w_N - c_N)^r}{s_N^r}\right) - E\left(\frac{(S_N - c_N)^r}{s_N^r}\right)\right\} = 0.$$

Proof. Since $S_N = \sum_{p \le a_N} X_p$, we have

$$E(S_N^r) = \sum_{k=1}^{r} \sum_{r_1 + \dots + r_k = r} \frac{r!}{r_1! \dots r_k!} \sum_{p_1 < \dots < p_k \le a_N} E(X_{p_1}^{r_1} \dots X_{p_k}^{r_k}),$$

but each of the random variables X_p takes only the values 0 and 1, so for any natural number m, we have $X_p^m = X_p$, which gives

$$E(X_{p_1}^{r_1} \ldots X_{p_k}^{r_k}) = E(X_{p_1} \ldots X_{p_k}) = \frac{1}{p_1 \ldots p_k}$$

and we obtain

(7.20) $$E(S_N^r) = \sum_{k=1}^{r} \sum_{r_1 + \ldots + r_k = r} \frac{r!}{r_1! \ldots r_k!} \sum_{p_1 < \ldots < p_k \leqslant a_N} \frac{1}{p_1 \ldots p_k}.$$

In an exactly analogous way we obtain

$$E_N(w_N^r) = \sum_{k=1}^{r} \sum_{r_1 + \ldots + r_k = r} \frac{r!}{r_1! \ldots r_k!} \sum_{p_1 < \ldots < p_k \leqslant a_N} E_N(e_{p_1}^{r_1} \ldots e_{p_k}^{r_k})$$

and

$$E_N(e_{p_1}^{r_1} \ldots e_{p_k}^{r_k}) = \frac{1}{N} \left[\frac{N}{p_1 \ldots p_k} \right]$$

and thus

(7.21) $$E_N(w_N^r) = \sum_{k=1}^{r} \sum_{r_1 + \ldots + r_k = r} \frac{r!}{r_1! \ldots r_k!} \sum_{p_1 < \ldots < p_k \leqslant a_N} \frac{1}{N} \left[\frac{N}{p_1 \ldots p_k} \right].$$

Considering the obvious inequality

$$\left| \frac{1}{p_1 \ldots p_k} - \frac{1}{N} \left[\frac{N}{p_1 \ldots p_k} \right] \right| < \frac{1}{N}$$

we obtain

$$|E(S_N^r) - E_N(w_N^r)| \leqslant \sum_{k=1}^{r} \sum_{r_1 + \ldots + r_k = r} \frac{r!}{r_1! \ldots r_k!} \sum_{p_1 < \ldots < p_k \leqslant a_N} \frac{1}{N}$$

$$= \frac{1}{N} \left(\sum_{p \leqslant a_N} 1 \right)^r \leqslant \frac{a_N^r}{N},$$

which in turn leads to

$$\left| E\left(\frac{(S_N - c_N)^r}{s_N^r} \right) - E_N\left(\frac{(w_N - c_N)^r}{s_N^r} \right) \right| = \left| \frac{1}{s_N^r} \sum_{k=0}^{r} \binom{r}{k} (E(S_N^k) - E_N(w_N^k))(-c_N)^{r-k} \right|$$

$$\leqslant \frac{1}{s_N^r} \sum_{k=0}^{r} \binom{r}{k} \frac{a_N^r c_N^{r-k}}{N} \leqslant \frac{(2a_N)^r}{N s_N^r}$$

$$= o\left(\frac{N^{\varepsilon r}}{N (\log \log N)^{\frac{1}{2} r}} \right) = o(1). \quad \blacksquare$$

It remains therefore to prove the following final lemma:

Lemma 7.6. *For* $r = 1, 2, \ldots$ *we have*

$$\lim_{N \to \infty} \mu_r \left(\frac{S_N - c_N}{s_N} \right) = \lim_{N \to \infty} E \left(\frac{(S_N - c_N)^r}{s_N^r} \right) = \mu_r^* \quad .$$

Proof. Applying (iii) to the sequence of independent random variables $X_p - E(X_p)$, we see that the distribution $\dfrac{S_N - c_N}{s_N}$ tends to Φ. In order to apply (ii), we must check that for all r and N

$$\left| E \left(\frac{(S_N - c_N)^r}{s_N^r} \right) \right| \leqslant C_r \quad ,$$

where C_r depends solely on r. For this purpose we take $Y_p = X_p - 1/p$ and note that $E(Y_p) = 0$, and hence

$$E \left((S_N - c_N)^r \right) = E \left(\left(\sum_{p \leqslant a_N} Y_p \right)^r \right)$$

$$= \sum_{k=1}^{r} \sum_{\substack{r_1 + \ldots + r_k = r \\ r_i \geqslant 2}} \frac{r!}{r_1! \ldots r_k!} \sum_{p_1 < \ldots < p_k \leqslant a_N} E(Y_{p_1}^{r_1}) \ldots E(Y_{p_k}^{r_k}).$$

Considering that $|Y_p| \leqslant 1$ we have for $r_i \geqslant 2$ the inequality

$$\left| E(Y_p^{r_i}) \right| \leqslant E(Y_p^2) \quad ,$$

and so

$$\sum_{p_1 < \ldots < p_k \leqslant a_N} E(Y_{p_1}^{r_1}) \ldots E(Y_{p_k}^{r_k}) \leqslant \sum_{p_1 < \ldots < p_k \leqslant a_N} E(Y_{p_1}^2) \ldots E(Y_{p_k}^2)$$

$$\leqslant \left(\sum_{p \leqslant a_N} E(Y_p^2) \right)^k = s_N^{2k}.$$

Since $r = r_1 + \ldots + r_k \geqslant 2k$, we have

$$|E \left((S_N - c_N)^r \right)| \leqslant \sum_{k=1}^{r} \sum_{\substack{r_1 + \ldots + r_k = r \\ r_i \geqslant 2}} \frac{r!}{r_1! \ldots r_k!} s_N^{2k}$$

$$\leqslant s_N^r \sum_{k=1}^{r} \sum_{\substack{r_1 + \ldots + r_k = r \\ r_i \geqslant 2}} \frac{r!}{r_1! \ldots r_k!} = C_r s_N^r$$

and

$$\left| E\left(\frac{(S_N - c_N)^r}{s_N^r} \right) \right| \leqslant C_r \; . \qquad \blacksquare$$

In their paper, P. Erdös and M. Kac proved Theorem 7.2 for all functions which are strongly additive and satisfy the condition $f(p) = O(1)$. I. P. Kubilius ([1]) and H. N. Shapiro ([2]) have obtained a better result by showing that Theorem 7.2 remains valid for all strongly additive functions $f(n)$ satisfying the conditions

$$B_N \to \infty, \qquad \sum_{\substack{p \leqslant N \\ |f(p)| > \epsilon B_N}} f^2(p) p^{-1} = o(B_N^2)$$

for every positive ϵ.

Exercises

1. Prove that in Theorem 7.2 the function $\omega(n)$ can be replaced by $\Omega(n)$.

2. Prove an analogue of Theorem 7.2 for the function $\omega_E(n) = \sum_{p \mid n, \, p \in E} 1$, where E is a given arithmetic progression containing infinitely many prime numbers.

3. From Theorem 7.2 deduce the theorem of Hardy and Ramanujan.

§3. Asymptotic distribution functions

1. If $f(n)$ is a given arithmetic function assuming real values, and N is a fixed natural number, then we can regard f as a random variable in the probability space Ω_N whose points are natural numbers in the interval $[0, N]$, and the probability of each of these points is identical and equal to $1/N$. We have already considered this probability space in the preceding section. It is clear that f has the distribution function $w_N(t)$ in Ω_N given by the formula

$$w_N(t) = N^{-1} \sum_{\substack{n \leqslant N \\ f(n) < t}} 1$$

and the question arises whether for N tending to infinity, $w_N(t)$ tends to a limit, say $w(t)$, which is also a distribution function. We say that a function $f(n)$ has an *asymptotic* (or *limit*) *distribution function* $w(t)$ if the function $w(t)$ is non-decreasing, satisfies

$$\lim_{t \to -\infty} w(t) = 0 \;, \qquad \lim_{t \to \infty} w(t) = 1 \quad,$$

and moreover at each point t of its continuity (and almost all points are such), the condition

$$w(t) = \lim_{N \to \infty} w_N(t)$$

is satisfied.

Not every function has an asymptotic distribution function because if we define f as follows:

$$f(n) = \begin{cases} 0 & \text{for} \quad n \in A \quad, \\ 1 & \text{for} \quad n \notin A \quad, \end{cases}$$

where A denotes the set of all natural numbers n that for some k satisfying the inequality

$$\frac{3^k}{2} < n \leqslant 3^k \quad,$$

then for $N = 3^r$ and $0 < t \leqslant 1$ we have

$$\sum_{\substack{n \leqslant N \\ f(n) < t}} 1 = \sum_{\substack{n \leqslant N \\ n \in A}} 1 = \sum_{k \leqslant r} \left(3^k - \left[\frac{1}{2} \, 3^k \right] \right) = \tfrac{1}{2} \sum_{k \leqslant r} 3^k + O(r)$$

$$= \tfrac{1}{4}(3^{1+r} - 1) + O(r) = \tfrac{3}{4} N + O(\log N),$$

and for $N = 3^r/2$ and $0 < t \leqslant 1$ we have

$$\sum_{\substack{n \leqslant N \\ f(n) < t}} 1 = \sum_{\substack{n \leqslant N \\ n \in A}} 1 = \sum_{\substack{n \leqslant 3^{r-1} \\ n \in A}} 1 = \tfrac{3}{4} \cdot 3^{r-1} + O(\log N) = \tfrac{1}{2} N + O(\log N),$$

and hence for $0 < t \leqslant 1$

$$\limsup_{N \to \infty} w_N(t) \geqslant \frac{3}{4}$$

and

$$\liminf_{N \to \infty} w_N(t) \leqslant \frac{1}{2} \quad,$$

and we see that $w_N(t)$ does not have a limit as $N \to \infty$.

Even in the case when a function behaves regularly, e.g. is strongly additive, the asymptotic distribution function may not exist. As an example let us consider the function $\omega(n)$. Then for $t > 1$ and $r = [t]$, we have

$$\sum_{\substack{n \leqslant x \\ \omega(n) < t}} 1 = \sum_{k \leqslant r} \sum_{\substack{n \leqslant x \\ \omega(n) = k}} 1 = \sum_{k \leqslant r} \left(\frac{1}{(k-1)!} \, x \, \frac{(\log \log x)^{k-1}}{\log x} + o \left(x \, \frac{(\log \log x)^{k-1}}{\log x} \right) \right)$$

$$= \left(\frac{1}{(r-1)!} + o(1) \right) x \, \frac{(\log\log x)^{r-1}}{\log x} = o(x),$$

and so for every t, $\lim\limits_{N\to\infty} w_N(t) = 0$.

However, if a function f is additive and both the series

$$\sum_p f^+(p) p^{-1} \quad , \qquad \sum_p (f^+(p))^2 p^{-1}$$

are convergent, where

$$f^+(p) = \begin{cases} f(p) & \text{for} \quad |f(p)| \leqslant 1 \quad , \\[2mm] 1 & \text{for} \quad |f(p)| > 1 \quad , \end{cases}$$

then, as P. Erdös ([3]) has shown, there exists an asymptotic distribution function for f. Furthermore, for additive functions the convergence of the above series is a necessary condition for the existence of an asymptotic distribution function. (P. Erdös, A. Wintner [1]). Here we shall prove the theorem of Erdös.

2.   Let us begin by recalling some fundamental facts about the distribution functions. Every non-decreasing function $w(t)$ continuous from the left, satisfying the conditions

$$\lim_{t\to -\infty} w(t) = 0 \, , \qquad \lim_{t\to\infty} w(t) = 1$$

is called a *distribution function*.

If $w_1(t)$, $w_2(t)$, ... is a sequence of distribution functions, then we say that it *converges* to the distribution function $w(t)$ if at every point t of continuity of $w(t)$, we have $\lim\limits_{n\to\infty} w_n(t) = w(t)$. Note that the function $w(t)$ is, as a monotonic function, continuous at every point with at most countably many exceptions.

Each distribution function $w(t)$ is associated with its *characteristic function* $\hat{w}(u)$ defined by the formula

$$\hat{w}(u) = \int_{-\infty}^{\infty} \exp(itu) \, dw(t) \quad .$$

We shall now prove a lemma concerning characteristic functions which will be useful in the sequel:

Lemma 7.7.

(i)   *If* $w_n(t)$ *are distribution functions for each* n, *and the sequence* w_n *is convergent to a distribution function* $w(t)$, *then for any real* u *we have*

$$\lim_{n\to\infty} \hat{w}_n(u) = \hat{w}(u) \quad .$$

(ii)   *If w_n are distribution functions for each n, and moreover if for any real u we have*

$$\lim_{n\to\infty} \hat{w}_n(u) = f(u) \quad ,$$

then the sequence w_n converges to some distribution function v, and $\hat{v} = f$. Furthermore, if $f = \hat{w}$, where w is a distribution function, then $w = v$.

Proof.   (i) Because w is a monotonic function, it has at most countably many discontinuities, so we can choose a number $a > 0$ so that the function w might be continuous at the points a and $-a$ and moreover $1 - w(a) < \frac{1}{4}\epsilon$ and $w(-a) < \frac{1}{4}\epsilon$, where ϵ is an arbitrarily fixed positive number. Now note that using integration by parts we obtain

$$\int_{-a}^{a} e^{itu}dw_n(t) = e^{iau}w_n(a) - e^{-iau}w_n(-a) - iu\int_{-a}^{a} e^{itu}w_n(t)dt \quad ,$$

$$\int_{-a}^{a} e^{itu}dw(t) = e^{iau}w(a) - e^{-iau}w(-a) - iu\int_{-a}^{a} e^{itu}w(t)dt$$

from these equalities, and noting that $\lim_{n\to\infty} w_n(a) = w(a)$, $\lim_{n\to\infty} w_n(-a) = w(-a)$ and the boundedness of the function $\exp(itu)w_n(t)$, we obtain

$$\lim_{n\to\infty}\int_{-a}^{a} e^{itu}dw_n(t) = \int_{-a}^{a} e^{itu}dw(t) \quad .$$

From this it follows that

$$\limsup_{n\to\infty} |\hat{w}_n(u) - \hat{w}(u)| \leqslant \limsup_{n\to\infty}\left|\int_{-\infty}^{-a} e^{itu}dw_n(t) - \int_{-\infty}^{-a} e^{itu}dw(t)\right|$$

$$+ \limsup_{n\to\infty}\left|\int_{a}^{\infty} e^{itu}dw_n(t) - \int_{a}^{\infty} e^{itu}dw(t)\right| \quad .$$

Furthermore,

$$\limsup_{n\to\infty}\left|\int_{a}^{\infty} e^{itu}dw_n(t) - \int_{a}^{\infty} e^{itu}dw(t)\right|$$

$$\leqslant \left|\int_{a}^{\infty} e^{itu}dw(t)\right| + \limsup_{n\to\infty}\left|\int_{a}^{\infty} e^{itu}dw_n(t)\right|$$

$$\leqslant 1 - w(a) + \limsup_{n\to\infty}(1 - w_n(a)) = 2(1 - w(a)) < \frac{1}{2}\epsilon$$

and similarly

$$\limsup_{n \to \infty} \left| \int_{-\infty}^{-a} e^{itu} dw_n(t) - \int_{-\infty}^{-a} e^{itu} dw(t) \right| < \frac{1}{2}\epsilon \quad .$$

Hence

$$\limsup_{n \to \infty} |\hat{w}_n(u) - \hat{w}(u)| < \epsilon \quad ,$$

which completes the proof of (i).

In order to prove (ii), let us first note that there exists a sequence $\{n_k\}$ possessing the property that at each rational point x, the sequence $\{ w_{n_k}(x) \}$ is convergent to some value, say $a(x)$, satisfying the condition $0 \leqslant a(x) \leqslant 1$. Actually, to prove this, it is sufficient to arrange rational numbers in a sequence $x_1, x_2, \ldots$ and apply the diagonal method of Cantor to the sequence $w_n(x_1), w_n(x_2), \ldots$. Moreover, note that by the monotonicity of w_n the inequality $c_1 < c_2$ implies $a(c_1) \leqslant a(c_2)$. This remark allows us to define a function $v(x)$ by the formula

$$v(x) = \sup_{\substack{y \in \mathscr{Q}, \\ y < x}} a(y).$$

Here we note that $\lim_{h \to 0} v(x + h) = \inf_{\substack{y \in \mathscr{Q}, \\ y > x}} a(y)$ and the function v is a distribution function. We show that at each point x_0 of its continuity the equality

$$(7.22) \qquad v(x_0) = \lim_{k \to \infty} w_{n_k}(x_0)$$

holds. In fact, for rationals $q > x_0$, we have

$$\limsup_{k \to \infty} w_{n_k}(x_0) \leqslant \limsup_{k \to \infty} w_{n_k}(q) = a(q) \quad ,$$

and therefore

$$\limsup_{k \to \infty} w_{n_k}(x_0) \leqslant \inf_{q > x_0} a(q) = \lim_{h \to \infty} v(x_0 + h) = v(x_0)$$

and analogously, considering the rationals $q < x_0$, we obtain

$$\liminf_{k \to \infty} w_{n_k}(x_0) \geqslant \sup_{q \leqslant x_0} a(q) = v(x_0) \quad ,$$

which leads to (7.22).

Now it remains to prove that for each point of continuity of the function $v(x)$ we have

$$(7.23) \qquad \lim_{n} w_n(x) = v(x) \quad ,$$

and also

$$(7.24) \qquad v(x) = w(x) \quad .$$

We shall show that from the equality of characteristic functions follows an equality of distributions, and from this the formulas (7.23) and (7.24) can be deduced by the following argument: if a subsequence $\{w_{m_k}(x)\}$ is convergent at each rational point x, then there exist a distribution function $v'(x)$ such that at each point of its continuity we have

(7.25) $\lim w_{m_k}(x) = v'(x)$,

and moreover, from each sequence $r_1 < r_2 < \ldots$ of natural numbers it is possible to choose a subsequence $\{m_k\}$ having this property. By (i) we have then

$$\hat{w}(x) = \lim_{k \to \infty} \hat{w}_{n_k}(x) = \hat{v}(x)$$

and

$$\hat{w}(x) = \lim_{k \to \infty} \hat{w}_{m_k}(x) = \hat{v}'(x) \quad .$$

Therefore

$$\hat{v}(x) = \hat{v}'(x) = \hat{w}(x)$$

and

(7.26) $v(x) = v'(x) = w(x)$.

Therefore the condition $\lim_{k \to \infty} w_{m_k}(x) = w(x)$ holds for any sequence $\{m_k\}$ with the property that at each rational point there exist the limit $\lim_{k \to \infty} w_{m_k}(x)$, and this gives the formula (7.23). The equality (7.24) now follows from (7.26).

Now we show that the equality of characteristic functions implies the equality of distribution functions. First observe that if $w(t), v(t)$ are distribution functions and $\hat{w}(t) = \hat{v}(t)$, then for any function f continuous and bounded on the real line we have

(7.27) $$\int_{-\infty}^{\infty} f(t)dw(t) = \int_{-\infty}^{\infty} f(t)dv(t) \quad .$$

To prove this we take an arbitrary function $f(t)$ continuous on the real line and satisfying the condition

$$\sup_{t \in \mathscr{R}} |f(t)| \leqslant M \quad ,$$

where M is some positive number. Moreover, let ϵ be an arbitrary positive number. Let us select $T > 0$ so that both the integrals

$$\int_{|t| > T} dw(t) \ , \qquad \int_{|t| > T} dv(t)$$

do not exceed ϵ. In the interval $[-T, T]$ we can approximate $f(t)$ uniformly by a finite linear combination of the functions of the form $\exp(\pi i t k / T)$ $(k \in \mathscr{Z})$ due to the theorem of Weierstrass concerning the approximation by trigonometrical polynomials.

Let $W(t)$ be such a linear combination, say

$$W(t) = \sum_{k=1}^{N} \lambda_k \exp(\pi i t n_k / T) \quad ,$$

for which

$$\sup_{|t| \leqslant T} |f(t) - W(t)| < \epsilon \quad .$$

Then

$$\sup_{|t| \leqslant T} |W(t)| \leqslant M + \epsilon \quad ,$$

but the function $W(t)$ is periodic with period $2T$, therefore

$$\sup_{t \in \mathcal{R}} |W(t)| \leqslant M + \epsilon \quad .$$

Using the assumption and the form of $W(t)$ we easily obtain the equality

$$\int_{-\infty}^{\infty} W(u) dw(u) = \int_{-\infty}^{\infty} W(u) dv(u) \quad ,$$

and so

$$\left| \int_{-\infty}^{\infty} f(u) dw(u) - \int_{-\infty}^{\infty} f(u) dv(u) \right| \leqslant \left| \int_{-\infty}^{\infty} \big(f(u) - W(u) \big) dw(u) \right.$$

$$\left. - \int_{-\infty}^{\infty} \big(f(u) - W(u) \big) dv(u) \right|$$

$$\leqslant \int_{|u| \geqslant T} |f(u) - W(u)| dw(u) + \int_{|u| \geqslant T} |f(u) - W(u)| dv(u)$$

$$+ \int_{-T}^{T} |f(u) - W(u)| dw(u) + \int_{-T}^{T} |f(u) - W(u)| dv(u)$$

$$\leqslant 2(2M + \epsilon)\epsilon + 4T\epsilon \quad ,$$

which, by the arbitrariness of ϵ, gives (7.27).

Now let $a < b$ be points of continuity of the functions w and v. We can approximate the characteristic function χ of the interval $[a, b]$ by a continuous function not exceeding unity, namely by the function f equal unity in this interval, vanishing outside $(a - 1/n, \ b + 1/n)$ and linear in the remaining intervals so that for sufficiently large n we may have

$$\int_{-\infty}^{\infty} |f(x) - \chi(x)| dw(x) = \int_{a-1/n}^{a} |f(x) - \chi(x)| dw(x) + \int_{b}^{b+1/n} |f(x) - \chi(x)| dw(x)$$

$$\leqslant w(a) - w(a - 1/n) + w(b + 1/n) - w(b) < \epsilon$$

and analogously

$$\int_{-\infty}^{\infty} |f(x) - \chi(x)| dv(x) < \epsilon \quad .$$

Then

$$|w(b) - w(a) - v(b) + v(a)| = \left| \int_{-\infty}^{\infty} \chi(x) dw(x) - \int_{-\infty}^{\infty} \chi(x) dv(x) \right|$$

$$\leqslant \int_{-\infty}^{\infty} |\chi(x) - f(x)| dw(x) + \int_{-\infty}^{\infty} |\chi(x) - f(x)| dv(x) < 2\epsilon \quad ,$$

and hence $w(b) - w(a) = v(b) - v(a)$. If now $a \to -\infty$, then we obtain

(7.28) $w(b) = v(b)$

at each point of continuity of these functions, but both of them are monotonic and continuous from the left, so from (7.28) it follows that $w = v$. ∎

3.   The following result beyond number theory used in this section is Bochner's theorem on positive definite functions, which we shall now prove:

Theorem 7.3 (S. Bochner [1]). *If $f(t)$ is a continuous bounded function defined on $\mathscr{R}$, $f(0) = 1$, and moreover f is positive definite, i.e. $f(-t) = \overline{f(t)}$ and for arbitrary reals $t_1, \ldots, t_k$ and complex $z_1, \ldots, z_k$ we have*

$$\sum_{1 \leqslant i, j \leqslant k} f(t_i - t_j) z_i \overline{z_j} \geqslant 0 \quad ,$$

then there exists a unique distribution function $w(x)$ for which $f(t)$ is the characteristic function, i.e.

$$f(t) = \int_{-\infty}^{\infty} e^{ixt} dw(x) \quad .$$

Proof. Let us consider first the case when $f(t)$ is integrable on the real line. Let $g(t)$ be any continuous, integrable function whose Fourier transform

$$\tilde{g}(x) = \int_{-\infty}^{\infty} g(t) e^{-ixt} dt$$

is also integrable. Then for any positive A we have

$$\int_{-A}^{A} \int_{-A}^{A} f(x-y)g(-x)\overline{g(-y)}\,dxdy \geqslant 0$$

since the corresponding Riemann's sums are non-negative. Passing from A to infinity, we obtain

(7.29)
$$\int_{-\infty}^{\infty} \int_{-\infty}^{\infty} f(x-y)g(-x)\overline{g(-y)}\,dxdy \geqslant 0 \quad .$$

By $\Phi(t)$ we denote the convolution of the functions $f(t)$, $g(t)$ and $g(-t)$, i.e.

(7.30)
$$(2\pi)^2\,\Phi(t) = \int_{-\infty}^{\infty} \int_{-\infty}^{\infty} f(t-x-y)g(x)\overline{g(-y)}\,dxdy$$

$$= \int_{-\infty}^{\infty} \int_{-\infty}^{\infty} f(t+x-y)g(-x)\overline{g(-y)}\,dxdy \quad .$$

Then for their Fourier transform we obtain

$$\tilde{\Phi} = |\tilde{g}|^2 \tilde{f}$$

and using the theorem on the inverse transform, we arrive at

$$\Phi(t) = \int_{-\infty}^{\infty} \exp(2\pi itx)\,|\tilde{g}(x)|^2\,\tilde{f}(x)\,dx$$

and

$$\Phi(0) = \int_{-\infty}^{\infty} |\tilde{g}(x)|^2\,\tilde{f}(x)\,dx \quad .$$

From formulas (7.29) and (7.30) we obtain $\Phi(0) \geqslant 0$, and hence we see that if g is continuous and integrable on the real line, and if its Fourier transform is also integrable, then

(7.31)
$$\int_{-\infty}^{\infty} |\tilde{g}(x)|^2\,\tilde{f}(x)\,dx \geqslant 0 \quad .$$

Since $\overline{\tilde{f}(-x)} = \overline{\tilde{f}(x)}$, and from our assumptions follows the equality $\overline{f(-x)} = f(x)$, therefore the Fourier transform $\tilde{f}$ has real values. Let us show its non-negativity. In fact, if in some interval $[a, b]$ we have $f(t) < 0$, and $g(t)$ is any continuous, integrable function with the integrable Fourier transform (one can take for g any function vanishing outside $[a, b]$ and having second continuous derivative), then we would have a contradiction with the formula (7.31).

Our next step is to show that the function $\tilde{f}$ is integrable. To do this let us consider for $n = 1, 2, \ldots$ the functions

$$f_n(x) = (\pi n)^{-1} \int_{-\infty}^{\infty} f(u) \frac{\sin^2(n(x-u))}{(x-u)^2} \, du \quad .$$

Taking into account the equality

$$\int_{-2n}^{2n} e^{ixt} \left(1 - \frac{|t|}{2n}\right) \tilde{f}(t) \, dt = \frac{1}{2\pi} \int_{-2n}^{2n} \left(1 - \frac{|t|}{2n}\right) e^{ixt} \int_{-\infty}^{\infty} f(u) e^{-itu} \, du \, dt$$

$$= \frac{1}{2\pi} \int_{-\infty}^{\infty} f(u) \, du \int_{-2n}^{2n} \left(1 - \frac{|t|}{2n}\right) e^{it(x-u)} \, dt$$

$$= \frac{1}{\pi} \int_{-\infty}^{\infty} f(u) \frac{2 \sin^2(n(x-u))}{2n(x-u)^2} \, du$$

$$= \frac{1}{n\pi} \int_{-\infty}^{\infty} f(u) \frac{\sin^2(n(x-u))}{(x-u)^2} \, du = f_n(x)$$

and

$$\tilde{f}(t) \geqslant 0 \quad ,$$

we obtain for $a < 2n$

$$\int_{-a}^{a} \left(1 - \frac{|t|}{2n}\right) \tilde{f}(t) \, dt \leqslant |f_n(0)| \leqslant \frac{1}{\pi n} \sup_{t \in \mathfrak{R}} |f(t)| \int_{-\infty}^{\infty} \frac{\sin^2 nu}{u^2} \, du$$

$$= \frac{1}{\pi} \sup_{t \in \mathfrak{R}} |f(t)| \int_{-\infty}^{\infty} \frac{\sin^2 u}{u^2} \, du = A ,$$

and therefore $\int_{-a}^{a} \tilde{f}(t) \, dt \leqslant A$ and $\int_{-\infty}^{\infty} \tilde{f}(t) \, dt \leqslant A$, which proves the integrability of $\tilde{f}$. Now it suffices to note that if f is continuous, integrable and has the integrable Fourier transform $\tilde{f}$, then

$$f(t) = \int_{-\infty}^{\infty} e^{iut} \, dw(u) = \hat{w}(t) \quad ,$$

where $w(t) = \int_{-\infty}^{t} \tilde{f}(u) \, du$. In fact it is sufficient to apply the theorem on inverse transform and note that

$$1 = f(0) = \int_{-\infty}^{\infty} dw(x) = \lim_{x \to \infty} w(x) \quad ,$$

and thus w is a distribution function.

Therefore in the case where f is continuous and integrable, the theorem has been proved. Now suppose that f is any continuous, positive definite and bounded function. Note that if $g(t)$ is non-negative and integrable, then the function

$$F(t) = f(t) \int_{-\infty}^{\infty} e^{ixt} g(x) dx$$

is also positive definite and

$$\sum_{1 \leqslant r,\, s \leqslant k} F(t_r - t_s) z_r \bar{z}_s = \int_{-\infty}^{\infty} \sum_{1 \leqslant r,\, s \leqslant k} \left(f(t_r - t_s) e^{it_r x} z_r \overline{e^{it_s x} z_s} \right) g(x) dx \geqslant 0 \quad .$$

From this it follows that for $n = 1, 2, \ldots$ the functions $g_n(x) = f(x) \exp(-x^2/n)$ are positive definite, and that from the boundedness of f follows their integrability, thus there exist distribution functions satisfying $g_n = \hat{w}_n$. It remains to apply the first part of Lemma 7.7 (ii). ∎

4.   Utilizing the results proved above, we can now give a necessary and sufficient condition for the existence of an asymptotic distribution, which in the case of additive functions, reduces the whole problem to that of the existence of the average value of multiplicative functions.

Theorem 7.4. *For the existence of an asymptotic distribution for a function f, it is necessary and sufficient that for all real t, the limit*

$$(7.32) \qquad \lim \frac{1}{N} \sum_{n \leqslant N} \exp\left(itf(n) \right) = M(\exp itf) = \varphi(t)$$

exists and the function $\varphi(t)$ is continuous.

Proof. Necessity: If f has an asymptotic distribution $w(t) = \lim_{N \to \infty} w_N(t)$, then applying Lemma 7.7 (i) in the case $g(t) = \exp(itx)$ we obtain on one hand

$$\lim_{N \to \infty} \int_{-\infty}^{\infty} \exp(itx) dw_N(x) = \int_{-\infty}^{\infty} \exp(itx) dw(x) \quad ,$$

but on the other hand we have

$$\int_{-\infty}^{\infty} \exp(itx) dw_N(x) = \frac{1}{N} \sum_{n \leqslant N} \exp\left(itf(n) \right) \quad ,$$

because w_N is a step function having finitely many steps. Hence

$$M\left(\exp(itf) \right) = \int_{-\infty}^{\infty} \exp(itx) dw(x) \quad .$$

Sufficiency: Suppose that the mean value $M\left(\exp(itf) \right) = \varphi(t)$ exists and is a continuous function. Then for any real $t_1, \ldots, t_k$ and complex $z_1, \ldots, z_k$ we have

$$\sum_{1 \le i,\, j \le k} \varphi(t_i - t_j) z_i \bar{z}_j = \sum_{1 \le i,\, j \le k} M\left(\exp\left(if(t_i - t_j)\right)\right) z_i \bar{z}_j$$

$$= M\left(\sum_{1 \le i,\, j \le k} \exp\left(if(t_i - t_j)\right) z_i \bar{z}_j\right) = M\left(\left|\sum_{j=1}^{k} \exp(it_j f) z_j\right|^2\right) \ge 0 \quad.$$

The function $\varphi(t)$ is therefore positive definite and from Theorem 7.3 it follows that

$$\varphi(t) = \int_{-\infty}^{\infty} \exp(itx) dw(x) \quad,$$

where $w(t)$ is a uniquely determined distribution function. Hence for all real t we have

$$\lim_{N \to \infty} \int_{-\infty}^{\infty} e^{itx} dw_N(x) = \int_{-\infty}^{\infty} e^{itx} dw(x) \quad,$$

whence by virtue of Lemma 7.7 (ii) the equality $\lim\limits_{N \to \infty} w_N(t) = w(t)$ at each point of continuity of the function $w_N(t)$ follows. ∎

In the case when the function f is additive, the problem of the existence of an asymptotic distribution reduces to the checking whether the multiplicative function $\exp(itf)$ has for each real t the mean value depending on t continuously. On account of this we must have at our disposal a result, assuring the existence of such a mean under mild conditions on the function f. It turns out that a theorem of H. Delange ([2]) is of use here. A. Rényi ([2]) found a simple proof of this result, which rests upon an application of the Turán-Kubilius inequality. We shall now give this proof for the case of strongly multiplicative functions.

Theorem 7.5.

(i) *If a function g is strongly multiplicative (i.e. in addition to the multiplicativity condition, it satisfies also the condition $g(p^k) = g(p)$ for all prime powers p^k). and moreover for all n we have $|g(n)| \le 1$ and the series*

$$\sum_{p} \frac{g(p) - 1}{p}$$

is convergent, then there exists the mean value of g equal to

$$M(g) = \prod_{p}\left(1 + \frac{g(p) - 1}{p}\right) \quad.$$

(ii) *If we have a given family $\{g_t(n): t \in \mathscr{R}\}$ of strongly multiplicative functions satisfying the assumptions of (i), and moreover for a fixed n, $g_t(n)$ is a continuous function of t and the series*

$$\sum_p \frac{g_t(p) - 1}{p}$$

is convergent almost uniformly with respect to t, *then the equality*

$$M(g_t) = \prod_p \left(1 + \frac{g_t(p) - 1}{p} \right)$$

holds and $M(g_t)$ *is a continuous function of* t.

Proof. The first step of the proof will be to show that without restricting the generality one can alter the values of the function $g_t(p)$ on a not too large set of prime numbers. This allows us to suppose in what follows that for all p we have $\text{Re} \, g_t(p) > \frac{1}{2}$. We shall be concerned next with the functions

$$g_t^{(N)}(n) = \prod_{\substack{p \leqslant \log N \\ p | n}} g_t(p) \quad,$$

and shall prove that they have the mean value and, finally, we shall show that their mean values approximate for large N some number which can be shown easily to be the mean value of g_t. Of course, in these arguments we have to be careful so that all the estimates would be almost uniform with respect to t.

Lemma 7.8. *If the families* $\{g_t\}$ *and* $\{h_t\}$ *satisfy the assumptions of the theorem, and moreover* $g_t(2) = h_t(2)$ *and the series*

$$\sum_{g_t(p) \neq h_t(p)} \frac{1}{p}$$

is convergent almost uniformly with respect to t *and the assertion (ii) is valid for the family* $\{g_t\}$, *then it is also valid for the family* $\{h_t\}$.

Proof. Since g_t and h_t are multiplicative, there exist a unique function f_t satisfying the condition $f_t * g_t = h_t$, where $*$ denotes the Dirichlet convolution. Furthermore, this function is multiplicative, and its values at the points p^k (p a prime number, $k = 1, 2, \dots$) can be determined recursively by the formula

$$(7.33) \qquad f_t(p^k) = h_t(p) - g_t(p) \sum_{j=0}^{k-1} f_t(p^j) \quad.$$

By simple induction, we obtain from this formula that in the case $g_t(p) = h_t(p)$ we have $f_t(p^k) = 0$ for all k, and when $g_t(p) \neq h_t(p)$ the estimate

$$|f_t(p^k)| \leqslant 2^k \qquad (k = 1, 2, \dots)$$

holds.

Therefore the product

$$(7.34) \qquad \prod_p \left(1 + \sum_{j=1}^{\infty} |f_t(p^j)| \, p^{-j} \right)$$

is convergent almost uniformly because if P_t is the set of those prime numbers p for which $g_t(p) \neq h_t(p)$, then for $p \in P_t$ we have

$$\sum_{j=1}^{\infty} |f_t(p^j)| p^{-j} \leqslant \sum_{j=1}^{\infty} \left(\frac{2}{p} \right)^j = \frac{2}{p-2} \leqslant \frac{6}{p} \quad ,$$

and the series $\displaystyle\sum_{p \in P_t} \frac{1}{p}$ is convergent almost uniformly by the assumption. In view of

$$\sum_{n=1}^{N} f_t(n) n^{-1} \leqslant \prod_{p \leqslant N} \left(1 + \sum_{j=1}^{\infty} |f_t(p^j)| p^{-j} \right)$$

we obtain now the absolute and almost uniform convergence of the series

$$\sum_{n=1}^{\infty} f_t(n) n^{-1}$$

Its sum $F(t)$ is continuous because from formula (7.33) it follows that the function $f_t(n)$ is continuous for $n = p^k$, and in the general case this is a consequence of the multiplicativity of f_t. Applying Theorem 2.19, we now obtain $M(h_t) = F(t) M(g_t)$, and hence $M(h_t)$ is a continuous function of t. It remains to show that

$$(7.35) \qquad M(h_t) = \prod_p \left(1 + \frac{h_t(p) - 1}{p} \right) .$$

To obtain this let us note that by the assumption we have

$$M(g_t) = \lim_{N \to \infty} \prod_{p \leqslant N} \left(1 + \frac{g_t(p) - 1}{p} \right) \quad ,$$

and hence

$$M(h_t) = F(t) M(g_t) = \lim_{N \to \infty} \prod_{p \leqslant N} \left(1 + \frac{g_t(p) - 1}{p} \right) \left(1 + \sum_{j=1}^{\infty} \frac{f_t(p^j)}{p^j} \right)$$

$$= \lim_{N \to \infty} \prod_{p \leqslant N} \left(1 - \frac{1}{p} \right) \left(1 + \sum_{j=1}^{\infty} \frac{g_t(p^j)}{p^j} \right) \left(1 + \sum_{j=1}^{\infty} \frac{f_t(p^j)}{p^j} \right)$$

$$= \lim_{N \to \infty} \prod_{p \leqslant N} \left(1 - \frac{1}{p} \right) \left(1 + \sum_{j=1}^{\infty} \frac{h_t(p^j)}{p^j} \right)$$

$$= \lim_{N \to \infty} \prod_{p \leqslant N} \left(1 + \frac{h_t(p) - 1}{p} \right) = \prod_{p} \left(1 + \frac{h_t(p) - 1}{p} \right). \qquad \blacksquare$$

Now we prove the theorem. Of course, it is sufficient to prove (ii). If $\{g_t\}$ is a family of functions satisfying the condition (ii), then the series

$$\sum_{p} \frac{1 - \operatorname{Re} g_t(p)}{p}$$

will be almost uniformly convergent, and because its terms are non-negative, the series

$$\sum_{\operatorname{Re} g_t(p) < \frac{1}{2}} \frac{1}{p}$$

will also be almost uniformly convergent. Using the preceding lemma, we can assume that for all prime numbers p we have

(7.36) $$\operatorname{Re} g_t(p) > \tfrac{1}{2} \ .$$

Let $r_t(p)$ denote the absolute value of $g_t(p)$ and $\vartheta_t(p)$ its argument with $-\pi < \vartheta_t(p) \leqslant \pi$. The condition (7.36) now takes the form

(7.37) $$r_t(p) \cos \vartheta_t(p) > \tfrac{1}{2} \ .$$

Lemma 7.9. *The series*

(7.38) $$\sum_{p} \frac{\log r_t(p)}{p} \ , \qquad \sum_{p} \frac{\log^2 r_t(p)}{p} \ , \qquad \sum_{p} \frac{\vartheta_t(p)}{p} \ , \qquad \sum_{p} \frac{\vartheta_t^2(p)}{p}$$

are almost uniformly convergent.

Proof. The series

$$\sum_{p} \frac{1 - r_t(p) \cos \vartheta_t(p)}{p} \qquad \text{and} \qquad \sum_{p} \frac{r_t(p) \sin \vartheta_t(p)}{p}$$

are almost uniformly convergent, and moreover we have

$$1 \geqslant r_t(p) \geqslant r_t(p) \cos \vartheta_t(p) > \tfrac{1}{2} \ ,$$

hence considering the inequality $\log(1/x) < 2(1 - x)$ valid for $1 > x > \tfrac{1}{2}$, we obtain the almost uniform convergence of the first of the series (7.38). In view of the estimate $\log^2 x \ll 1 - x$ valid for $1 > x \geqslant \tfrac{1}{2}$, we obtain

$$\log^2 r_t(p) \ll 1 - r_t(p) \ll 1 - r_t(p) \cos \vartheta_t(p) \ ,$$

and this gives the almost uniform convergence of the second of the series (7.38).

Now consider the fourth series. The formula (7.36) implies the positivity of $\cos \vartheta_t(p)$, therefore the equality $\sin \vartheta_t(p) = 0$ can occur only in the case when $\vartheta_t(p) = 0$. Hence

$$\vartheta_t(p) \ll |\sin \vartheta_t(p)|$$

and

$$\frac{\vartheta_t^2(p)}{p} \ll \frac{\sin^2 \vartheta_t(p)}{p} \quad ,$$

where the constants implied by Vinogradov's symbol depend neither on t nor on p. Now it remains to note that the series

$$\sum_p \frac{\sin^2 \vartheta_t(p)}{p}$$

is almost uniformly convergent due to the inequality

$$\frac{\sin^2 \vartheta_t(p)}{1 - r_t(p) \cos \vartheta_t(p)} \leq \frac{\sin^2 \vartheta_t(p)}{1 - \cos \vartheta_t(p)} = 2 \cos^2 \frac{\vartheta_t(p)}{2} \leq 2 \ ,$$

and this gives the almost uniform convergence of the fourth of the series (7.38).

We are left with the series $\sum_p \dfrac{\vartheta_t(p)}{p}$. In view of

$$\vartheta_t(p) = \sin \vartheta_t(p) + O\left(\vartheta_t^2(p)\right)$$

and the convergence of the fourth series, it suffices to show that the series $\sum_p \dfrac{\sin \vartheta_t(p)}{p}$ is almost uniformly convergent. Now

$$\sum_p \frac{\sin \vartheta_t(p)}{p} = \sum_p \frac{r_t(p) \sin \vartheta_t(p)}{p} + \sum_p \frac{\left(1 - r_t(p)\right) \sin \vartheta_t(p)}{p} \quad ,$$

but the first of these series is almost uniformly convergent by our assumption, and so is the second in view of the estimate

$$\left| \frac{\left(1 - r_t(p)\right) \sin \vartheta_t(p)}{p} \right| \leq \frac{1 - r_t(p)}{p} \quad . \quad \blacksquare$$

Now we shall define the function $g_t^{(N)}(n)$ approximating $g_t(n)$ by the formula

$$g_t^{(N)}(n) = \prod_{\substack{p|n \\ p \leqslant \log N}} g_t(p)$$

and prove the following result:

Lemma 7.10. *Almost uniformly with respect to t, we have*

$$\lim_{N \to \infty} \frac{1}{N} \sum_{n \leqslant N} g_t^{(N)}(n) = \prod_p \left(1 + \frac{g_t(p) - 1}{p} \right).$$

Proof. Let $G_t^{(N)} = \mu * g_t^{(N)}$. Then for $p \leqslant \log N$, we have $G_t^{(N)}(p^k) = g_t(p) - 1$, and for all higher prime powers p^k we have

$$G_t^{(N)}(p^k) = \sum_{i=0}^{k} \mu(p^i) g_t^{(N)}(p^{k-i}) = g_t^{(N)}(p^k) - g_t^{(N)}(p^{k-1}) \quad.$$

Since for $p \leqslant \log N$, $k \geqslant 2$ we have

$$g_t^{(N)}(p^k) - g_t^{(N)}(p^{k-1}) = g_t(p) - g_t(p) = 0 \quad,$$

and for $p > \log N$, $k \geqslant 1$ the equality

$$g_t^{(N)}(p^k) - g_t^{(N)}(p^{k-1}) = 1 - 1 = 0$$

holds, thus for $k \geqslant 2$ and for all primes p we have $G_t^{(N)}(p^k) = 0$ and also $G_t^{(N)}(p) = 0$ for $p > \log N$, therefore for sufficiently large n we have $G_t^{(N)}(n) = 0$, and from Euler's identity it follows that

$$(7.39) \qquad \sum_{n=1}^{\infty} \frac{G_t^{(N)}(n)}{n} = \prod_{p \leqslant \log N} \left(1 + \frac{g_t(p) - 1}{p} \right) \quad.$$

Here $g_t^{(N)} = 1 * G_t^{(N)}$, and so

$$\frac{1}{N} \sum_{n \leqslant N} g_t^{(N)}(n) = \frac{1}{N} \sum_{d \leqslant N} G_t^{(N)}(d) \left[\frac{N}{d} \right]$$

$$= \sum_{d=1}^{\infty} \frac{G_t^{(N)}(d)}{d} + O\left(\sum_{d > N} \frac{|G_t^{(N)}(d)|}{d} \right) + O\left(\frac{1}{N} \sum_{d \leqslant N} |G_t^{(N)}(d)| \right)$$

$$= \prod_{p \leqslant \log N} \left(1 + \frac{g_t(p) - 1}{p} \right) + O\left(\frac{1}{N} \sum_{d=1}^{\infty} |G_t^{(N)}(d)| \right)$$

in view of (7.39), but

$$\frac{1}{N} \sum_{d=1}^{\infty} \left| G_t^{(N)}(d) \right| = \frac{1}{N} \prod_{p \leqslant \log N} \left(1 + |g_t(p) - 1| \right) \leqslant \frac{1}{N} \, 2^{\log N} = N^{-\delta} \quad ,$$

where

$$\delta = 1 - \log 2 > 0 \quad . \quad \blacksquare$$

Corollary. *The product* $\displaystyle\prod_p \left(1 + \frac{g_t(p) - 1}{p} \right)$ *is a continuous function of* t. $\quad \blacksquare$

The last step in the proof of Delange's theorem is the following lemma:

Lemma 7.11. *If*

$$\Delta_t(N) = \frac{1}{N} \sum_{n \leqslant N} \left(g_t(n) - g_t^{(N)}(n) \right) \quad ,$$

then $\Delta_t(N)$ *converges almost uniformly to zero as* N *tends to infinity.*

Proof. Since

$$g_t(n) = \prod_{p | n} g_t(p) \qquad \text{and} \qquad g_t^{(N)}(n) = \prod_{\substack{p | n \\ p \leqslant \log N}} g_t(p) \quad ,$$

$$\Delta_t(N) = \frac{1}{N} \sum_{n \leqslant N} g_t^{(N)}(n) \left(\prod_{\substack{p | n \\ p > \log N}} g_t(p) - 1 \right) \quad .$$

Now we define a strongly additive function $\psi_t^{(N)}$ by the formula

$$\psi_t^{(N)}(p) = \begin{cases} 0 & \text{for } p \leqslant \log N \quad , \\[2mm] \log g_t(p) & \text{for } p > \log N \quad , \end{cases}$$

where we fix the value $\log g_t(p)$ in such a way that firstly, for a fixed p it is a continuous function of t which can be done due to (7.36), and secondly, for some constant M independent of p nor t, we have

$$(7.40) \qquad |\log g_t(p)| \leqslant M \quad .$$

Then

$$|\Delta_t(N)| \leqslant \frac{1}{N} \sum_{n \leqslant N} |\exp \psi_t^{(N)}(n) - 1| \leqslant \frac{1}{N} \sum_{n \leqslant N} |\psi_t^{(N)}(n)|$$

taking into account the inequality

$$|e^z - 1| = \left| \int_0^z \exp t \, dt \right| \leqslant |z|$$

valid for $\operatorname{Re} z \leqslant 0$.

Now if we take

$$A_t^{(N)} = \sum_{p \leqslant N} \psi_t^{(N)}(p) p^{-1} \quad ,$$

then from Cauchy's inequality we obtain

$$|\Delta_t(N)|^2 \leqslant \frac{1}{N} \sum_{n \leqslant N} |\psi_t^{(N)}(n)|^2 \leqslant \frac{2}{N} \sum_{n \leqslant N} |\psi_t^{(N)}(n) - A_t^{(N)}|^2 + 2|A_t^{(N)}|^2 \quad ,$$

and from Theorem 7.1 it follows that almost uniformly with respect to t, we have

$$\frac{2}{N} \sum_{n \leqslant N} |\psi_t^{(N)}(n) - A_t^{(N)}|^2 \ll \sum_{p^k \leqslant N} |\psi_t^{(N)}(p^k)|^2 p^{-k}$$

$$= \sum_{p \leqslant N} |\psi_t^{(N)}(p)|^2 \sum_{\substack{k \geqslant 1 \\ p^k \leqslant N}} p^{-k} \ll \sum_{p \leqslant N} \frac{|\psi_t^{(N)}(p)|^2}{p}$$

$$= \sum_{\log N < p \leqslant N} \frac{|\log g_t(p)|^2}{p}$$

$$= \sum_{\log N < p \leqslant N} \frac{\log^2 r_t(p) + \vartheta_t^2(p)}{p} \quad .$$

The last sum converges almost uniformly to zero by virtue of Lemma 7.9, and so does

$$|A_t^{(N)}|^2 = \left| \sum_{\log N < p \leqslant N} \frac{\log r_t(p)}{p} \right|^2 + \left| \sum_{\log N < p \leqslant N} \frac{\vartheta_t(p)}{p} \right|^2 \quad . \qquad \blacksquare$$

Using Lemmas 7.10 and 7.11 we now obtain

$$M(g_t) = \lim_{N \to \infty} \frac{1}{N} \sum_{n \leqslant N} g_t(n) = \prod_p \left(1 + \frac{g_t(p) - 1}{p} \right) \quad ,$$

and the continuity of $M(g_t)$ is contained in Corollary to Lemma 7.10. $\qquad \blacksquare$

    5.   Utilizing the theorem of Delange just proved, we can give a proof of the theorem of Erdös which gives a sufficient condition for the existence of an asymptotic

distribution of a strongly additive function. This theorem is true for any additive function; moreover, the condition contained in it is also necessary (the Erdös-Wintner theorem), but we shall not prove this here.

Theorem 7.6. *If a function $f(n)$ is strongly additive, real-valued, and if the series*

$$\sum_p f^+(p)/p \qquad \text{and} \qquad \sum_p \left(f^+(p)\right)^2/p \ ,$$

where

$$f^+(p) = \begin{cases} f(p), & \text{if} \quad |f(p)| \leqslant 1 \ , \\[2mm] 1, & \text{if} \quad |f(p)| > 1 \ , \end{cases}$$

are convergent, then the asymptotic distribution function $w(t)$ for the function f exists such that

$$\hat{w}(t) = \int_{-\infty}^{\infty} \exp(itx)\, dw(x) = \prod_p \left(1 - \frac{1}{p}\right)\left(1 + \frac{\exp(itf(p)) - 1}{p}\right) .$$

Proof. In view of Theorem 7.4 it is necessary only to show that the function $g_t(n) = \exp\left(itf(n)\right)$ has a mean value $M(g_t)$ which is a continuous function of t. For this purpose we note that the series

$$\sum_{\substack{p \\ |f(p)| > 1}} \frac{1}{p} \ , \qquad \sum_{\substack{p \\ |f(p)| \leqslant 1}} \frac{f^2(p)}{p} \ , \qquad \sum_{\substack{p \\ |f(p)| \leqslant 1}} \frac{f(p)}{p}$$

are convergent. Actually, the convergence of the first two of them follows from the equality

$$\sum_p \frac{\left(f^+(p)\right)^2}{p} = \sum_{\substack{p \\ |f(p)| \leqslant 1}} \frac{f^2(p)}{p} + \sum_{\substack{p \\ |f(p)| > 1}} \frac{1}{p} \ ,$$

and the convergence of the third is a consequence of the equality

$$\sum_{\substack{p \\ |f(p)| \leqslant 1}} \frac{f(p)}{p} = \sum_p \frac{f^+(p)}{p} - \sum_{\substack{p \\ |f(p)| > 1}} \frac{1}{p} \ .$$

Therefore the series

$$\sum_{\substack{p \\ |f(p)| > 1}} \frac{1 - \exp(itf(p))}{p}$$

is absolutely and uniformly convergent because it is majorized by the series

$$\sum_{\substack{p \\ |f(p)| > 1}} \frac{2}{p} .$$

In the case where $|f(p)| < 1$ we have

$$\frac{1 - \exp(itf(p))}{p} = -it \frac{f(p)}{p} + \frac{1 - \cos(tf(p))}{p} + i\left(\frac{tf(p)}{p} - \frac{\sin tf(p)}{p}\right)$$

and, since for $|t| < T$ we have, on noting that $1 - \cos a \ll a^2$,

$$0 \leqslant \frac{1 - \cos(tf(p))}{p} \ll \frac{t^2 f^2(p)}{p} \ll T^2 \frac{f^2(p)}{p} ,$$

and also, by $a - \sin a \ll a^3$,

$$\left| \frac{tf(p)}{p} - \frac{\sin(tf(p))}{p} \right| \ll T^3 \frac{|f^3(p)|}{p} \leqslant T^3 \frac{|f(p)|^2}{p} ,$$

so the series

$$-it \sum_{\substack{p \\ |f(p)| < 1}} \frac{f(p)}{p} , \quad \sum_{\substack{p \\ |f(p)| < 1}} \frac{1 - \cos(tf(p))}{p} , \quad \sum_{\substack{p \\ |f(p)| < 1}} \frac{tf(p) - \sin(tf(p))}{p}$$

are almost uniformly convergent and we see that the same holds for the series

$$\sum_{\substack{p \\ |f(p)| < 1}} \frac{1 - \exp(itf(p))}{p} .$$

Finally we obtain the almost uniform convergence of the series

$$\sum_{p} \frac{1 - \exp(itf(p))}{p}$$

and from Theorem 7.5 (ii) our assertion follows, with

$$\hat{w}(t) = M(g_t) = \prod_{p}\left(1 + \frac{g_t(p) - 1}{p}\right) = \prod_{p}\left(1 + \frac{\exp(itf(p)) - 1}{p}\right)$$

Exercises

1.   Determine the characteristic function for the asymptotic distribution function of

$\log \dfrac{\varphi(n)}{n}$.

2.   In probability theory the following theorem of P. Lévy is proved :
If F_1 , F_2 , . . . *is a sequence of step distribution function, and if the infinite convolution*
$F_1 * F_2 * \ldots$ *is convergent to* F, *then for* F *to be continuous it is necessary and sufficient that the series*

$$\sum_{k=1}^{\infty} (1 - d_k)$$

is divergent, where d_k *denotes the maximal jump of the function* F_k. *Otherwise* F *is a step function.*

Utilizing this theorem prove that the asymptotic distribution function of an additive function f satisfying the condition of Theorem 7.1 is continuous if and only if the series

$$\sum_{f(p) \neq 0} \frac{1}{p}$$

is divergent.

CHAPTER VIII

DIOPHANTINE APPROXIMATION

In this chapter we shall present fundamental results concerning the approximation of real numbers by rational numbers and we shall also speak about problems of distribution of sequences of numbers in the interval $[0, 1)$.

§1. Continued fractions

1. It is a well-known fact that each real number can be approximated by rational numbers with arbitrary exactitude. In fact, taking sufficiently many digits of the decimal expansion, we obtain a rational approximation with arbitrary accuracy. In this section we shall try to find the approximation of real numbers by rational numbers with not too large denominators. For this purpose decimal expansions are of little use because e.g. approximating $\pi = 3.14159\ldots$ by $3.1 = 31/10$, we commit the error exceeding 0.04, and approximating it by the number $22/7$, we obtain, with a smaller denominator, the error less than 0.02. Of course, approximating a given number by a fraction with a given denominator N, we cannot guarantee an approximation better than $1/2N$, but it happens that by a skilful choice of N we can get an error less than $1/N^2$. This result is a consequence of the Corollary to Theorem 5.6 but it can easily be proved in a direct manner: Let α be an arbitrary real number and N a given natural number. Let us consider the numbers $\{r\alpha\}$ $(r = 0, 1, 2, \ldots, N)$. There are $N+1$ of them, and all of them lie in the interval of length 1, so that we can select two among them which differ from each other by a number less than $1/N$, say $\{r_1\alpha\}$ and $\{r_2\alpha\}$ $(0 \leqslant r_1 < r_2 \leqslant N)$. Then there exist an integer P such that for $Q = r_1 - r_2$ we have

$$\left|\alpha - \frac{P}{Q}\right| = \frac{1}{Q}\,|Q\alpha - P| = \frac{1}{Q}\,|(r_2 - r_1)\alpha - P| < \frac{1}{QN} \leqslant \frac{1}{Q^2}\ .$$

One can raise the question of how the fractions giving such a good approximation can be found. For this purpose we shall now introduce the method of continued fractions.

A finite sequence $\langle a_0; a_1, \ldots, a_n \rangle$ of real numbers is called a *continued fraction* if the numbers $a_1, \ldots, a_n$ are positive. We call the number n the *length* of such a fraction and the number

$$a_0 + \cfrac{1}{a_1 + \cfrac{1}{a_2 + \cfrac{1}{a_3 + \cdots}}} \\ {\scriptstyle \cdots\cdots} \\ {\scriptstyle + \cfrac{1}{a_n}}$$

its *value*. We shall denote the value of the continued fraction $\langle a_0; \ldots, a_n \rangle$ by $[a_0; a_1, \ldots, a_n]$. We call the numbers $[a_0; a_1, \ldots, a_k]$ with $k = 0, 1, \ldots, n$ the *k-th partial quotient* of this fraction and the numbers $a_1, \ldots, a_n$ its *denominators*. Elementary properties of the notions introduced here are contained in the following lemma:

Lemma 8.1. *If* $\langle a_0; a_1, \ldots, a_n \rangle$ *is a finite continued fraction, then*

(i) $[a_0; a_1, \ldots, a_n] = a_0 + [0; a_1, \ldots, a_n]$,

(ii) $[0; a_1, \ldots, a_n] = \dfrac{1}{[a_1; \ldots, a_n]}$,

(iii) $[a_0; a_1, \ldots, a_n] = a_0 + \dfrac{1}{[a_1; \ldots, a_n]}$,

(iv) $[a_0; a_1, \ldots, a_n] = [a_0; a_1, \ldots, a_{k-1}, [a_k; a_{k+1}, \ldots, a_n]]$ $(k = 1, \ldots, n)$,

(v)   *if for* $i = 1, \ldots, n$ *we have* $a_i \geqslant 1$, *then* $0 < [0; a_1, \ldots, a_n] \leqslant 1$ *and moreover the equality* $[0; a_1, \ldots, a_n] = 1$ *can occur only in the case* $n = 1$, $a_1 = 1$.

Proof. Formulas (i) - (iv) follow directly from the definition, so we shall give a proof of (v) only. Actually it is obvious that

$$0 < [0; a_1, \ldots, a_n] = \cfrac{1}{a_1 + \cfrac{1}{a_2 + \cdots}} \leqslant \frac{1}{a_1} \leqslant 1, \\ {\scriptstyle \cdots\cdots} \\ {\scriptstyle + \cfrac{1}{a_n}}$$

and if $n \geqslant 2$, then

$$[0; a_1, \ldots, a_n] = \frac{1}{a_1 + [0; a_2, \ldots, a_n]} < 1.$$

Therefore $n = 1$, $1 = [0; a_1] = \dfrac{1}{a_1}$ and $a_1 = 1$. ∎

2. For an easier calculation of the partial quotients of a given continued fraction, which will be useful to us in the following section, we now define recursively two sequences of polynomials with non-negative integer coefficients:

We put

$$P_{-1} = 1, \quad Q_{-1} = 0, \quad P_0(x_0) = x_0, \quad Q_0(x_0) = 1.$$

$$P_{k+1}(x_0, \ldots, x_{k+1}) = x_{k+1} P_k(x_0, \ldots, x_k) + P_{k-1}(x_0, \ldots, x_{k-1}),$$
(8.1)
$$Q_{k+1}(x_0, \ldots, x_{k+1}) = x_{k+1} Q_k(x_0, \ldots, x_k) + Q_{k-1}(x_0, \ldots, x_{k-1})$$

for $k = 0, 1, \ldots$.

Before demonstrating the utility of these polynomials (see Theorem 8.1 below) we prove a simple lemma concerning them:

Lemma 8.2. *For $k = -1, 0, 1, \ldots$ we have the identities*

$$P_{k+1} Q_k - Q_{k+1} P_k = (-1)^k,$$
$$P_{k+2} Q_k - Q_{k+2} P_k = (-1)^k x_{k+2}.$$

Proof. The first identity is trivial in the case $k = -1$ and if we assume that it is true for index k, then in view of (8.1) we obtain

$$P_{k+2} Q_{k+1} - Q_{k+2} P_{k+1} = (x_{k+2} P_{k+1} + P_k) Q_{k+1} - (x_{k+2} Q_{k+1} + Q_k) P_{k+1}$$
$$= P_k Q_{k+1} - Q_k P_{k+1} = -(-1)^k = (-1)^{k+1}.$$

The second identity is a consequence of the first because

$$P_{k+2} Q_k - Q_{k+2} P_k = (x_{k+2} P_{k+1} + P_k) Q_k - (x_{k+2} Q_{k+1} + Q_k) P_k$$
$$= x_{k+2}(P_{k+1} Q_k - Q_{k+1} P_k) = (-1)^k x_{k+2}. \quad ∎$$

Theorem 8.1. *If $\langle a_0; a_1, \ldots, a_n \rangle$ is a finite continued fraction, then for $k = 0, 1, \ldots, n$ the equality*

$$[a_0; a_1, \ldots, a_k] = \frac{P_k(a_0, \ldots, a_k)}{Q_k(a_0, \ldots, a_k)}.$$

holds.

In particular, the value of a continued fraction $\langle a_0; a_1, \ldots, a_n \rangle$ is the number $P_n(a_0, \ldots, a_n)/Q_n(a_0, \ldots, a_n)$.

Proof. Let us apply induction. For $k = 0$ the assertion is obvious and for $k = 1$ we have

$$[a_0; a_1] = a_0 + \frac{1}{a_1} = \frac{a_0 a_1 + 1}{a_1} = \frac{P_1(a_0, a_1)}{Q_1(a_0, a_1)}.$$

If we assume that the assertion is true for some $k \geqslant 1$ and all continued fractions, then we obtain, using Lemma 8.1 (iv), the equality

$$[a_0; a_1, \ldots, a_{k+1}] = \left[a_0; a_1, \ldots, a_k + \frac{1}{a_{k+1}}\right] = \frac{P_k\left(a_0, a_1, \ldots, a_k + \frac{1}{a_{k+1}}\right)}{Q_k\left(a_0, a_1, \ldots, a_k + \frac{1}{a_{k+1}}\right)}$$

$$= \frac{P_{k-1}(a_0, a_1, \ldots, a_{k-1})\left(a_k + \frac{1}{a_{k+1}}\right) + P_{k-2}(a_0, a_1, \ldots, a_{k-2})}{Q_{k-1}(a_0, a_1, \ldots, a_{k-1})\left(a_k + \frac{1}{a_{k+1}}\right) + Q_{k-2}(a_0, a_1, \ldots, a_{k-2})}$$

$$= \frac{P_k(a_0, \ldots, a_k) a_{k+1} + P_{k-1}(a_0, \ldots, a_{k-1})}{Q_k(a_0, \ldots, a_k) a_{k+1} + Q_{k-1}(a_0, \ldots, a_{k-1})}$$

$$= \frac{P_{k+1}(a_0, \ldots, a_{k+1})}{Q_{k+1}(a_0, \ldots, a_{k+1})}. \quad \blacksquare$$

3.   Now we introduce *infinite continued fractions*. We shall call a sequence $\langle a_0; a_1, \ldots \rangle$ of real numbers in which the elements a_i are positive for $i > 1$ an infinite continued fraction. If moreover the limit

(8.2)
$$\lim_{n \to \infty} [a_0; a_1, \ldots, a_n]$$

exists, then we say that the fraction $\langle a_0; a_1, \ldots \rangle$ is *convergent*, call this limit the *value* of this fraction and denote it by

$$[a_0; a_1, \ldots]$$

In the following we shall exclusively be concerned with convergent continued fractions. The numbers $a_1, a_2, \ldots$ are called *denominators* of the continued fraction and the number $[a_0; a_1, \ldots, a_k]$ its *k-th partial quotient* for $k = 0, 1, 2, \ldots$. It is quite obvious that Theorem 8.1 is true also for infinite continued fractions.

We shall prove now a lemma which gives a simple sufficient condition for the existence of the limit (8.2):

Lemma 8.3. *If for $k \geqslant 1$ we have $a_k \geqslant 1$, then the limit (8.2) exists.*

Proof. Using Lemma 8.1 and Theorem 8.1 we can write with the notation $r_n = [a_0; a_1, \ldots, a_n]$

$$(8.3) \qquad |r_{n+1} - r_n| = \left| \frac{P_{n+1}(a_0, \ldots, a_{n+1})}{Q_{n+1}(a_0, \ldots, a_{n+1})} - \frac{P_n(a_0, \ldots, a_n)}{Q_n(a_0, \ldots, a_n)} \right|$$

$$= \left| \frac{P_{n+1} Q_n - P_n Q_{n+1}}{Q_n Q_{n+1}} \right| = \frac{1}{Q_n Q_{n+1}}.$$

We now note that by our assumptions we have $Q_k \geqslant k$. Actually, for $k = 0, 1$ this is obvious and if this inequality holds for $k \leqslant m$, with $m \geqslant 1$, then

$$Q_{m+1}(a_0, \ldots, a_{m+1}) = a_{m+1} Q_m(a_0, \ldots, a_m) + Q_{m-1}(a_0, \ldots, a_{m-1}) \geqslant 2m - 1 \geqslant m + 1.$$

Using (8.3), we now obtain the inequality

$$|r_{n+1} - r_n| \leqslant \frac{1}{n(n+1)},$$

hence for any $N > n$ we have

$$|r_N - r_n| \leqslant \sum_{j=n}^{N-1} |r_{j+1} - r_j| \leqslant \sum_{j=n}^{\infty} \frac{1}{j(j+1)} < \varepsilon,$$

if only n is sufficiently large.

Corollary. *Each continued fraction whose denominators are natural numbers is convergent.*

For infinite continued fractions the following lemma, analogous to Lemma 8.1 holds.

Lemma 8.4. *If $\langle a_0; a_1, \ldots \rangle$ is a convergent continued fraction, then*

(i) $[a_0; a_1, \ldots] = a_0 + [0; \alpha_1, \ldots]$,

(ii) $[0; a_1, \ldots] = \dfrac{1}{[a_1; a_2, \ldots]}$,

(iii) $[a_0; a_1, \ldots] = a_0 + \dfrac{1}{[a_1; a_2, \ldots]}$,

(iv) $[a_0; a_1, \ldots] = [a_0; a_1, \ldots, a_{k-1}, [a_k; a_{k+1}, \ldots]]$ $\quad (k = 1, 2, \ldots, n)$.

(v) *if for $i > 1$ we have $a_i \geqslant 1$, then $0 < [0; a_1, \ldots] \leqslant 1$, and the equality holds solely in the case where the fraction $\langle 0; a_1, \ldots \rangle$ is finite, has length 1 and $a_1 = 1$.*

Proof. Equalities (i) and (iv) follow, by passage to the limit, from Lemma 8.1. We obtain equalities (ii) and (iii) in the same way if only we check that $[a_1; a_2, \ldots] \neq 0$. In order to get this, we note that for any n we have

$$[a_1; a_2, \ldots, a_n] = a_1 + [0; a_2, \ldots, a_n] > a_1 > 0$$

in view of part (v) of Lemma 8.1, hence

$$[0; a_1, \ldots] \geqslant 0, \qquad [a_1; a_2, \ldots] \geqslant a_1 .$$

We obtain the inequality $[0; a_1, \ldots] \leqslant 1$ from the corresponding one of Lemma 8.1 by passage to the limit just as we obtained the inequality

$$[0; a_1, \ldots] \geqslant 0.$$

The equality $[0; a_1, \ldots] = 0$ would contradict (ii), and in order to determine all the cases in which the equality

$$[0; a_1, \ldots] = 1,$$

holds, we note that if our fraction is infinite, then

$$[0; a_1, \ldots] = \lim_{n \to \infty} [0; a_1, \ldots, a_n],$$

but for $n \geqslant 3$ we have

$$[0; a_1, \ldots, a_n] = \cfrac{1}{a_1 + \cfrac{1}{[a_2; a_3, \ldots, a_n]}} = \cfrac{1}{a_1 + \cfrac{1}{a_2 + [0; a_3, \ldots, a_n]}} \leqslant \cfrac{1}{a_1 + \cfrac{1}{a_2 + 1}} .$$

Hence our fraction is finite and we can apply part (v) of Lemma 8.1. ■

From now on we shall restrict ourselves to those continued fractions whose denominators are natural numbers and a_0 is an integer, and we shall not state this restriction in the assumptions of the following theorems and lemmas.

The following lemma describes the relative position of successive partial quotients of a continued fraction on the real line.

Lemma 8.5. *If* r_n *denotes the n-th partial quotient of some continued fraction (infinite or not), then for each* n *we have the inequality* $r_{2n} < r_{2n+2}$ *and* $r_{2n+1} < r_{2n-1}$, *and furthermore, for all* m, n *we have* $r_{2m} < r_{2n+1}$.

Proof. From Theorem 8.1 and Lemma 8.2 we obtain for $k = 0, 1, \ldots$ the equality

$$r_{k+2} - r_k = \frac{P_{k+2}}{Q_{k+2}} - \frac{P_k}{Q_k} = \frac{(-1)^k a_{k+2}}{Q_k Q_{k+2}}$$

and in view of the fact that $Q_k, Q_{k+2}, a_{k+2} \geqslant 1$ we obtain for even k the inequality $r_{k+2} > r_k$, and for odd k the inequality $r_{k+2} < r_k$, which proves the first part of the lemma. The second part of the lemma follows at once from the first, considering that the sequences r_{2k} and r_{2k+1} have a common limit.

Corollary. *If* a *is the value of a continued fraction* $\langle a_0; a_1, \ldots \rangle$, *then for any* k *we have* $r_{2k} \leqslant a \leqslant r_{2k+1}$.

4.   Let us now turn to the approximation of real numbers by rational numbers. We begin by an example. Let us consider the continued fraction $\langle 1; 1, \ldots \rangle$ and denote its value by a. By Lemma 8.4 (iii) we can write $a = 1 + 1/a$ and, solving this equation,

we see that $a = \dfrac{1+\sqrt{5}}{2}$ because the other root of this equation is negative while the number a must be possible by Lemma 8.4 (i), (v). If r_n is the n-th partial quotient of our fraction, then we have $r_{n+1} = 1 + 1/r_n$ and $r_0 = 1$, hence $r_1 = 2$, $r_2 = 1.5$, $r_3 = 1.66\ldots$, $r_4 = 1.6$, $r_5 = 1.625$ and $r_6 = 1.61538\ldots$. Now $a = 1.61803\ldots$ and so we see that the successive partial quotients give good approximations. It happens that this is not an exceptional case. In fact, the following lemma holds:

Lemma 8.6. *If a is the value of a continued fraction $\langle a_0 ; a_1, \ldots \rangle$, and $r_n = P_n(a_0, \ldots, a_n)/Q_n(a_0, \ldots, a_n)$ its n-th partial quotient, then*

$$\frac{1}{2Q_n Q_{n+1}} < \left| a - \frac{P_n}{Q_n} \right| < \frac{1}{Q_n Q_{n+1}} < \frac{1}{Q_n^2},$$

and, besides, the fraction P_n/Q_n is irreducible, i.e. $(P_n, Q_n) = 1$.

Proof. From the Corollary to Lemma 8.5 and Lemma 8.2 follows

$$\left| a - \frac{P_n}{Q_n} \right| \leqslant \left| \frac{P_{n+1}}{Q_{n+1}} - \frac{P_n}{Q_n} \right| = \frac{1}{Q_n Q_{n+1}} < \frac{1}{Q_n^2},$$

because

$$Q_{n+1}(a_0, \ldots, a_{n+1}) = a_{n+1} Q_n(a_0, \ldots, a_n) + Q_{n-1}(a_0, \ldots, a_n) > Q_n(a_0, \ldots, a_n).$$

Similarly,

$$\left| a - \frac{P_n}{Q_n} \right| = \left| \frac{P_{n+2}}{Q_{n+2}} - \frac{P_n}{Q_n} \right| = \frac{a_{n+2}}{Q_n Q_{n+2}} = \frac{a_{n+2}}{Q_n(a_{n+2}Q_{n+1} + Q_n)}$$

$$\geqslant \frac{1}{Q_n(Q_n + Q_{n+1})} > \frac{1}{2Q_n Q_{n+1}}$$

The last half of the assertion follows from Lemma 8.2. ∎

In order to be able to make use of the proved lemma to obtain good approximations of reals by rationals, it is necessary to show that each real number is the value of some continued fraction and to give a method for finding such a fraction. This is the content of the following theorem:

Theorem 8.2.

(i) *Every rational number is the value of two continued fractions. Both of them are finite and of the form*

$$\langle a_0 ; a_1, \ldots, a_{N+1} \rangle, \quad \langle a_0 ; a_1, \ldots, a_{N+1} - 1, 1 \rangle \quad (a_{N+1} \neq 1) .$$

(ii) *Every irrational real number is the value of exactly one continued fraction. This fraction is an infinite fraction.*

(iii) *We can find successive denominators of a continued fraction with a given value a from the recurrence formulas:*

$$a_0 = a, \qquad a_{k+1} = \frac{1}{a_k - [a_k]}, \qquad a_k = [a_k]$$

(where the square bracket [ ] means the integral part, not the value of a continued fraction of length zero). *In the case of a rational a we arive at the fraction* $[a_0; a_1, \ldots, a_N]$ *with* $a_N \neq 1$ *in this way.*

Proof. Let a be an arbitrary real number. Let us first prove that it is a value of at least one continued fraction whose denominator can be found in the way given in (iii). If $a \in \mathscr{Z}$, then a is the value of the continued fraction $\langle a \rangle$ of length zero and (iii) holds

clearly. We may therefore suppose further that $a \notin \mathscr{Z}$. Let $\alpha_0 = a$, $\alpha_{k+1} = \dfrac{1}{\alpha_k - [\alpha_k]}$.

This recurrence definition makes sense as long as α_k is not an integer. If a is an irrational number, then this case will never happen and taking $a_n = [\alpha_n]$ $(n = 0, 1, \ldots)$ we obtain

$$a = \left[a_0; a_1, \ldots, a_n + \frac{1}{a_{n+1}} \right] \qquad (n = 1, 2, \ldots).$$

To get the equality $a = [a_0; a_1, \ldots]$ it now suffices to prove the formula

$$(8.4) \qquad \lim_{n \to \infty} \left(\left[a_0; a_1, \ldots, a_n + \frac{1}{a_{n+1}} \right] - [a_0; a_1, \ldots, a_n] \right) = 0,$$

For this we turn to the following lemma:

Lemma 8.7. *If* a_0 *is an arbitrary real number, and the numbers* $a_1, \ldots, a_n, a_n'$ *are not less than* 1 *(we do not assume them to be integers), then*

$$(8.5) \qquad |[a_0; a_1, \ldots, a_{n-1}, a_n] - [a_0; a_1, \ldots, a_{n-1}, a_n']| \leqslant \frac{1}{a_1 \ldots a_n + n - 1} .$$

Proof. First we note that for positive t, x the function)n

$$f_x(t) = [a_0; x] - [a_0; x+t] = \frac{t}{x(x+t)}$$

is an increasing function of a variable t for a fixed x, bounded by $1/x$, and so

$$|[a_0; a_1] - [a_0; a_1']| \leqslant \frac{1}{a_1},$$

which proves the assertion for $n = 1$. Now suppose the truth of (8.5) for an index $N \geqslant 1$. Then

$$|[a_0; a_1, \ldots, a_{N+1}] - [a_0; a_1, \ldots, a_{N+1}']|$$
$$= |[a_0; a_1 + [0; a_2, \ldots, a_{N+1}]] - [a_0; a_1 + [0; a_2, \ldots, a_{N+1}']]|$$
$$= f_b(|[0; a_2, \ldots, a_{N+1}] - [0; a_2, \ldots, a_{N+1}']|)$$

(where $b = a_1 + [0; a_2, \ldots, a_{N+1}]$), and this does not exceed by

$$f_b\left(\frac{1}{a_2 \ldots a_{N+1} + N - 1}\right) = \frac{\dfrac{1}{a_2 \ldots a_{N+1} + N - 1}}{b(b + a_2 \ldots a_{N+1} + N - 1)}$$

$$\leqslant \frac{\dfrac{1}{a_2 \ldots a_{N+1} + N - 1}}{a_1(a_1 + a_2 \ldots a_{N+1} + N - 1)} \leqslant \frac{1}{a_1 a_2 \ldots a_{N+1} + N} \; . \quad \blacksquare$$

From the proved·lemma the inequality

$$\left|\left[a_0; a_1, \ldots, a_n + \frac{1}{a_{n+1}}\right] - [a_0; a_1, \ldots, a_n]\right| \leqslant \frac{1}{a_1 \ldots a_n + n - 1} \leqslant \frac{1}{n},$$

follows, which implies (8.4).

In the case of a rational a we consider the index n such that the number α_n is not an integer. It must of course be a rational number, say $\alpha_n = r/s$, where $(r, s) = 1$. Then

$$0 < \alpha_n - [\alpha_n] = \frac{r - s[\alpha_n]}{s} < 1,$$

i.e. $r - s[\alpha_n] < s$ and the number $\alpha_{n+1} = \dfrac{s}{r - s[\alpha_n]}$ has its denominator less than that

of α_n. Therefore the denominators of the α_n's will decrease and we must arrive in the end at an index N such that the number α_N will be an integer. Putting $a_j = [\alpha_j]$ for $j \leqslant N$ we then get the equality

$$a = [a_0; a_1, \ldots, a_N].$$

If we had $a_N = \alpha_N = 1$ here, then the number

$$a_{N-1} = [a_{N-1}] + \frac{1}{a_N} = [a_{N-1}] + 1$$

would already be an integer, contrary to the choice of N. $\quad \blacksquare$

Now we shall determine all the continued fractions with a given value and we shall appeal to the following lemma in this connection:

Lemma 8.8. *If* $[a_0; a_1, \ldots] = [b_0; b_1, \ldots]$ *(the continued fractions appearing here can be finite) and* $a_0 < b_0$, *then the fraction* $\langle b_0; b_1, \ldots \rangle$ *has length zero and the fraction* $\langle a_0; a_1, \ldots \rangle$ *has length* 1, *and moreover* $a_0 = b_0 - 1$, $a_1 = 1$.

Proof. Since

$$a_0 + [0; a_1, \ldots] = [a_0; a_1, \ldots] = [b_0; b_1, \ldots] = b_0 + [0; b_1, \ldots],$$

we have

$$[0; a_1, \ldots] - [0; b_1, \ldots] \in \mathscr{Z}.$$

By Lemma 8.1 (v) this can hold solely when $[0; a_1, \ldots] = 1$ and the fraction $[0; b_1, \ldots]$ has length 0 and so the length of $[0; a_1, \ldots]$ must be equal to 1 and $a_1 = 1$, which implies $a_0 = b_0 - 1$.

Let us now suppose that $[a_0; a_1, \ldots] = [b_0; b_1, \ldots]$ and let $a_0 \leqslant b_0$. If $a_0 < b_0$, then from the lemma it follows that the value of the fraction under consideration is b_0, and the above equality reduces to $[b_0] = [b_0 - 1; 1]$ which agrees with the assertion of the theorem. On the other hand, if $a_0 = b_0$, then we denote by N the largest index for which we have $a_j = b_j$ for $j = 0, 1, \ldots, N$. If one of the continued fractions under consideration has length N, and if we suppose without loss of generality that it is $\langle a_0; a_1, \ldots \rangle$, then the other is either infinite or has length greater than N and we may write in any case,

$$(8.6) \qquad [a_0; a_1, \ldots, a_N] = [a_0; a_1, \ldots, a_N, b_{N+1}, \ldots]$$
$$= [a_0; a_1, \ldots, a_{N-1}, a_N + [0; b_{N+1}, \ldots]].$$

If, on the other hand, both of the fractions have length greater than N (or they are infinite), then $a_{N+1} \neq b_{N+1}$ and we can write

$$(8.7) \quad [a_0; a_1, \ldots, a_N, [a_{N+1}; a_{N+2}, \ldots]] = [a_0; a_1, \ldots, a_N, [b_{N+1}; b_{N+2}, \ldots]].$$

The function $[a_0; a_1, \ldots, a_k, x]$ is monotonic and so has different values for $k = 0, 1, \ldots$. Hence applying this fact to (8.6) (with $k = N - 1$) we obtain

$$a_N = a_N + [0; b_{N+1}, \ldots],$$

which contradicts Lemma 8.1. Similarly, we get in the second case

$$[a_{N+1}; a_{N+2}, \ldots] = [b_{N+1}; b_{N+2}, \ldots],$$

which proves by Lemma 8.8 that the value of the fraction $\langle a_0; a_1, \ldots \rangle$ is a rational number, and has itself one of the two forms given in (i).

5.   Now we prove a result characterizing the partial quotients of a continued fraction with a given value. We say that a rational number a/b ($b > 0$, $(a, b) = 1$) is a best approximation to a real number α if for all rational numbers $a'/b' \neq a/b$ with a denominator $0 < b' \leqslant b$, we have

$$(8.8) \qquad\qquad b \left| a - \frac{a}{b} \right| < b' \left| a - \frac{a'}{b'} \right|.$$

Then evidently the inequality

$$\left| a - \frac{a}{b} \right| < \left| a - \frac{a'}{b'} \right|.$$

also holds.

Theorem 8.3. *Let* $\langle a_0 ; a_1 , \ldots \rangle$ *be a continued fraction with the value* α, *where in the case of a fraction with finite length* N *we assume* $a_N \neq 1$. *Then every rational number that is a best approximation to* α *is equal to some partial quotient of this fraction and conversely, for* $k \geqslant 1$ *the* k-*th partial quotient* r_k *of this fraction is a best approximation to* α.

Proof. Let a/b be a best approximation to α and let $\alpha = [a_0 ; a_1 , \ldots]$. We denote the successive partial quotients of this continued fraction by $r_0 , r_1 , \ldots$. If a/b is none of them, then *a priori* (according to Lemma 8.5), there can occur the following three possibilities:

(i) $\dfrac{a}{b} < r_0$,

(ii) $\dfrac{a}{b} > r_1$,

(iii)   for some n the number a/b lies between r_n and r_{n+2} .

Let us show that all these possibilities lead to a contradiction. If (i) holds, then by the Corollary to Lemma 8.5 we have

$$\frac{a}{b} < r_0 = \frac{a_0}{1} \leqslant \alpha \quad ,$$

and hence $b|\alpha - a/b| > |\alpha - a_0 /1|$, contrary to (8.8).

In the case (ii) we have

$$\frac{a}{b} > r_1 = \frac{P_1}{Q_1} \geqslant \alpha,$$

and so

$$\left| \alpha - \frac{a}{b} \right| > \left| \frac{P_1}{Q_1} - \frac{a}{b} \right| \geqslant \frac{1}{bQ_1} .$$

From Theorem 8.2 (iii) we know that $a_1 = [\alpha_1]$, $\alpha_1 = \dfrac{1}{\alpha - [\alpha]}$, which gives

$$b\left| \alpha - \frac{a}{b} \right| \geqslant \frac{1}{Q_1} = \frac{1}{a_1} \geqslant \left| \alpha - \frac{[\alpha]}{1} \right|$$

contrary to the inequality (8.8).

There remains the case (iii). If $r_n = P_n /Q_n$, $r_{n+1} = P_{n+1}/Q_{n+1}$, $r_{n+2} = P_{n+2}/Q_{n+2}$, then

$$\frac{1}{bQ_n} \leqslant \left| \frac{a}{b} - r_n \right| < |r_n - r_{n+1}| = \frac{1}{Q_n Q_{n+1}},$$

and so $b > Q_{11}$. By virtue of Lemma 8.6 we have

$$Q_{n+1}|r_{n+1} - \alpha| \leqslant \frac{1}{Q_{n+2}} \leqslant b \left| \frac{P_{n+2}}{Q_{n+2}} - \alpha \right| \leqslant b \left| \alpha - \frac{a}{b} \right|,$$

which contradicts (8.8). In this way we have proved the first part of the theorem. In order to prove the second part we consider first the case $k = 1$. If r_1 were not a best approximation to the number α, then there would exist integers a' and b' satisfying $a'/b' \neq r_1$ and $0 < b' \leqslant a_1$ for which the inequality

$$b' \left| \frac{a'}{b'} - \alpha \right| \leqslant a_1 |\alpha - r_1|$$

holds.

Choose the smallest of such b'. Then a'/b' is a best approximation for α, and so, by virtue of that part of the theorem already proved, $a'/b' = r_0$, i.e. $a' = a_0$, $b' = 1$. But this leads to the inequality $|a_0 - \alpha| \leqslant |\alpha a_1 - a_0 a_1 - 1|$, or

$$1 \leqslant \left| a_1 - \frac{1}{\alpha - a_0} \right|,$$

which is impossible in view of $a_1 = \left[\dfrac{1}{\alpha - a_0} \right]$.

Therefore for $k = 1$ the theorem is true. Let us suppose that it is true for all indices not exceeding $k - 1$. If our fraction has length k, then r_k is of course a best approximation. Otherwise let Q be the smallest number with the following properties:

(i) $\quad Q > Q_k$,

(ii) $\quad$ there exists an integer P such that

$$Q \left| \alpha - \frac{P}{Q} \right| < Q_k \left| \alpha - \frac{P_k}{Q_k} \right|.$$

Then P/Q is a best approximation for α and on the basis of that part of the theorem already proved we must have for some m the equality

$$\frac{P}{Q} = r_m.$$

From Lemma 8.6 it follows that

$$Q_{k+1} \left| \alpha - \frac{P_{k+1}}{Q_{k+1}} \right| < \frac{1}{Q_{k+2}},$$

and moreover

$$\left| \alpha - \frac{P_k}{Q_k} \right| > \left| \frac{P_{k+2}}{Q_{k+2}} - \frac{P_k}{Q_k} \right| \geqslant \frac{1}{Q_k Q_{k+2}},$$

which gives

$$Q_{k+1}\left|a-\frac{P_{k+1}}{Q_{k+1}}\right| < Q_k\left|a-\frac{P_k}{Q_k}\right|$$

and we see that the number Q_k satisfies (i) and (ii), and therefore $Q_{k+1} \geqslant Q = Q_m > Q_k$, i.e. $Q_{k+1} = Q$. It remains to note that $P = P_{k+1}$ since otherwise the number $\dfrac{P}{Q_{1+k}}$ would lie farther from α than $\dfrac{P_{k+1}}{Q_{k+1}}$.

Remark. In the case of a continued fraction of length N in which the N-th partial quotient is equal to unity, the assertion of the theorem ceases to be true as is seen for example from the continued fraction $[0; 1, 1, 1]$ with the value $\frac{2}{3}$, whose successive partial quotients are the numbers 1, $\frac{1}{2}$, $\frac{2}{3}$, but $\frac{1}{2}$ is not the best approximation, for

$$2\cdot|\tfrac{2}{3}-\tfrac{1}{2}| = \tfrac{1}{3} = 1\cdot|\tfrac{2}{3}-1|$$

and the condition (8.8) is not fulfilled for $a' = b' = 1 < 2$.

6.   We shall close this section by proving a theorem refining, in some cases, the Corollary to Theorem 5.6 as well as Lemma 8.6.

Theorem 8.4 (A. Hurwitz, 1891). *If a is any irrational number, then for infinitely many rational numbers P/Q $(P, Q \in \mathscr{Z}$, $(P, Q) = 1)$ we have*

(8.9)
$$\left|a-\frac{P}{Q}\right| < \frac{1}{\sqrt{5}\,Q^2},$$

and each of them is a partial quotient of the continued fraction with the value a. The number $\sqrt{5}$ in (8.9) cannot be replaced by any larger number.

Proof. By P_k/Q_k $(P_k, Q_k \in \mathscr{Z}$, $.(P_k, Q_k) = 1)$ denote the k-th partial quotient of the continued fraction $\langle a_0 ; a_1 , \ldots \rangle$ having the value a.

Lemma 8.9. *If*

$$A_k = Q_{k-2}/Q_{k-1}, \qquad B_k = A_k+[a_k; a_{k+1}, \ldots] \qquad (k = 1, 2, \ldots),$$

and moreover for some $n \geqslant 2$ we have $B_{n-1} \leqslant \sqrt{5}$, $B_n \leqslant \sqrt{5}$, then $A_n > \dfrac{\sqrt{5}-1}{2}$.

Proof. Note that by Lemma 8.1, Theorem 8.1 and (8.1) we have

$$\frac{1}{A_{k+1}}+\frac{1}{[a_{k+1}; a_{k+2}, \ldots]} = \frac{Q_k}{Q_{k-1}}+[a_k; a_{k+1}, \ldots]-a_k$$

$$= a_k+A_k+[a_k; a_{k+1}, \ldots]-a_k = B_k,$$

therefore

$$\frac{1}{A_n} + \frac{1}{[a_n; a_{n+1}, \ldots]} \leqslant \sqrt{5},$$

and

$$A_n + [a_n; a_{n+1}, \ldots] \leqslant \sqrt{5}$$

which give altogether

$$(\sqrt{5} - A_n)\left(\sqrt{5} - \frac{1}{A_n}\right) \geqslant 1 \quad \text{or} \quad 5 - \sqrt{5}\left(A_n + \frac{1}{A_n}\right) \geqslant 0.$$

Here the equality sign cannot hold because $\sqrt{5}$ is not a rational number, and so $A_n + A^{-1} > \sqrt{5}$, which implies immediately the assertion of the lemma.

Lemma 8.10 (Borel's theorem). *If $k \geqslant 2$, then the inequality*

$$\left| a - \frac{P_j}{Q_j} \right| < \frac{1}{\sqrt{5} Q_j^2}$$

holds for at least one of the numbers $j = k - 2$, $k - 1$, k.

Proof. If this were not true, then we would have for some $k \geqslant 2$

$$\left| a - \frac{P_j}{Q_j} \right| \geqslant \frac{1}{\sqrt{5} Q_j^2} \qquad (j = k-2, k-1, k).$$

From Theorem 8.1 and Lemma 8.1 we obtain

$$a = [a_0; a_1, \ldots] = [a_0; a_1, \ldots, a_j, [a_{j+1}; \ldots]]$$

$$= \frac{P_{j+1}(a_0; \ldots, a_j, [a_{j+1}; \ldots])}{Q_{j+1}(a_0; \ldots, a_j, [a_{j+1}; \ldots])} = \frac{[a_{j+1}; a_{j+2}, \ldots] P_j + P_{j-1}}{[a_{j+1}; a_{j+2}, \ldots] Q_j + Q_{j-1}}.$$

and so

$$\left| a - \frac{P_j}{Q_j} \right| = \left| \frac{[a_{j+1}; \ldots] P_j + P_{j-1}}{[a_{j+1}; \ldots] Q_j + Q_{j-1}} - \frac{P_j}{Q_j} \right| = \frac{1}{B_{j+1} Q_j^2},$$

which gives $B_{j+1} \leqslant \sqrt{5}$ $(j = k-2, k-1, k)$.

From the preceding lemma it follows that

$$A_k > \frac{\sqrt{5} - 1}{2}, \quad A_{k+1} > \frac{\sqrt{5} - 1}{2},$$

and hence

$$a_k = \frac{1}{A_{k+1}} - A_k < 1,$$

which is impossible.

From this lemma the first part of the theorem follows. In order to prove the second part it is enough, by Theorem 8.3, to show that each number P/Q satisfying (8.9) is a best approximation to a. Now, if

$$B \left| a - \frac{A}{B} \right| \leqslant Q \left| a - \frac{P}{Q} \right| < \frac{1}{\sqrt{5}\,Q} < \frac{1}{2Q},$$

then

$$\left| a - \frac{A}{B} \right| < \frac{1}{2BQ}$$

and

$$\frac{1}{BQ} \leqslant \left| \frac{P}{Q} - \frac{A}{B} \right| \leqslant \left| a - \frac{P}{Q} \right| + \left| a - \frac{A}{B} \right| < \frac{1}{Q^2} + \frac{1}{2BQ},$$

whence $B > Q$.

There remains the last part of the theorem to be proved. If for the number

$$a = \frac{\sqrt{5}+1}{2} = [1; 1, 1, \dots]$$ and infinitely many rational numbers P/Q the inequality

$$\left| a - \frac{P}{Q} \right| < \frac{1}{cQ^2}$$

holds for some $c > \sqrt{5}$, then the numbers P/Q were partial quotients of the continued fraction $\langle 1; 1, 1, \dots \rangle$. But, if P_k/Q_k is such a partial quotient, then, as we have seen above,

$$\left| a - \frac{P_k}{Q_k} \right| = \frac{1}{B_{k+1} Q_k^2},$$

where

$$B_{k+1} = \frac{Q_{k-2}}{Q_{k-1}} + [a_k; a_{k+1}, \dots] = \frac{Q_{k-2}}{Q_{k-1}} + \frac{\sqrt{5}+1}{2},$$

and so

(8.10) $$c \leqslant \frac{Q_{k-2}}{Q_{k-1}} + \frac{\sqrt{5}+1}{2}.$$

Now we evaluate the limit $\lim\limits_{k \to \infty} \dfrac{Q_{k-2}}{Q_{k-1}}$, and for that purpose the following lemma will be useful:

Lemma 8.11. *If* $k \geqslant 1$ *and* $P_k/Q_k = [a_0; a_1, \dots, a_k]$, *then* $Q_k/Q_{k-1} = [a_k; a_{k-1}, \dots, a_0]$.

Proof. For $k = 1$ the assertion is trivial, and supposing its truth for some k we have

$$\frac{Q_{k+1}}{Q_k} = a_{k+1} + \frac{1}{[a_k; a_{k-1}, \dots]} = [a_{k+1}; a_k, \dots].$$

In our case $a_j = 1$ $(j = 0, 1, 2, \ldots)$, and hence

$$\lim_{k \to \infty} \frac{Q_k}{Q_{k-1}} = [1; 1, 1, \ldots] = \frac{\sqrt{5}+1}{2}.$$

Since the inequality (8.10) holds for infinitely many k, we have

$$c \leqslant \frac{2}{\sqrt{5}+1} + \frac{\sqrt{5}+1}{2} = \sqrt{5},$$

contrary to the assumption.   ∎

Exercises

1.   Prove that an infinite continued fraction with positive denominators is convergent iff the series formed by its denominators is divergent.

2.   Let a be the value of an infinite continued fraction whose denominators form a sequence which is periodic from some term on. Show that there exist integers A, B, C $(A \neq 0)$ such that $Aa^2 + Ba + C = 0$. (The converse is also true and was proved by Lagrange).

3.   Show that for every sequence $c_1, c_2, \ldots$ of positive numbers there exists an irrational number a such that the inequality

$$\left| a - \frac{P}{Q} \right| < c_Q$$

has infinitely many solutions $P \in \mathcal{Z}$, $Q \in \mathcal{N}$.

4.   Show that if $a = [a_0; a_1, \ldots]$ is an irrational number, and the sequence $\{a_n\}$ is bounded, then there exists a positive constant c depending only on a such that for any rational P/Q the inequality

$$\left| a - \frac{P}{Q} \right| \geqslant \frac{c}{Q^2}.$$

holds.

5.   Let α, β be irrational numbers and let

$$\alpha = [a_0; a_1, \ldots], \qquad \beta = [b_0; b_1, \ldots].$$

Prove that for the equality $a_n = b_n$ to hold for sufficiently large n it is necessary and sufficient that there exist integers a, b, c, d satisfying $ad \neq bc$ for which

$$\beta = \frac{a\alpha + c}{b\alpha + d}.$$

§2. Uniform distribution

1. We shall say that a sequence $a_1, a_2, \ldots$ of numbers in the intervals $[0, 1)$ is *uniformly distributed (has equidistribution)* if for any subinterval (a, b) of this interval, the number $A_N(a, b)$ of terms a_k with $k \leqslant N$ which lie in (a, b) satisfies

(8.11) $A_N(a, b) = (b - a + o(1))N$

as N tends to infinity. Such sequences "measure" subintervals of $[0, 1)$ in a sense.

If $a_1, a_2, \ldots$ is an arbitrary sequence of real numbers, then we say that it *is uniformly distributed* (mod 1) if the sequence $\{a_n\}$ formed from the fractional parts of the numbers a_n is uniformly distributed.

These notions were introduced by H. Weyl [1] who also proved the following theorem giving a necessary and sufficient condition for a given sequence to be uniformly distributed (mod 1).

Theorem 8.5. *For a sequence* $a_1, a_2, \ldots$ *of real numbers to be uniformly distributed* (mod 1) *it is necessary and sufficient that for each non-zero integer* m *the formula*

$$(8.12) \qquad \lim_{N \to \infty} \frac{1}{N} \sum_{k=1}^{N} \exp(2\pi i m a_k) = 0.$$

holds.

Proof. Since the left-hand side of formula (8.12) does not undergo a change when we replace the numbers a_k by their fractional parts, we may suppose that for $k = 1, 2, \ldots$ we have $0 \leqslant a_k < 1$.

Suppose that the condition (8.11) is satisfied. We prove that for every complex-valued function $F(t)$ which is Riemann-integrable in the interval $[0, 1]$, the formula

$$(8.13) \qquad \lim_{N \to \infty} \frac{1}{N} \sum_{n=1}^{N} F(a_n) = \int_0^1 F(t) \, dt.$$

holds. Obviously it suffices to prove it for real-valued functions. In the case when $F(t)$ is a characteristic function of any interval, this formula is a rewriting of formula (8.11) in another form, whence it follows the validity of (8.13) for any step function, viz. for any linear combination of characteristic functions of intervals. If $F(t)$ is now an arbitrary real function, Riemann-integrable in the interval $[0, 1]$, then for any given positive ϵ we can find two step functions $f_1(t)$ and $f_2(t)$ satisfying

$$f_1(t) \leqslant F(t) \leqslant f_2(t)$$

and

$$\left| \int_0^1 F(t) \, dt - \int_0^1 f_i(t) \, dt \right| < \epsilon \qquad (i = 1, 2).$$

Then

$$\limsup_{N \to \infty} \frac{1}{N} \sum_{n=1}^{N} F(a_n) \leqslant \lim_{N \to \infty} \frac{1}{N} \sum_{n=1}^{N} f_2(a_n) = \int_0^1 f_2(t) \, dt \leqslant \int_0^1 F(t) \, dt + \epsilon$$

and similarly

$$\liminf_{N \to \infty} \frac{1}{N} \sum_{n=1}^{N} F(a_n) \geqslant \int_0^1 F(t) \, dt - \epsilon,$$

whence (8.13) follows by arbitrariness of ϵ. Taking $F(t) = \exp(2\pi imt)$ $(m \in \mathscr{Z}$, $m \neq 0)$ we obtain the necessity of the condition (8.12).

In order to prove the sufficiency of the condition (8.12) we shall argue in an analogous way, proving that (8.13) follows from (8.12) for all functions, continuous in $[0, 1]$. Since evidently

$$\lim_{N \to \infty} \frac{1}{N} \sum_{k=1}^{N} 1 = 1,$$

(8.13) holds for $F(t) = \exp(2\pi imt)$ with any $m \in \mathscr{Z}$, and so for all trigonometrical polynomials of the form

$$(8.14) \qquad\qquad W(t) = \sum_{m=-k}^{k} c_m \exp\{2\pi imt\}.$$

If $F(t)$ is continuous in $[0, 1]$, then for every $\epsilon > 0$ there exists a $W(t)$ of the form (8.14) satisfying

$$|W(t) - F(t)| < \epsilon \qquad\qquad (0 \leqslant t \leqslant 1) .$$

If $\operatorname{Re} W(t) = u(t)$, $\operatorname{Re} F(t) = \varphi(t)$, then $|u(t) - \varphi(t)| < \epsilon$, and since

$$\lim_{N \to \infty} \frac{1}{N} \sum_{n=1}^{N} u(a_n) = \int_0^1 u(t)\, dt,$$

it follows that

$$-\epsilon + \int_0^1 u(t)\, dt \leqslant \liminf_{N \to \infty} \frac{1}{N} \sum_{n=1}^{N} \varphi(a_n) \leqslant \limsup_{N \to \infty} \frac{1}{N} \sum_{n=1}^{N} \varphi(a_n) \leqslant \epsilon + \int_0^1 u(t)\, dt.$$

But

$$\left| \int_0^1 u(t)\, dt - \int_0^1 \varphi(t)\, dt \right| < \epsilon,$$

which leads to

$$-2\epsilon + \int_0^1 \varphi(t)\, dt \leqslant \liminf_{N \to \infty} \frac{1}{N} \sum_{n=1}^{N} \varphi(a_n)$$

$$\leqslant \limsup_{N \to \infty} \frac{1}{N} \sum_{n=1}^{N} \varphi(a_n) \leqslant 2\epsilon + \int_0^1 \varphi(t)\, dt .$$

By arbitrariness of ϵ we obtain

$$\lim_{N \to \infty} \sum_{n=1}^{N} \varphi(a_n) = \int_0^1 \varphi(t)\, dt$$

and in just the same way we arrive at

$$\lim_{N\to\infty} \sum_{n=1}^{N} \psi(a_n) = \int_0^1 \psi(t)\,dt,$$

where $\psi(t) = \operatorname{Im} F(t)$. From these the validity of (8.13) for continuous functions follows immediately. The theorem will be proved if we show that (8.13) holds also for characteristic functions of an interval $(a, b) \subset [0, 1)$. If $F(t)$ is such a function, then for given $\epsilon > 0$ we shall construct two continuous functions $f(t)$ and $g(t)$ in the following way: For $t \leqslant a$ and $t \geqslant b$ we define $f(t) = 0$. In the interval $[a + \epsilon, b - \epsilon]$ we define $f(t) = 1$, and in the intervals $(a, a + \delta)$ and $(b - \epsilon, b)$, the function will be linear. Similarly we define $g(t) = 0$ for $t \leqslant a - \epsilon$ and $t \geqslant b + \epsilon$, $g(t) = 1$ in the interval $[a, b]$, and in the intervals $(a - \epsilon, a)$ and $(b, b + \epsilon)$ the function $g(t)$ will be a linear function. Then

$$f(t) \leqslant F(t) \leqslant g(t)$$

and

$$-\epsilon + \int_0^1 g(t)\,dt \leqslant \int_0^1 F(t)\,dt \leqslant \int_0^1 f(t)\,dt + \epsilon.$$

Hence

$$\limsup_{N\to\infty} \frac{1}{N} \sum_{n=1}^{N} F(a_n) \leqslant \lim_{N\to\infty} \frac{1}{N} \sum_{n=1}^{N} g(a_n) = \int_0^1 g(t)\,dt \leqslant \int_0^1 F(t)\,dt + \epsilon$$

and

$$\liminf_{N\to\infty} \frac{1}{N} \sum_{n=1}^{N} F(a_n) \geqslant \lim_{N\to\infty} \frac{1}{N} \sum_{n=1}^{N} f(a_n) = \int_0^1 f(t)\,dt \geqslant \int_0^1 F(t)\,dt - \epsilon.$$

From arbitrariness of ϵ we obtain

$$\lim_{N\to\infty} \frac{1}{N} \sum_{\substack{n\leqslant N \\ a_n\in(a,b)}} 1 = \lim_{N\to\infty} \frac{1}{N} \sum_{n=1}^{N} F(a_n) = \int_0^1 F(t)\,dt = b - a. \quad \blacksquare$$

Corollary. *If a is an irrational number, then the sequence $a, 2a, \ldots, na, \ldots$ is uniformly distributed (mod 1).*

Proof. It is sufficient to check that for integral and non-zero m we have

$$\sum_{n=0}^{N} \exp(2\pi i m n a) = o(N),$$

but this follows from the identity

$$\sum_{n=0}^{N} \exp(2\pi i m n a) = \frac{\exp(2\pi i m (N+1)a) - 1}{\exp(2\pi i m a) - 1} = O(1).$$

2.   Now let us consider the case when $a_n = W(n)$, and $W(t)$ is a polynomial with real coefficients among which at least one, different from the constant term, is an irrational number. This sequence turns out to be uniformly distributed (mod 1), but here the direct application of Theorem 8.5 is not easy. In order to avoid it, we apply a method due to van der Corput.

Lemma 8.12. *If* $z_1, \ldots, z_N$ *are arbitrary complex numbers and* $n \leqslant N$, *then*

$$n^2 \left| \sum_{k=1}^{N} z_k \right|^2 \leqslant n(N+n-1) \sum_{k=1}^{N} |z_k|^2 + 2(N+n-1) \sum_{l=1}^{n-1} (n-l) \left| \sum_{k=1}^{N-l} \bar{z}_k z_{k+l} \right|.$$

Proof. We take $z_k = 0$ for $k \leqslant 0$ and $k \geqslant N+1$. Then

$$\sum_{l=1}^{N+n-1} \sum_{k=0}^{n-1} z_{l-k} = \sum_{l=1}^{N+n-1} \sum_{r=l-n+1}^{l} z_r = \sum_{r=2-n}^{N+n-1} z_r \sum_{l=r}^{r+n-1} 1$$

$$= n \sum_{r=2-n}^{N+n-1} z_r = n \sum_{r=1}^{N} z_r,$$

and therefore

$$n^2 \left| \sum_{r=1}^{N} z_r \right|^2 \leqslant (N+n-1) \sum_{l=1}^{N+n-1} \left| \sum_{k=0}^{n-1} z_{l-k} \right|^2 = (N+n-1) \sum_{l=1}^{N+n-1} \sum_{k=0}^{n-1} \sum_{k'=0}^{n-1} z_{l-k} \bar{z}_{l-k'}$$

$$= (N+n-1) \sum_{l=1}^{N+n-1} \sum_{k=0}^{n-1} |z_{l-k}|^2 + (N+n-1) \sum_{l=1}^{N+n-1} \sum_{\substack{0 \leqslant k, k' \leqslant n-1 \\ k \neq k'}} z_{l-k} \overline{z_{l-k'}}.$$

Since

$$\sum_{l=1}^{N+n-1} \sum_{\substack{0 \leqslant k, k' \leqslant n-1 \\ k \neq k'}} z_{l-k} \bar{z}_{l-k'} = \sum_{l=1}^{N+n-1} \sum_{\substack{l-n+1 \leqslant r, s \leqslant l \\ r \neq s}} z_r \bar{z}_s,$$

and in this sum the term $z_r \bar{z}_{r+h}$ or $\bar{z}_r z_{r+h}$ can occur for $h > 0$ only when $h \leqslant n-1$ in which case it appears $n-h$ times, we have

$$n^2 \left| \sum_{r=1}^{N} z_r \right|^2 \leqslant (N+n-1) \sum_{l=1}^{N+n-1} \sum_{k=0}^{n-1} |z_{l-k}|^2$$

$$+ (N+n-1) \sum_{h=1}^{n-1} (n-h) \left| \sum_{r=1}^{N-h} z_r \bar{z}_{r+h} + \bar{z}_r z_{r+h} \right|$$

$$\leqslant (N+n-1) n \sum_{k=1}^{N} |z_k|^2 + 2(N+n-1) \sum_{h=1}^{n-1} (n-h) \left| \sum_{r=1}^{N-h} \bar{z}_r z_{r+h} \right| \quad \blacksquare$$

Theorem 8.6. *If* a_n *is a sequence of real numbers with the property that for every natural number* l, *the sequence* $b_n = a_{n+l} - a_n$ *is uniformly distributed* (mod 1), *then* a_n *is also uniformly distributed* (mod 1).

Proof. We take

$$z_k = \exp(2\pi i m a_k)$$

in the lemma just proved, where m is an arbitrary non-zero integer, and we let n be any natural number. Then for $N > n$ we shall have

$$\frac{1}{N^2} \left| \sum_{k=1}^{N} \exp(2\pi i m a_k) \right|^2$$

$$\leqslant \frac{N+n-1}{nN^2} \sum_{k=1}^{N} 1 + 2\frac{N+n-1}{N^2} \sum_{l=1}^{n-1} (n-l) \left| \sum_{k=1}^{N-l} \exp(2\pi i m(a_{k+l} - a_k)) \right|,$$

but the right-hand side tends to $\dfrac{1}{n}$ as $N \to \infty$, and hence

$$\limsup_{N\to\infty} \frac{1}{N} \left| \sum_{k=1}^{N} \exp(2\pi i m a_k) \right| \leqslant \frac{1}{n},$$

which gives the assertion in view of the arbitrariness of n and of Theorem 8.5. ∎

Corollary. *If* $W(x) = a_n x^n + \ldots + a_0$ *is a polynomial with real coefficients, and the coefficient* a_n *is irrational, then the sequence* $W(1)$, $W(2)$, ... *is uniformly distributed* (mod 1).

Proof. Let us apply induction on the degree n of the polynomial $W(x)$. For $n = 1$ the assertion follows directly from Theorem 8.5, and if it is true for all polynomials of degree at most equal to $n-1$, then (for $m \in \mathscr{Z}$, $m \neq 0$) the polynomial

$$V_m(x) = W(x+m) - W(x) = ma_n x^{n-1} + \cdots$$

satisfies the assumption and we can apply Theorem 8.6.

Exercises

1.   Prove that if the sequence $a_1, a_2, \ldots$ is uniformly distributed (mod 1), then in formula (8.11) the term $o(1)$ tends to zero uniformly in a, b.

2.   Let $\varphi(t)$ be a distribution on the interval $[0, 1]$. Prove that for a sequence $a_1, a_2, \ldots$ of the numbers in this interval to satisfy

$$\lim_{N\to\infty} \frac{\sum_{n\leqslant N,\, a_n \leqslant t} 1}{N} = \varphi(t),$$

it is necessary and sufficient that for every non-zero integer m, the formula

$$\lim_{N\to\infty} \frac{1}{N} \sum_{n=1}^{N} \exp(2\pi i m a_n) = \int_0^1 \exp(2\pi i m t)\, d\varphi(t).$$

holds.

3.   We say that a sequence $a_1, a_2, \ldots$ is uniformly distributed (mod M) if for every r we have

$$\lim_{N \to \infty} \frac{1}{N} \sum_{\substack{n \leqslant N \\ a_n \equiv r \,(\mathrm{mod}\, M)}} 1 = \frac{1}{M}.$$

Find a necessary and sufficient condition, analogous to Theorem 8.5, for a given sequence to be uniformly distributed (mod M).

CHAPTER IX

ALGEBRAIC NUMBERS AND p-ADIC NUMBERS

§1. Algebraic numbers and algebraic integers

1. The last chapter of this book is devoted to the generalization of the concept of rational integers. We shall consider two such generalizations — in §1 and §2 we shall be concerned with algebraic integers, and in §3 we shall discuss p-adic integers.

Let us begin by recalling the concept of algebraic numbers familiar in elementary algebra: we call a complex number a an *algebraic number* if there exists a non-zero polynomial with rational coefficients of which a is a root. We shall call the least of degrees of such polynomials the *degree* of a and denote it by deg a. We shall call each of the polynomials $F(x) \in \mathcal{Q}[x]$ satisfying the conditions: deg $F = $ deg a, $F(a) = 0$, the *minimal polynomial* for a.

Note that in the above definitions, the condition of rationality of coefficients of the polynomial can be replaced by the condition of their integrality, for it is enough to multiply the given polynomial with rational coefficients by the l.c.m. of the coefficients in order to get a polynomial with coefficients in $\mathcal{Z}$ and with the same roots.

Minimal polynomials for a given number are determined only up to constant rational factors. Choosing a suitable factor, we can always obtain a minimal polynomial of the shape $a_0 + a_1 x + \ldots + a_n x^n$, where $n = $ deg a, $a_i \in \mathcal{Z}$, $a_n > 0$ and $(a_0, a_1, \ldots, a_n) = 1$. The polynomial satisfying these conditions is determined uniquely by a and we shall call it the *normalized minimal polynomial of* a.

Note that each rational number is an algebraic number of degree 1 because $a \in \mathcal{Q}$ is a root of the polynomial $x - a \in \mathcal{Q}[x]$, and if m is a rational number which is not a square of any rational number, then $m^{1/2}$ is an algebraic number of degree 2. In fact, it is a root of the polynomial $x^2 - m$, and not a root of any polynomial of degree 1. These examples can be generalized. For this purpose we prove the following simple lemma.

Lemma 9.1.

(i)   *Every minimal polynomial of an algebraic number is irreducible over* $\mathscr{Q}$.

(ii) *If* a *is a root of an irreducible polynomial* F *with rational coefficients and* deg $F = N$, *then* a *is an algebraic number of degree* N, *and* F *is its minimal polynomial. Moreover, every polynomial in* $\mathscr{Q}[x]$ *for which* a *is a root must be divisible by* F.

Proof.

(i)   If a minimal polynomial $G(x)$ of a were reducible over $\mathscr{Q}$, then a would be a root of some of its factors, contrary to the minimality of G.

(ii) Suppose that G is a minimal polynomial of a and let F be any polynomial with rational coefficients for which a is a root. Then deg $G \leqslant$ deg F, and hence we may write $F = AG + B$ with suitable polynomials $A, B \in \mathscr{Q}[x]$, where deg $B <$ deg G or $B = 0$. Substituting $x = a$ in this equality, we obtain $B(a) = 0$, and so $B = 0$ in view of minimality of G. Moreover, if F is an irreducible polynomial of degree N, then we must have $F = cG$ and F is also a minimal polynomial of a, whence deg $a =$ deg $F = N$.   ∎

Making use of appropriate criterions for irreducibility of polynomials one can determine the degrees of various algebraic numbers. As an example let us prove a theorem on the degree of roots of natural numbers.

Theorem 9.1. *If* N *is a natural number, and* $m > 1$ *is a natural number not being d-th power of any natural number for* d *dividing* N *and* $d \neq 1$, *then the number* $m^{1/N}$ *is an algebraic number of degree* N.

Proof. Let $m^{1/N} = t$. This is a root of the polynomial $x^N - m$. In view of Lemma 9.1 it suffices to show the irreducibility of this polynomial. For that let us suppose that we have a decomposition

$$x^N - m = P(x) \, Q(x) \quad ,$$

where by Gauss' lemma we may suppose that the coefficients of P, Q are integers. If we denote by z_N any primitive N-th root of unity, e.g. $z_N = \exp(2\pi i/N)$, then

$$x^N - m = \prod_{j=0}^{N-1} (x - z_N^j t) \ .$$

We may therefore express the set $\{0, 1, \ldots, N-1\}$ of indices in the form of a union of disjoint sets A and B such that

$$P(x) = \prod_{j \in A} (x - z_N^j t) \ , \qquad Q(x) = \prod_{j \in B} (x - z_N^j t) \ .$$

If we denote the cardinality of the sets A and B by r and s respectively, then for some natural numbers R, S we shall have

$$P(0) = (-1)^r \, t^r z_N^R \, , \qquad Q(0) = (-1)^s \, t^s z_N^S \ ,$$

and hence t^r and t^s are natural numbers. Let i denote the least natural index for which the number t^i is rational. Note that if j is a natural number and $t^j \in \mathcal{Q}$, then i divides j, because otherwise we would have $j = ai + b$ with some $a, b \in \mathcal{Q}$ satisfying the condition $0 < b < i$, but the number $t^b = t^j t^{-ia}$ is rational and b is less than i, which implies $b = 0$.

Hence all the numbers r, s, N are divisible by i. Note that t^i is a natural number. Indeed, from $(t^i)^{N/i} = m$ and $t^i = A/B$ ($(A, B) = 1$) we obtain $mB^{N/i} = A^{N/i}$, and so each prime factor of B divides A, which, in view of $(A, B) = 1$, implies $B = 1$. Thus we see that $m = (t^i)^{N/i}$, $t^i \in \mathcal{N}$, and $N/i \neq 1$, which contradicts our assumption. ∎

2. Here it is worth mentioning complex numbers which are not algebraic numbers. We call such numbers *transcendental numbers*. Their existence follows from the following reason: since the set of all non-zero polynomials with integral coefficients is countable, and each of them has finitely many roots, the set of all such roots is countable, i.e. the set of all algebraic numbers is countable and there must exist transcendental numbers because the set of complex numbers is uncountable.

This argument does not enable us to give any example of a transcendental number. In order to give such an example, we prove the following theorem of Liouville:

Theorem 9.2. *If a is a real algebraic number of degree $N \neq 1$, then there exists a constant $C = C(a)$ such that for all integers A, B $(B > 0)$, the inequality*

$$\left| a - \frac{A}{B} \right| \geq \frac{C}{B^N}$$

holds.

Proof. Let $F(x) = a_N x^N + \ldots + a_0$ $(a_j \in \mathcal{Q})$ be a minimal polynomial for a. Then $F(x) = (x - a) G(x)$, where G has real coefficients. Since the polynomial F is irreducible, it cannot have multiple roots, and so $G(a) \neq 0$. Therefore there exist positive numbers ϵ, δ such that for x in the interval $(a - \epsilon, a + \epsilon)$, we have $0 < |G(x)| \leq \delta$.

Let A/B be a rational number in this interval $(A, B \in \mathcal{Q}, (A, B) = 1)$. Then $G(A/B) \neq 0$, and hence $F(A/B) \neq 0$ and we obtain

$$\left| a - \frac{A}{B} \right| = \left| \frac{F\left(\dfrac{A}{B}\right)}{G\left(\dfrac{A}{B}\right)} \right| = \left| \sum_{k=0}^{N} a_k A^k B^{N-k} \right| B^{-N} |G(A/B)|^{-1} \geq \frac{1}{\delta B^N} \ ,$$

because $\displaystyle\sum_{k=0}^{N} a_k A^k B^{N-k}$ is a non-zero element of $\mathcal{Q}$. If, on the other hand, A/B

$(A, B \in \mathcal{Q}, (A, B) = 1)$ lies outside the interval $(a - \epsilon, a + \epsilon)$, then

$$\left| a - \frac{A}{B} \right| \geqslant \epsilon \geqslant \frac{\epsilon}{B^N}$$

and, taking $C = \min(\epsilon, \delta^{-1})$ we obtain the assertion. ∎

Corollary. *The number* $\displaystyle\sum_{n=1}^{\infty} 2^{-n!}$ *is transcendental.*

Proof. If the number $a = \displaystyle\sum_{n=1}^{\infty} 2^{-n!}$ were algebraic of degree N, then for some

constant C we would have for all natural numbers k

$$\sum_{n=k+1}^{\infty} 2^{-n!} = \left| a - \sum_{n=1}^{k} 2^{-n!} \right| \geqslant \frac{C}{2^{Nk!}} \quad ,$$

because the fraction $\displaystyle\sum_{n=1}^{k} 2^{-n!}$ has the denominator $2^{-k!}$. On the other hand, we

have

$$\sum_{n=1+k}^{\infty} 2^{-n!} \leqslant \frac{2}{2^{(1+k)!}} \quad ,$$

and therefore

$$\frac{C}{2^{Nk!}} \leqslant \frac{2}{2^{(1+k)!}} \quad ,$$

i.e.

$$2^{k!(1+k-N)} \leqslant \frac{2}{C} \quad ,$$

which is impossible for sufficiently large k. ∎

Liouville's theorem shows that algebraic numbers cannot be very well approximated by rational numbers, and its formulation suggests that the degree of approximation depends upon that of a given number. However, this is not the case, as showed in 1955 by K. F. Roth.

If a is a real algebraic number, $a \in \mathcal{2}$, then for any $\epsilon > 0$ there exists a positive constant $C = C(a, \epsilon)$ such that for any $A \in \mathcal{2}$, $B \in \mathcal{N}$ we have

$$\left| a - \frac{A}{B} \right| \geqslant CB^{-2-\epsilon} \quad .$$

(Cf. e.g. J. W. S. Cassels [1]).

This result was extended to approximations of a system of n algebraic numbers by fractions with the same denominator by W. M. Schmidt [1] who has proved the following theorem:

If $a_1, \ldots, a_n$ are real algebraic numbers, and moreover the system $\{1, a_1, \ldots, a_n\}$ is linearly independent over $\mathscr{Q}$, then for each $\epsilon > 0$ the system of inequalities

$$\left| a_i - \frac{A_i}{B} \right| < B^{-1-1/n-\epsilon}$$

has at most a finite number of solutions $A_1, \ldots, A_n, B \in \mathscr{Z}$, $B > 0$.

The reader can easily check that for $n = 1$, Schmidt's result coincides with Roth's theorem.

In 1873, C. Hermite proved that the number e is transcendental, and in 1882, F. Lindemann showed the transcendency of the number π, thus solving negatively the classical problem of the quadrature of a circle.

In 1934, A. O. Gel'fond and T. Schneider showed independently that if $a \neq 0, 1$ is an algebraic number and b is an irrational algebraic number, then a^b is transcendental. This result was strengthened by A. Baker [1] who proved that if $a_1, \ldots, a_n$ are non-zero algberaic numbers $b_1, \ldots, b_n$ are algebraic numbers and both the systems $\{1, b_1, \ldots, b_n\}$ and $\{2\pi i, \log a_1, \ldots, \log a_n\}$ are linearly independent over $\mathscr{Q}$, then the product

$$a_1^{b_1} \ldots a_n^{b_n}$$

is transcendental, and moreover the numbers $\log a_1, \ldots, \log a_n$ are linearly independent over the field of all algebraic numbers. The method employed by Baker gives also quantitative results and enables one to estimate from below the modulus of the linear combination

$$\sum_{j=1}^{n} c_j \log a_j \ ,$$

where a_j satisfy the assumptions of the above theorem and c_j are any algebraic numbers not simultaneously vanishing. Such estimations are useful in the theory of Diophantine equations. Let us cite here only one of the results of this type (A. Baker [2]):

If

$$f(x, y) = \sum_{j=0}^{n} a_j x^j y^{n-j}$$

is a non-degenerate form of degree $n \geqslant 3$ with integral coefficients and m is a given natural number, then all the integer solutions of the equation

$$f(X, Y) = m$$

satisfy

$$\log \{ \max(|x|, |y|) \} \leqslant C \ ,$$

where $C = C(m, n, H)$ is a constant and $H = \max\limits_{0 \leqslant j \leqslant n} \left\{ |a_j| \right\}$. Moreover, one may take

$$C = (nH)^{(10n)^5} + (\log m)^{2n+2}$$

One can get acquainted with the actual state of transcendence theory due to the following survey works: N. I. Fel'dman, A. B. Shidlovskii [1] and S. Lang [1]. M. Waldschmidt's book [1] presents a survey of the most important results together with their proofs.

3.   We shall call a number an *algebraic integer* when it is a root of some non-zero polynomial with coefficients in $\mathscr{Z}$ and with leading term equal to 1. Almost directly from the definition we obtain the lemma:

Lemma 9.2. *The following conditions are equivalent:*
(i)   *a is an algebraic integer.*
(ii)   *a is an algebraic number and its normalized minimal polynomial has the leading coefficient equal to one.*

Proof.   The implication (ii) → (i) is trivial, and the reverse implication is a consequence of Gauss' lemma.   ∎

Corollary. *If a is a rational number which is an algebraic integer, then $a \in \mathscr{Z}$. Conversely, each number in $\mathscr{Z}$ is an algebraic integer.*

Proof.   It suffices to note that the normalized minimal polynomial for a rational number A/B $(A, B \in \mathscr{Z}, (A, B) = 1, B > 0)$ is $Bx - A$.   ∎

In the rest of this chapter we shall call elements of $\mathscr{Z}$ rational integers and call algebraic integers simply *integers*. By the above corollary this cannot lead to a misunderstanding. Note that using this terminology, we can formulate the above corollary as follows: *every rational integer is an integer which is rational, and vice versa.* The notion of an integer is a special case of the notion of an integral element over an arbitrary integral domain R, i.e. over a commutative ring with unit element and without zero divisors. Namely, we say that an element a of a field $K \supset R$ is integral over R if there exists a non-zero polynomial P with coefficients in R and with leading coefficient equal to unity such that $P(a) = 0$. Note that for $R = \mathscr{Z}$, $K = \mathfrak{Z}$ we obtain in this way the notion of an algebraic integer, and for $R = \mathscr{Q}$, $K = \mathfrak{Z}$ we obtain the notion of an algebraic number. This remark allows us to prove at the same time that the set of all algebraic numbers is a field, and the set of all algebraic integers is a ring.

Theorem 9.3. *If R is an integral domain and K is a field containing R, then the set S of elements of K, integral over R, is a ring. If R is a field, then S also is a field.*

Proof.   We use a lemma which is of interest in itself:

Lemma 9.3. *If R is an integral domain and K is a field containing R, then the following properties of an element $a \in K$ are equivalent:*

(i)   The element a is integral over R.

(ii) The ring $R[a] \subset K$ generated by R and a is finitely generated R-module.

(iii) There exists a non-zero, finitely generated R-module $M \subset K$ satisfying $aM \subset M$.

Proof. (i) → (ii). Let $F(x) = x^n + a_{n-1} x^{n-1} + \ldots + a_0$ be a polynomial with coefficients in R of which a is a root. We shall prove that the elements $1, a, a^2, \ldots, a^{n-1}$ generate $R[a]$ as an R-module. They are clearly contained in $R[a]$, so it suffices to show that for any N, the element a^N is a linear combination of $1, a, \ldots, a^{n-1}$ with coefficients in R. We prove this by induction. For $N \leqslant n-1$ there is nothing to prove, so let $N \geqslant n$ and suppose that all the powers a^j with $j < N$ are such combinations. Then

$$a^N = a^{N-n} a^n = a^{N-n}(-a_0 - a_1 a - \ldots - a_{n-1} a^{n-1})$$

$$= -a_0 a^{N-n} - a_1 a^{N-n+1} - \ldots - a_{n-1} a^{N-1}$$

is also such a combination.

The implication (ii) → (iii) is trivial because we can take $R[a]$ as M. Finally, to prove the implication (iii) → (i), let us denote by $z_1, \ldots, z_n$ the generators of the module M. From the assumption we have $z_i a \in M$ for $i = 1, 2, \ldots, n$, so that with suitably chosen $c_{ij} \in R$ we may write

$$az_i = \sum_{j=1}^{n} c_{ij} z_j \qquad (i = 1, 2, \ldots, n) .$$

Not all the elements z_i can vanish, and hence the determinant

$$\begin{vmatrix} c_{11} - x & c_{12} & \cdots & c_{1n} \\ c_{21} & c_{22} - x & \cdots & c_{2n} \\ \vdots & & & \\ c_{n1} & c_{n2} & \cdots & c_{nn} - x \end{vmatrix}$$

must vanish for $x = a$, which after expanding turns out to be a polynomial with coefficients in R, of degree n and with leading coefficient equal to $(-1)^n$. Multiplying it by $(-1)^n$ we obtain the assertion. ∎

Now let $a, b \in K$ be integral and M, N be non-zero, finitely generated R-modules satisfying $aM \subset M$, $bN \subset N$. Then the module MN is also finitely generated and non-zero, and moreover $(a \pm b) MN \subset MN$ and $abMN \subset MN$. Hence $a + b$, $a - b$, ab are R-integral and we see that S is a ring. Furthermore if R is a field and $a \in S$ is a non-zero element and

$$a^n + a_{n-1} a^{n-1} + \ldots + a_0 = 0 \qquad (a_i \in R) ,$$

then

$$1 + a_{n-1} a^{-1} + \ldots + a_0 a^{-n} = 0 \ .$$

The polynomial $a_0 x^n + \ldots + a_{n-1} x + 1$ is clearly non-zero and after dividing it by its leading coefficient we obtain a polynomial, with leading coefficient equal to unity, of which $1/a$ is a root. Hence $1/a \in S$.   ■

The ring which was referred to in the preceding theorem is called the *integral closure* of R in K.

Corollary 1. *The sum, difference, product and quotient of algebraic numbers are algebraic, and the sum, difference, product and quotient of integers are integers.*   ■

Corollary 2. *If K is an arbitrary subfield of the field of complex numbers, then the set of all albegraic numbers contained in K is a field, and the set of all integers contained in K is a ring.*   ■

4.   The ring of all integers is too large to construct in it in a simple way any theory analogous to that in the ring $\mathscr{Z}$. We shall therefore restrict ourselves to the ring of all integers contained in finite extensions of the field of rational numbers. Here we shall suppose that the reader is familiar with elementary facts concerning such extensions contained in the book of van der Waerden ([1]) which we shall denote simply by vdW in what follows. For convenience we shall give below fundamental definitions and theorems of this theory, restricting ourselves to subfields of the field of complex numbers.

If K is a field and L is a field containing K, then we say that L is an *extension* of the field K. We denote such an extension by L/K. We shall call it a *finite extension* if the dimension of L as a linear space over K is finite. In this case we call the dimension the *degree of extension* L/K and denote it by $[L:K]$. If $K \subset L$ and $a \in L$, then we denote the smallest subfield of L containing K and a by $K(a)$. Here, the following facts are valid:

(i)   *If the extensions L/K and M/L are finite, then so is the extension M/K and moreover*

$$[M:K] = [M:L][L:K] \quad \text{(vdW, Chap. IV, §41, Theorem on degrees)}$$

(ii)   *If a is an algebraic number of degree N, then the field $\mathscr{Q}(a)$ coincides with the set*

$$\{m_0 + m_1 a + \ldots + m_{N-1} a^{N-1} : m_j \in \mathscr{Q}\}$$

and $[\mathscr{Q}(a):\mathscr{Q}] = N$. (vdW, Chap. IV, §39, d)).

As an immediate corollary to (i) and (ii) we find that if a is an algebraic number of degree N and b lies in $\mathscr{Q}(a)$, then b is also an algebraic number and its degree divides N.

Now let a be an algebraic number of degree N and with its minimal polynomial $F(x)$ and let $a_1 = a$, $a_2, \ldots, a_N$ be all the roots of this polynomial. We call the numbers a_i the *conjugates* of a and the fields $K_i = \mathscr{Q}(a_i)$ the *conjugate fields* to $K = \mathscr{Q}(a)$.

(iii) *For* $i = 1, 2, \ldots, N$ *we have*

$$K_i = \left\{ m_0 + m_1 a_i + \ldots + m_{N-1} a_i^{N-1} : m_j \in \mathcal{2} \right\} \quad ,$$

and the mappings $\varphi_i : K \to K_i$ *given by*

$$\varphi_i(m_0 + \ldots + m_{N-1} a^{N-1}) = m_0 + \ldots + m_{N-1} a_i^{N-1}$$

are isomorphisms. Moreover, every isomorphism of K *in the field of complex numbers is of this form.*

Proof. The fact that φ_i are isomorphisms is contained in the proof of a proposition in Chap. IV, §39. To show that each isomorphism $\varphi : K \to \mathcal{3}$ is of such a form it suffices to note that $b = \varphi(a)$ must also be a root of F in view of

$$0 = \varphi(0) = \varphi(F(a)) = F(\varphi(a)) = F(b) \quad ,$$

and hence $b = a_i$ with suitable i and we obtain at once the required shape of φ. ∎

Corollary. *The isomorphisms* φ_i *do not depend on the choice of* a, *i.e. if* $\mathcal{2}(a) = \mathcal{2}(b)$ *and we construct the isomorphisms* φ_i *using the number* b *in place of* a, *then we obtain the same isomorphisms.* ∎

(iv) *Abel's theorem on the primitive element.*

If K *is a subfield of the field of complex numbers and* L/K *is its finite extension, then there exists an element* $a \in L$ *such that* $L = K(a)$. (vdW, Chap. IV, 46, the first proposition).

(v) *A theorem on the extension of an isomorphism.*

If M_1, M_2 *are subfields of the field of complex numbers,* $\varphi : M_1 \to M_2$ *is an isomorphism onto* M_2, $F(x) = \sum_{k=0}^{n} a_k x^k$ $(a_k \in M_1)$ *is an irreducible polynomial over* M_1 *and*

$$G(x) = \sum_{k=0}^{n} b_k x^k \quad ,$$

where $b_k = \varphi(a_k) \in M_2$ *and finally* $F(a) = G(b) = 0$, $a, b \in \mathcal{3}$, *then there exists an isomorphism* ψ *of the field* $M_1(a)$ *onto* $M_2(b)$ *with the following properties:*

For $c \in M_1$ *we have* $\varphi(c) = \psi(c)$, *and moreover* $\psi(a) = b$. (vdW, Chap. IV, §41, the second proposition).

Recall also a fundamental theorem on symmetric functions (vdW, Chap. V, §33, The fundamental theorem on symmetric functions).

(vi) *If* $A(x_1, \ldots, x_n)$ *is a rational function with coefficients in a field* K *which remains unchanged under any permutation of variables, then there exists a rational function* $B(x_1, \ldots, x_n)$ *such that*

$$A(x_1, \ldots, x_n) = B(t_1(x_1, \ldots, x_n), \ldots, t_n(x_1, \ldots, x_n)) \quad ,$$

where $t_1, \ldots, t_n$ *are fundamental symmetric polynomials.*

From the results (iii) and (v) we now deduce a simple, but useful corollary:

Lemma 9.4. *If* a *is an algebraic number of degree* N, $K = \mathcal{Q}(a)$ *and* $\{\varphi_1, \ldots, \varphi_N\} = \Phi$ *is the set of all isomorphisms of* K *into the field of complex numbers, then every number of the form* $\varphi_i(a)$ *is a conjugate of* a *and conversely, every number, conjugate to* a *is of this form.*

Proof. The first part is a consequence of (iii), and for the proof of the second, it suffices to apply (v) in the case where $M_1 = M_2 = \mathcal{Q}$ to the minimal polynomial of a.   ∎

Corollary. *Isomorphisms of the fields* $K_1 \to K_2$ *preserve integrity.*

5.   We shall now make use of the notation of Lemma 9.4 to define two important notions. We call the product

$$\prod_{\varphi \in \Phi} \varphi(b)$$

the *norm* of an element $b \in K$ in this field and denote it by $N_K(b)$, and if K is fixed, then we simply write $N(b)$. We call the sum

$$\sum_{\varphi \in \Phi} \varphi(b)$$

the *trace* of an element b in the field K and denote it by $T_K(b)$ or $T(b)$. The following theorem describes the fundamental properties of the norm and the trace:

Theorem 9.4.

(i)   *For* $b_1, b_2 \in K$ *the following equalities hold:*

$$N_K(b_1 b_2) = N_K(b_1) \cdot N_K(b_2), \qquad T_K(b_1 b_2) = T_K(b_1) + T_K(b_2) .$$

(ii)   *For* $b \in K$ *we have* $N_K(b) \in \mathcal{Q}$ *and* $T_K(b) \in \mathcal{Q}$ *, and if* b *is an integer, then* $N_K(b) \in \mathcal{Z}$ *and* $T_K(b) \in \mathcal{Z}$ *.*

Proof. Since $\varphi \in \Phi$ are isomorphisms, we have

$$N_K(b_1 b_2) = \prod_{\varphi \in \Phi} \varphi(b_1 b_2) = \prod_{\varphi \in \Phi} \varphi(b_1) \prod_{\varphi \in \Phi} \varphi(b_2) = N_K(b_1) N_K(b_2)$$

and similarly for the trace.

For the proof of (ii) we use the following remark:

The polynomial

$$\prod_{\varphi \in \Phi} \left(X - \varphi(b) \right)$$

has rational coefficients.

To prove this we write $\Phi = \{\varphi_1, \ldots, \varphi_N\}$ and

$$\prod_{\varphi \in \Phi} \left(X - \varphi(b) \right) = \sum_{j=0}^{N} A_{N-j} X^j \quad ,$$

where

$$A_{N-j} = (-1)^{N-j} \sum_{1 \leqslant i_1 < \ldots < i_j \leqslant N} \varphi_{i_1}(b) \ldots \varphi_{i_j}(b) \quad ,$$

but

$$b = \sum_{k=0}^{N-1} c_k a^k \qquad (c_k \in \mathcal{Z}) \quad ,$$

and so

$$A_{N-j} = (-1)^{N-j} \sum_{1 \leqslant i_1 < \ldots < i_j \leqslant N} \sum_{k_1=0}^{N-1} \ldots \sum_{k_j=0}^{N-1} c_{k_1} \ldots c_{k_j} a_{i_1}^{k_1} \ldots a_{i_k}^{k_j} \quad ,$$

where $a_i = \varphi_i(a)$.

Hence we obtain

$$A_{N-j} = (-1)^{N-j} \sum_{k_1=0}^{N-1} \ldots \sum_{k_j=0}^{N-1} c_{k_1} \ldots c_{k_j} \sum_{1 \leqslant i_1 < \ldots < i_j \leqslant N} a_{i_1}^{k_1} \ldots a_{i_j}^{k_j} \quad ,$$

but

$$\sum_{1 \leqslant i_1 < \ldots < i_j \leqslant N} a_{i_1}^{k_1} \ldots a_{i_j}^{k_j}$$

is a symmetric function in $a_1, \ldots, a_N$, so that it can be expressed in terms of fundamental symmetric polynomials of these numbers. These are the coefficients of the minimal polynomial for a with leading coefficient equal to 1, when multiplied by ± 1 suitably, hence lie in $\mathcal{Z}$. Therefore A_{N-j} belong to $\mathcal{Z}$. It remains to note that

$$N_K(b) = (-1)^N A_N \qquad \text{and} \qquad T_K(b) = -A_1 \quad .$$

If b is an integer, then $N_K(b)$ and $T_K(b)$ are integers and lie in $\mathcal{Z}$, and so must belong to $\mathcal{Z}$. ∎

6.   In order to illustrate the notions introduced above we shall now discuss quadratic fields, i.e. fields of degree 2 over the rational field, which arise from $\mathcal{Z}$ by the adjunction of a root of an irreducible polynomial of degree 2. The description of such fields is contained in the following theorem:

Theorem 9.5. *If* $[K: \mathcal{Z}] = 2$, *then there exists a rational integer* D *distinct from* 1 *and not divisible by any square of a prime such that*

$$K = \mathcal{Z}(D^{\frac{1}{2}})$$

and conversely, every extension $K/\mathcal{Z}$ *of this form is of degree* 2.

Proof. The second part of the theorem is clear, hence we shall be concerned exclusively with the first part. Let $K = \mathcal{Z}(a)$ and let a be a root of a quadratic polynomial $X^2 + Ax + B$ with rational coefficients, irreducible over $\mathcal{Z}$. This last condition means that the discriminant $\varDelta = A^2 - 4B$ of our polynomial is not a square of any rational number. Noting that

$$a = \frac{-A \pm \varDelta^{\frac{1}{2}}}{2} \in \mathscr{Q}\!\left(\varDelta^{\frac{1}{2}}\right)$$

we obtain $\mathscr{Q}(a) \subset \mathscr{Q}\!\left(\varDelta^{\frac{1}{2}}\right)$, and that $\mathscr{Q}\!\left(\varDelta^{\frac{1}{2}}\right)$ is also a quadratic extension of $\mathscr{Q}$, therefore by (i) we have $\left[\mathscr{Q}\!\left(\varDelta^{\frac{1}{2}}\right) : K\right] = 1$, so that $K = \mathscr{Q}\!\left(\varDelta^{\frac{1}{2}}\right)$. If $\varDelta$ is an integer, then we may take for D the product of $\operatorname{sgn}\varDelta$ by the product of all prime numbers dividing $\varDelta$ in odd powers, and obtain the assertion. Otherwise, we write

$$A = \frac{M}{N} \qquad ((M, N) = 1, \; M, N \in \mathscr{Z})$$

and observe that

$$\mathscr{Q}\!\left(\varDelta^{\frac{1}{2}}\right) = \mathscr{Q}\!\left((MN)^{\frac{1}{2}}\right) \qquad ,$$

which results from

$$(MN)^{\frac{1}{2}} = N\varDelta^{\frac{1}{2}} \in \mathscr{Q}\!\left(\varDelta^{\frac{1}{2}}\right)$$

and

$$\varDelta^{\frac{1}{2}} = \frac{1}{N}(MN)^{\frac{1}{2}} \in \mathscr{Q}\!\left((MN)^{\frac{1}{2}}\right) \qquad .$$

Now, if

$$MN = \epsilon \prod_{i=1}^{r} p_i^{2\alpha_i + 1} \prod_{i=1+r}^{s} p_i^{2\alpha_i}$$

($\epsilon = \pm 1$, $p_1, \ldots, p_s$ prime numbers) is the canonical decomposition of MN, then $D = \epsilon p_1 \ldots p_s$ satisfies our assertion.   ∎

Using the theorem proved above, we can give an explicit form for the norm and the trace in quadratic extensions:

Corollary. *If D satisfies the conditions of the theorem and $K = \mathscr{Q}\!\left(D^{\frac{1}{2}}\right)$, then for $a, b \in \mathscr{Q}$ we have $N_K\!\left(a + bD^{\frac{1}{2}}\right) = a^2 - b^2 D$ and $T_K\!\left(a + bD^{\frac{1}{2}}\right) = 2a$.*

Proof. It suffices to prove that

$$\varphi_1\!\left(a + bD^{\frac{1}{2}}\right) = a + bD^{\frac{1}{2}} \qquad \text{and} \qquad \varphi_2\!\left(a + bD^{\frac{1}{2}}\right) = a - bD^{\frac{1}{2}}$$

are the only isomorphisms of K into the field of complex numbers.   ∎

In the case of a positive D we call $\mathscr{Q}\!\left(D^{\frac{1}{2}}\right)$ a *real quadratic field*, and in the case of a negative D we call this an *imaginary quadratic field*.

Now we shall describe algebraic integers in quadratic fields:

Theorem 9.6. *If D is a square-free integer, $K = \mathcal{Q}\left(D^{\frac{1}{2}}\right)$ and $D \equiv 2, 3 \pmod 4$, then the expression*

$$(9.1) \qquad a = x + yD^{\frac{1}{2}} \qquad (x, y \in \mathscr{Z})$$

gives all algebraic integers of K.

If $D \equiv 1 \pmod 4$, then all algebraic integers of K are given by the expression

$$(9.2) \qquad a = \frac{x + yD^{\frac{1}{2}}}{2} \qquad (x, y \in \mathscr{Z}, \ x \equiv y \pmod 2) \quad.$$

In both cases the numbers x, y are uniquely determined by a.

Proof. First note that the expressions (9.1) and (9.2) give exclusively integers. Indeed, $D^{\frac{1}{2}}$ is integral as a root of the polynomial $x^2 - D$, and so is $x + yD^{\frac{1}{2}}$ $(x, y \in \mathscr{Z})$ as a sum of integers. We argue similarly in the case of x, y divisible by 2 in formula (9.2). Now, if $D \equiv 1 \pmod 4$ and $x \equiv y \equiv 1 \pmod 2$, then

$$\frac{x + yD^{\frac{1}{2}}}{2} = \frac{1 + D^{\frac{1}{2}}}{2} + \frac{x - 1}{2} + \frac{y - 1}{2} D^{\frac{1}{2}}$$

and we obtain the integrality of this element from the fact that $\dfrac{1 + D^{\frac{1}{2}}}{2}$ is a root of the polynomial $X^2 - X + \frac{1}{4}(1 - D)$.

Now we show that our formulas give all integers in quadratic fields. Since the elements $1, D^{\frac{1}{2}}$ are linearly independent over $\mathscr{Q}$, every element x of K can be uniquely written in the form

$$x = a + bD^{\frac{1}{2}}$$

with rationals a, b. If $b = 0$, then x is a rational integer and evidently is of the form (9.1) resp. (9.2). Hence we may suppose that $b \neq 0$. If $x^2 + Ax + B$ $(A, B \in \mathscr{Z})$ is a minimal polynomial for x, then

$$(a^2 + b^2D + Aa + B) + (2ab + Ab)D^{\frac{1}{2}} = 0 \quad,$$

so that

$$a^2 + b^2D + Aa + B = 0 ,$$
$$2ab + Ab = 0$$

which in turn lead to $A + 2a = 0$, i.e. $a = \frac{1}{2}m$ with some $m \in \mathscr{Z}$, and $b^2D - a^2 + B = 0$, i.e.

$$(9.3) \qquad b^2D - a^2 \in \mathscr{Z} \ .$$

Writing $b = P/Q$, where $(P, Q) = 1, P, Q \in \mathscr{Z}$, we can rewrite the above condition in the form

$$4P^2 D - m^2 Q^2 \equiv 0 \pmod{4Q^2} \ .$$

This in turn implies $4P^2 D \equiv 0 \pmod{Q^2}$, hence $4D \equiv 0 \pmod{Q^2}$, but, since D is a product of distinct primes, it follows that $Q = 1$ or $Q = 2$. In any case we may write $b = \frac{1}{2} n \ (n \in \mathscr{Z})$, and then (9.3) gives

$$n^2 D - m^2 \equiv 0 \pmod{4} \ .$$

Now, if $D \equiv 1 \pmod{4}$, then $n^2 \equiv m^2 \pmod{4}$, so that $n \equiv m \pmod{2}$ and if $D \equiv 2 \pmod{4}$ or $D \equiv 3 \pmod{4}$, then we obtain $2n^2 \equiv m^2 \pmod{4}$ or $3n^2 \equiv m^2 \pmod{4}$ respectively, which lead to $m \equiv n \equiv 0 \pmod{2}$. Therefore, in all cases, x is of the form given in formulas (9.1), (9.2).   ∎

Corollary. *If D satisfies the conditions of the theorem and if we put*

$$w = \begin{cases} D^{\frac{1}{2}} & \text{if } D \equiv 2, 3 \pmod{4} \ , \\ \\ \dfrac{1 + D^{\frac{1}{2}}}{2} & \text{if } D \equiv 1 \pmod{4} \ , \end{cases}$$

then the set of all integers of the field $\mathscr{Q}\left(D^{\frac{1}{2}}\right)$ coincides with the set of the elements of the form $a + bw \ (a, b \in \mathscr{Z})$.

7.   We shall close this section with a result generalizing Corollary to Theorem 9.6 in the case of an arbitrary finite extension of the field of rational numbers. Let K be such an extension and let R_K denote the ring composed of all integers contained in K. We shall call a system of numbers $w_1, \ldots, w_n$ an *integral basis* if every element of R_K can be expressed uniquely in the form

$$a_1 w_1 + \ldots + a_n w_n$$

with $a_1, \ldots, a_n \in \mathscr{Z}$. That there always exists such a basis forms the contents of the following theorem:

Theorem 9.7. *Every field K of degree n over $\mathscr{Q}$ has an integral basis consisting of n elements.*

Proof. Choose any system of elements $v_1, \ldots, v_n \in K$ forming a basis of K, as a linear space over $\mathscr{Q}$. Here we may suppose that these elements lie in R_K, because the following lemma holds:

Lemma 9.5. *If v is an algebraic number, then there exists a natural number N such that Nv is an integer.*

Proof. The number v is a root of a polynomial, say, of

$$\sum_{k=0}^{m} A_k x^k \qquad (a_i \in \mathscr{Z}, \ A_m \neq 0)$$

Then the number $A_m v$ is a root of the polynomial

$$x^m + \sum_{k=1}^{m} A_{m-k} A_m^{k-1} x^{m-k} \quad ,$$

and so we may take $N = |A_m|$.   ∎

Now we show that there exists a rational integer $M \neq 0$ with the property that for any $a \in R_K$, the number Ma is expressed in the form

$$Ma = \lambda_1 v_1 + \ldots + \lambda_n v_n \qquad (\lambda_i \in \mathcal{Z}) \quad ,$$

i.e. it lies in the additive group generated by $v_1, \ldots, v_n$. It suffices to show this for $a \neq 0$. For this purpose we write

$$a = \frac{1}{c_0} (c_1 v_1 + \ldots + c_n v_n) \quad ,$$

where $c_0, \ldots, c_n \in \mathcal{Z}$, $c_0 \neq 0$, $(c_0, c_1, \ldots, c_n) = 1$. Let $\varphi_1, \ldots, \varphi_n$ be all the isomorphisms of K into the field of complex numbers and let

$$a^{(j)} = \varphi_j(a) , \qquad v_i^{(j)} = \varphi_j(v_i)$$

$(i, j = 1, \ldots, n)$. Then for $j = 1, \ldots, n$ we have

$$a^{(j)} = \sum_{i=1}^{n} \frac{c_j}{c_0} v_i^{(j)} \quad ,$$

so that the system of linear equations

(9.4) $$a^{(j)} = \sum_{i=1}^{n} X_i v_i^{(j)} . \qquad (j = 1, \ldots, n)$$

with determinant $D = \det[v_i^{(j)}]$ has a solution $X_i = c_i/c_0$. Note that $D \neq 0$. For if $K = \mathcal{Q}(a)$, then with suitable $\lambda_{ik} \in \mathcal{Q}$ we have

$$v_i = \sum_{k=0}^{n} \lambda_{ik} a^k \qquad (i = 1, \ldots, n) \quad ,$$

$$v_i^{(j)} = \varphi_j(v_i) = \sum_{k=0}^{n} \lambda_{ik} \varphi_j(a)^k \qquad (i = 1, \ldots, n) \quad ,$$

and so

$$D = \det[v_i^{(j)}] = \det[\lambda_{ik}] \det[\varphi_j(a)^k] \neq 0 \quad ,$$

since $\det[\varphi_j(a)^k]$ is the Vandermonde determinant and both the systems $\{v_1, \ldots, v_n\}$ and $\{1, a, \ldots, a^{n-1}\}$ form a basis of K as a linear space over $\mathcal{Q}$, which implies that the matrix $[\lambda_{ik}]$ is non-singular.

Moreover,

$$D^2 = \det[v_i^{(j)}]\,\det[v_i^{(j)}] \;=\; \det[v_i^{(j)}]\,\det[v_i^{(j)}]^T$$

$$= \det \sum_{k=1}^{n} v_i^{(k)} v_j^{(k)} \;=\; \det[T_K(v_i v_j)] \in \mathscr{Z}$$

by virtue of Theorem 9.4, and hence D is an integer, not necessarily lying in the field K. Applying Cramer's formula to the system (9.4), we obtain the integrality of the numbers $A_j = Dc_j/c_0$. From this it follows that the numbers $DA_j = D^2 A_j/D$ are rational integers, for DA_j is integral and D^2 and A_j/D are rational. Hence $D^2 c_j/c_0$ is a rational integer, and c_0 must divide D^2. Therefore we may take $M = D^2$.

Hence we see that the additive group of the ring R_K is a subgroup of the additive group generated by $v_1, \ldots, v_n$ which is isomorphic with $\mathscr{Z}^n$. From Theorem 1.12 follows the existence of elements $w_1, \ldots, w_k$ such that the additive group R_K is isomorphic with $\bigoplus_{j=1}^{k} \mathscr{Z} w_j$, but R_K contains n linearly independent elements over $\mathscr{Z}$, hence $k = n$. ∎

If $w_1, \ldots, w_n$ is an integral basis of the field K, then we call the number

(9.5) $\det[\varphi_j(w_i)]^2 \in \mathscr{Z}$

the *discriminant of the field* K and denote it by $d(K)$. For this definition to be meaningful it is necessary to show that $d(K)$ does not depend on the choice of an integral basis, but it is clear that the change of a basis is accomplished with the aid of an invertible matrix with integer coefficients, and hence with determinant equal to ± 1, the number (9.5) does not depend upon the choice of a basis.

Integral bases of quadratic fields were found in Corollary to Theorem 9.6. An easy calculation shows that in notation of that theorem we have for the field $K = \mathscr{Q}\left(D^{\frac{1}{2}}\right)$

$$d(K) = \begin{cases} D, & \text{if } D \equiv 1 \pmod 4 , \\[2mm] 4D, & \text{if } D \not\equiv 1 \pmod 4 . \end{cases}$$

Exercises

1 (L. K ronecker). Show that if a is an integer all of whose conjugates lie in the unit circle $|z| \leqslant 1$, then a is a root of unity.

2 (L. K ronecker). Show that if a is an integer all of whose conjugates lie in the interval $[-2, 2]$, then there exists a rational number r such that $a = 2\cos\pi r$.

3.   Give a sufficient condition for the number $\sum_{n=1}^{\infty} q^{-a_n}$ $(q, a_n \in \mathscr{Z})$ to be transcendental.

4. Estimate from below the constant $C(a)$ appearing in Theorem 9.2.

5. Describe all integral bases of the field $\mathcal{Q}(i)$.

6 (L. Stickelberger). Prove that no discriminant of any field is congruent to 2, 3 (mod 4).

7. Porve that number of integers in $\mathcal{Q}(i)$ with norm not exceeding x is equal to $\pi x + o(x)$.

8. Prove that if r_1, r_2 are given natural numbers, then there exists an integer having r_1 real and $2r_2$ complex conjugates.

9. Determine the sign of the discriminant of the field $\mathcal{Q}(a)$ depending on the number of real and non-real conjugates of a.

10. Find an integral base and the discriminant of the field $\mathcal{Q}(m^{1/3})$, for $m \in \mathcal{Z}$.

11. Show that in the field $\mathcal{Q}(\zeta_p) \left(\zeta_p = \exp\left(\dfrac{2\pi i}{p}\right) \right)$, the elements $1, \zeta_p, \ldots, \zeta_p^{p-2}$ form an integral basis, and evaluate the discriminant of this field.

12. Prove that no field has its discriminant equal to 1 or -1.

Hint. Use Minkowski's theorem on convex bodies.

§2. Ideals in the rings of algebraic integers

1. In this section we shall be more closely concerned with the algebraic structure of the ring R_K consisting of all integers of the field K, a finite extension of $\mathcal{Q}$, and in particular, with its ideals.

In the case where K is the rational field $\mathcal{Q}$, we have clearly $R_K = \mathcal{Z}$, and there every ideal is principal. Similarly, in the Gauss ring consisting of the elements $a + bi$, with rational integrals a, b, every ideal is principal, which follows from the existence of Euclid's algorithm in this ring. This ring forms the ring of integers of the field $\mathcal{Q}(i)$ (Theorem 9.6). We shall give later examples of fields K for which the rings R_K are not principal ideal domains.

This problem is related to the question when in the ring R_K, the unique factorization theorem holds. For fields $\mathcal{Q}$ and $\mathcal{Q}(i)$ the answer is positive, but this situation is rather exceptional. It happens, however, that if we introduce the product of ideals in some natural way, then in any ring R_K every non-zero ideal is uniquely expressed in the form of a product of irreducible ideals, which in our case will prove to be simply prime ideals, and even maximal ideals. An integral ring in which this unique factorization theorem for ideals holds is called a *Dedekind ring*. We begin by considering the fundamental properties of this class of rings. For methodical reasons, we shall first give a somewhat different definition which will be proved later to be equivalent to that given above.

Let R be any integral domain (i.e., commutative, with unity and without zero divisors) and let K be its field of fractions. If M is a non-zero R-module contained in K, then we say that M is a *fractional ideal*, provided that there exists a non-zero element $a \in R$ such that $aM \subset R$. Evidently, if $M \subset R$, then M is a fractional ideal iff M is an ideal of the ring R. From this it follows that if M is a fractional ideal and $aM \subset R$, then aM is an ideal in R, so that we may express every fractional ideal in the form $a^{-1}I$, where I is an ideal in R and a is a non-zero element of R. This fact explains the name "fractional ideal".

If M is a fractional ideal, then by M^* we shall denote the set

$$\{c \in K \colon cM \subset R\} \ .$$

It is clearly a non-zero R-module, and moreover it is a fractional ideal. Indeed, if $r \in R$ is so chosen that $r \neq 0$ and $rM \subset R$, and $m \in M$, $m \neq 0$, then $rmM^* \subset MM^* \subset R$ and $rm \in R$, $rm \neq 0$.

Now we define the product of fractional ideals: If M_1, M_2 are fractional ideals, then under their *product* $M_1 M_2$ we shall understand the smallest R-module containing all products $m_1 m_2$ $(m_i \in M_i$, $i = 1, 2)$. It is not difficult to check that $M_1 M_2$ is a fractional ideal and that it coincides with the set of all sums

$$\sum_{i=1}^{N} m_1^{(i)} m_2^{(i)} \qquad (m_1^{(i)} \in M_1, \ m_2^{(i)} \in M_2) \ .$$

As an example we take $R = \mathscr{Z}$. Then $K = \mathscr{Q}$ and we see that the fractional ideals are the sets of the form $r\mathscr{Z}$, where $r \in \mathscr{Q}$, $r \neq 0$ (we may assume that $r > 0$, since $r\mathscr{Z} = (-r)\mathscr{Z}$), and the product of fractional ideals $r_1 \mathscr{Z}$ and $r_2 \mathscr{Z}$ is the fractional ideal $r_1 r_2 \mathscr{Z}$. We see that in this case the set of all fractional ideals is a group with respect to multiplication, isomorphic with the multiplicative group of positive numbers of the field $\mathscr{Q}$. The unit element of this group is $\mathscr{Z}$, and moreover

$$(r\mathscr{Z})^* = r^{-1}\mathscr{Z} \ .$$

This example suggests the following definition: We shall call a fractional ideal M *invertible* if $MM^* = R$. In this case we shall also write M^{-1} in place of M^*. The following simple lemma holds:

Lemma 9.6. *Every principal fractional ideal* (i.e. *a fractional ideal of the form* aR $(a \in K$, $a \neq 0)$) *is invertible, and the set of all invertible fractional ideals form a group under multiplication.*

Proof. The first part follows from the remark that $(aR)^* = a^{-1}R$, and hence $(aR)(aR)^* = R$. To prove the second, we first note that $MR = M$ for all fractional ideals M. Furthermore, from the equality $M_1 M_2 = R$ follows $M_2 = M_1^*$. Indeed, $M_2 \subset M_1^*$, so that

$$R = M_1 M_2 \subset M_1 M_1^* \subset R \ ,$$

whence $M_1 M_1^* = R$ and $M_1^* = M_1^* R = M_1^* M_1 M_2 = RM_2 = M_2$.

Finally, the invertibility of a product of invertible fractional ideals and the equality $M = (M^{-1})^{-1}$ follow immediately from the definition and our assertion results. ∎

We can now give a definition of a *Dedekind ring*: It is an integral domain every fractional ideal of which is invertible. From this it follows at once that every principal ideal domain is a *Dedekind ring*. In particular, the ring $\mathscr{Z}$ of rational integers is such a ring and one of the aims of the present section is to show that every ring R_K is also a Dedekind ring. First let us prove some fundamental properties of a Dedekind ring:

Theorem 9.8. *If R is a Dedekind ring, then the following three conditions are satisfied:*

(i)   *R is a Noetherian domain, i.e. every ideal of R has a finite set of generators.*

(ii) *Every non-zero prime ideal in R is a maximal ideal.*

(iii) *The ring R is its own integral closure in its field K of fractions.*

(The following theorem shall show that every integral domain satisfying (i), (ii) and (iii) is a Dedekind ring, and this will allow us to show that R_K is such a ring).

Proof.

(i)   Let I be a non-zero ideal in R. Since $II^{-1} = R$, we can find elements $a_i \in I$, $b_i \in I^{-1}$ satisfying

$$\sum_{i=1}^{n} a_i b_i = 1 \ .$$

If now $x \in I$, then

$$x = \sum_{i=1}^{n} a_i (x b_i) \ ,$$

and $x b_i \in R$, hence I is generated by $a_1, \ldots, a_n$. Thus R is a Noetherian domain.

(ii) Let p be any non-zero prime ideal in R. Then there exists a maximal ideal P containing p. This fact, which follows immediately from Zorn-Kuratowski's lemma, can be accounted for in our case, without this lemma, by arguing as follows: If there were no such maximal ideals P, we would find an infinite sequence $p \subset I_1 \subset I_2 \subset \ldots$ of distinct ideals, and then their union would be an ideal without a finite set of generators, contrary to (i). The fractional ideal $pP^{-1} \subset PP^{-1} = R$ is an ideal of the ring R. By the equality $(pP^{-1})P = p$ we must have either $pP^{-1} \subset p$ or $P \subset p$, for p is a prime ideal. In the first case we would have $P^{-1} \subset p^{-1}pP^{-1} \subset p^{-1}p = R$, and so $P^{-1} = R$ and $P = R$, which is impossible. Therefore the inclusion $P \subset p$ must occur implying $p = P$.

(iii) Let x be any element of K integral over R. Then, by Lemma 9.3 the ring $R[x]$ is a finitely generated R-module. It is even a fractional ideal, for if $c_1, \ldots, c_n$ are its generators and a non-zero element $b \in R$ is so chosen that $bc_i \in R$ for $i = 1, \ldots, n$, then clearly $bR[x] \subset R$. Since $R[x]$ is a ring with unity, $R[x]^2 = R[x]$, so that

$$R[x] = RR[x] = R[x]\,R[x]^{-1}\,R[x] = R[x]^2\,R[x]^{-1} = R[x]\,R[x]^{-1} = R$$

and we see that $x \in R$. ∎

2.   Let us now show that the conditions (i), (ii) and (iii) of the preceding theorem characterize Dedekind rings among all integral domains.

Theorem 9.9. *If R is an integral domain satisfying the conditions* (i), (ii), (iii) *of Theorem 9.8, then R is a Dedekind ring.*

Proof. First we shall show that every non-zero prime ideal in R is invertible, and next, that every non-zero ideal of this ring, distinct from R, is a product of invertible ideals. Since a product of invertible ideals is invertible, it will follow that every non-zero ideal in R is invertible (the ideal R is invertible in view of $RR = R$), and from this and the remark that every fractional ideal is of the form aI with $a \in K$, $I \subset R$, follows the invertibility of any fractional ideal.

Lemma 9.7. *If R is a Noetherian domain and I its ideal $\neq \{0\}$, R, then we can find prime ideals $p_1, \ldots, p_r \subset R$ such that*

(9.6) $$p_1 p_2 \cdots p_r \subset I \subset p_1 \cap p_2 \cap \ldots \cap p_r \ .$$

Proof. Denote by I a maximal element of the set of all ideals that do not have this property, supposing that, contrary to our assertion, this set is non-empty. I cannot be a prime ideal, for every prime ideal satisfies the condition (9.6). Therefore we can find elements $a, b \notin I$ whose product ab lies in I. Now let $A = I + aR$, $B = I + bR$. Then $AB \subset I \subset A \cap B$. The ideals A, B are different from R, because e.g. from $A = R$ it would follow that $I = B \ni b$. Hence A, B satisfies (9.6), but from this follows that this condition is satisfied by I, contrary to the choice of I. This contradiction obtained shows that the set of ideals without the property (9.6) is in fact empty. ∎

Lemma 9.8. *If an integral domain R satisfies the conditions* (i), (ii), (iii), *then every non-zero prime ideal of R is invertible.*

Proof. Select any prime ideal $p \neq 0$ in R and choose a non-zero element a in it so that the ideal aR contains a product $p_1 \cdots p_k$ of as few prime ideals as possible. This is possible according to the preceding lemma. Since $p_1 \cdots p_k \subset p$, e.g. the ideal p_1 is contained in p and by (ii) we must have $p_1 = p$. The product $p_2 \cdots p_k$ is not contained in aR, therefore we can find $b \in p_2 \cdots p_k \backslash aR$. For this b we now obtain

$$bp \subset pp_2 \cdots p_k \subset aR, \qquad ba^{-1}p \subset R \ ,$$

i.e. $ba^{-1} \subset p^*\backslash R$, whence $R \subset p^*$, $R \neq p^*$.

The product pp^* is an ideal in R, and moreover

$$p = pR \subset pp^* \subset R \ .$$

We wish to show that $pp^* = R$. If this is not so, then by the maximality of p we must have $pp^* = p$ and by easy induction $p(p^*)^n = p$ for $n = 1, 2, \ldots$. Now let $x \in p$, $y \in p^* \setminus R$. For every natural n we have now $xy^n \in p \subset R$, and so $xR[y] \subset R$, hence $xR[y]$ is an ideal in R. By (i) it has a finite set of generators, say $a_1, \ldots, a_n$, so that $R[y]$ has a finite number of generators $a_1 x^{-1}, \ldots, a_n x^{-1}$. In view of Lemma 9.3, y is integral over R and by (iii) it must belong to R, contrary to the assumption. ∎

For the completion of the proof of the theorem it remains to prove the following lemma:

Lemma 9.9. *If R is an integral domain satisfying* (i), (ii), (iii), *then every ideal different from $\{0\}$ and R is a product of non-zero prime ideals.*

Proof. Let I be an ideal which is not such a product and contains a product with the least possible prime ideal factors, say $p_1 \ldots p_k$. Let p be a maximal ideal containing I. Then, analogously as in the proof of the preceding lemma, we obtain e.g. $p = p_1$, and

$$p_2 \cdots p_k \subset p^{-1} I \subset p^{-1} p = R \quad ,$$

which means that $p^{-1} I$ is an ideal in R. By the choice of I we must have

$$p^{-1} I = q_1 \ldots q_s \quad ,$$

where $q_1, \ldots, q_s$ are prime ideals, and this gives $I = p q_1 \ldots q_s$, contrary to the hypothesis. ∎

Using the theorem proved above, we can now prove the fundamental theorem concerning multiplicative structure of the set of ideals in a Dedekind ring R:

Theorem 9.10. *If R is a Dedekind ring, then its every ideal different from $\{0\}$ and R can be expressed uniquely as a product of prime ideals.*

Proof. The existence of such an expression results from Theorem 9.8 and Lemma 9.9. It remains to show its uniqueness. Suppose that there exists an ideal I having two distinct expressions as a product of prime ideals, say

$$(9.7) \qquad I = p_1 \ldots p_r = q_1 \ldots q_s \qquad (r \leqslant s)$$

and assume that I is so chosen that the number r is as small as possible. One of the ideals q_i, say q_1, is contained in p_1, and so $q_1 = p_1$, which gives

$$I p_1^{-1} = p_2 \cdots p_r = q_2 \cdots q_s \quad .$$

The ideal $I p_1^{-1}$ has, by the choice of I, only one decomposition into a product of prime ideals, so that $r = s$ and changing the ordering, if necessary, we may write $p_i = q_i$ ($i = 2, \ldots, r$). Hence the decompositions in (9.7) coincide. ∎

Corollary. *Every ideal I of a Dedekind ring R, different from $\{0\}$ and R itself, can be expressed uniquely in the form*

$$I = \prod_p p^{a(p)} \quad ,$$

where p runs through all the non-zero prime ideals of R, $a(p) \in \mathscr{Z}$, are non-negative, and are different from zero for only finitely many p. ∎

    3.   Using Theorem 9.10 we can now construct the theory of divisibility in the set of ideals of a Dedekind ring R. We say that an ideal $I \subset R$ is *divisible by* an ideal $J \subset R$ and write $J | I$ if there exists an ideal $A \subset R$ for which $I = AJ$' Note that a non-zero ideal in R has only finitely many divisors (this follows from Theorem 9.10). We shall call the *greatest common divisor* of ideals A and B an ideal C which divides A and B, and has the property that every divisor of A and B is a divisor of C. Of course, we must prove the existence of such an ideal C:

    Lemma 9.10. *Every pair of non-zero ideals A, B of a Dedekind ring R has exactly one greatest common divisor.*

    Proof. From the preceding theorem it follows that we may write

$$A = \prod_p p^{a(p)} , \qquad B = \prod_p p^{b(p)} ,$$

where p runs through all prime ideals of R, and $a(p)$, $b(p)$ are non-negative rational integers which are different from zero for only finitely many p. Observe that if

$$I = \prod_p p^{i(p)} ,$$

then I divides A iff for all p we have $i(p) \leqslant a(p)$, so that I divides A and B exactly when

$$i(p) \leqslant \min \{a(p), b(p)\} = c(p) .$$

Hence the ideal $C = \prod p^{c(p)}$ is a greatest common divisor of A, B. To prove the uniqueness suppose that C' is also a greatest common divisor of A and B. Then $C | C'$ and $C' | C$, so that $C = IC'$, $C' = JC$, and hence we see that $C = IJC$, where I, J are ideals in R. But this gives $IJ = R$ and $C = C'$. ∎

    We denote by (A, B) the greatest common divisor of ideals A and B. In the case $(A, B) = R$ we say that ideals A and B are *relatively prime*. The simplest properties of the greatest common divisor are contained in the following theorem:

    Theorem 9.11. *If R is a Dedekind ring, then*

(i)   *If $(A, B) = R$ and $A | BC$, then $A | C$.*

(ii)   *If A, B are ideals in R, then the inclusion $A \subset B$ holds iff B divides A.*

(iii) *If A and B are relatively prime ideals in R, then $AB = A \cap B$.*

(iv) *For any ideals A, B in R we have $(A, B) = A + B$.*

    Proof.

(i)   Follows immediately from Theorem 9.10.

(ii)   If $A = BC$, where C is an ideal in R, then obviously we have $A \subset B$ by the definition of the product of ideals. Conversely, if $A \subset B$, then

        $AB^{-1} \subset BB^{-1} = R$, so AB^{-1} is an ideal in R, and $B(AB^{-1}) = A$, so that B divides A.

(iii) By virtue of (ii), $A|A \cap B$ and $B|A \cap B$, hence by $(A, B) = R$ and (i), we have $AB|A \cap B$, i.e. $A \cap B \subset AB$. On the other hand, $AB \subset A \cap B$ from the definition of the product.

(iv) By virtue of (ii), $A + B$ divides (A, B); but on the other hand, $A + B$ is the smallest ideal containing A and B, and hence it must be divisible by (A, B). ∎

Corollary. *If $a \in R$, $a \neq 0$, then only finitely many ideals $I \subset R$ can contain a.*

Proof. If $a \in I$, $a \neq 0$, then $aR \subset I$, so by (ii) we have $I|aR$, and hence for I we have only a finite number of possibilities. ∎

4. We shall finish our study of Dedekind rings by sketching the theory of congruences in them, which is a generalization of the theory of congruences in the ring of rational integers, discussed in Chapter I.

If R is a Dedekind ring and $I \subset R$ is its ideal, then we shall write $a \equiv b \pmod{I}$ to mean that $a - b \in I$. Since the relation $a \equiv b \pmod{I}$ is a relation of equality of cosets in the quotient ring R/I, we may apply to congruences such operations as addition, subtraction or multiplication in the same way as in $\mathscr{Z}$.

The problem of solving a linear congruence is settled by the following lemma:

Lemma 9.11. *If R is a Dedekind ring, I its ideal and a, $b \in R$, then the congruence*

$$ax \equiv b \pmod{I}$$

has a solution iff the element b lies in the ideal $I + aR$.

Proof. If x satisfies this congruence, then $ax - b \in I$, and so $b \in I + aR$, which proves the necessity of the condition. The sufficiency is also easy to verify, for from $b \in I + aR$ follows the existence of $x \in R$ satisfying $b - ax \in I$, and so $b \equiv ax \pmod{I}$. ∎

Corollary 1. *If p is a prime ideal in a Dedekind ring R and $a \in R \setminus p$, then for any natural N the congruence*

$$ax \equiv b \pmod{p^N}$$

has a solution for any $b \in R$.

Proof. It suffices to observe that under our assumptions we have $p^N + aR = R$. ∎

Corollary 2 (Chinese remainder theorem). *If $I_1, \ldots, I_m$ are ideals in a Dedekind ring, such that for $i \neq j$ we have $(I_i, I_j) = R$ and the elements $a_1, \ldots, a_m \in R$ are given, then there exists a solution of the system of congruences*

(9.8) $$x \equiv a_i \pmod{I_i} \qquad (i = 1, 2, \ldots, n) \quad .$$

Proof. If

$$I_i = \prod_{j=1}^{k_i} p_{ij}^{a_{ij}} \quad ,$$

then the system (9.8) is equivalent to

$$x \equiv a_i \pmod{p_{ij}^{a_{ij}}} \qquad (i = 1, \ldots, m; \, j = 1, \ldots, k_i) \, ,$$

so that it is sufficient to prove that the system of congruences

$$x \equiv a_i \pmod{p_{ij}^N}$$

has a solution, where $N = \max\limits_{i,\,j} a_{ij}$.

For this purpose choose for $1 \leqslant r \leqslant m$, $1 \leqslant s \leqslant k_r$,

$$b_{rs} \in \prod_{\langle i,\,j \rangle \,\neq\, \langle r,\,s \rangle} p_{ij}^N \setminus p_{rs}$$

and denote by x_{rs} any solution of the congruence

$$b_{rs} x_{rs} \equiv a_r \pmod{p_{rs}^N} \quad .$$

Then the element

$$\sum_{i,\,j} b_{ij} x_{ij}$$

is a solution of (9.8).   ∎

Making use of the last corollary, we can prove a theorem on the structure of the quotient ring R/I:

Theorem 9.12. *If I is an ideal in a Dedekind ring R and*

$$I = \prod_p p^{a(p)}$$

is its decomposition into the product of powers of prime ideals, then

$$R/I \simeq \bigoplus_p R/p^{a(p)} \quad .$$

Proof. By f denote the homomorphism

$$R \rightarrow \bigoplus_p R/p^{a(p)}$$

given by

$$f : x \rightarrow \langle x \pmod{p^{a(p)}} \rangle_p \quad .$$

From the Chinese remainder theorem it follows that f is an epimorphism, and if x lies in its kernel $\operatorname{Ker} f$, then $x \equiv 0 \pmod{p^{a(p)}}$, and hence $x \in I$, whence $\operatorname{Ker} f \subset I$. Since the inclusion $I \subset \operatorname{Ker} f$ is trivial, hence $I = \operatorname{Ker} f$ and

$$R/I \simeq \bigoplus_p R/p^{a(p)} \quad . \qquad ∎$$

5. At present, the only example that we have seen of a Dedekind ring was the ring $\mathcal{Z}$ of rational integers, and from Lemma 9.6 we may conclude that every principal ideal domain is also a Dedekind ring. We shall now prove a theorem showing that all the rings R_K, i.e. the rings of integers in finite extensions K of the rational field are also Dedekind rings.

Theorem 9.13. *Let K be a finite extension of the rational field and R_K be the ring consisting of all integers contained in K. Then R_K is a Dedekind ring.*

Proof. It is necessary to check that the conditions (i), (ii), (iii) of Theorem 9.8 are satisfied. The easiest is the verification of the condition (i). Indeed, by Theorem 9.7 the additive group R_K is isomorphic with $\mathcal{Z}^n$, and so its every subgroup is finitely generated. Hence the ideals of R_K are finitely generated and R_K is a Noetherian domain.

To show that the condition (ii) is satisfied, we prove:

Lemma 9.12. *If P is a prime ideal in R_K, then the ideal $P \cap \mathcal{Z}$ of the ring $\mathcal{Z}$ is also a prime ideal. Moreover, if $P \subset P_1$, where P_1 is a prime ideal and $P \cap \mathcal{Z} = P_1 \cap \mathcal{Z}$, then $P = P_1$.*

Proof. Observe that the image of the ideal $P \cap \mathcal{Z}$ under the embedding of $\mathcal{Z}$ into R_K lies in P, and so this embedding induces a homomorphism $\mathcal{Z}/P \cap \mathcal{Z} \to R_K/P$ which is also an embedding. But R_K/P does not have any zero divisors, so that $\mathcal{Z}/P \cap \mathcal{Z}$ does not have them, either, and we see that $P \cap \mathcal{Z}$ is a prime ideal. If $P \subset P_1$, $P \neq P_1$ and $P \cap \mathcal{Z} = P_1 \cap \mathcal{Z}$, then let x be any element of $P_1 \backslash P$ and let

$$X^n + a_{n-1} X^{n-1} + \ldots + a_0$$

be the minimal polynomial for x over $\mathcal{Q}$ with coefficients in $\mathcal{Z}$. If all the coefficients a_i lie in $P \cap \mathcal{Z}$, then by

$$x^n = -(a_{n-1} x^{n-1} + \ldots + a_0) \in P$$

we would have $x \in P$, contrary to the hypothesis. Let j be the smallest index such that $a_j \notin P \cap \mathcal{Z}$. Then

$$x^j(x^{n-j} + a_{n-1} x^{n-j-1} + \ldots + a_j) \in P \ ,$$

and so

$$x^{n-j} + a_{n-1} x^{n-j-1} + \ldots + a_j \in P \subset P_1 \quad ,$$

but the element x lies in P_1, whence we obtain

$$x^{n-j} + a_{n-1} x^{n-j-1} + \ldots + a_{j+1} x \in P_1$$

and we see that $a_j \in P_1 \cap \mathcal{Z} = P \cap \mathcal{Z}$ contrary to the choice of j. $\blacksquare$

Now let P be any prime ideal in R_K. If it is not a maximal ideal, then a maximal ideal P_1 containing P would be different from P. Then $P \cap \mathscr{Z} \subset P_1 \cap \mathscr{Z}$ would be two prime ideals in $\mathscr{Z}$, but this situation is possible only in the case $P \cap \mathscr{Z} = P_1 \cap \mathscr{Z}$, which contradicts the lemma 9.12. Hence P is a maximal ideal and the condition (ii) is satisfied.

It remains to check that the condition (iii) is satisfied. Here again we shall need a simple lemma concerning integral domains:

Lemma 9.14. *If* R *is an integral domain contained in the field* K, L *an extension of* K, *and* S *is the integral closure of* R *in* L, *then* S *is an integrally closed domain.*

Proof. Let c be any element of L, integral over S and let $a_0 + a_1 x + \ldots + a_{n-1} x^{n-1} + x^n$ be a polynomial with coefficients in S of which c is a root. Put $R_{-1} = R$, and for $i = 0, 1, \ldots, n-1$ let R_i be the ring generated by R and $a_0, \ldots, a_i$. Note that R_i is a finitely generated R_{i-1}-module, for $i = 0, 1, \ldots, n-1$. Indeed, we have $R_i = R_{i-1}[a_i]$, $a_i \in S$, so that a_i is an element integral over R and a fortiori over R_i. From the implication (i) $\to$ (ii) of Lemma 9.3 it follows that R_{n-1} is a finitely generated R-module, for, if $A \subset B \subset C$ are rings, B is finitely generated as an A-module and C is finitely generated as a B-module, then C is finitely generated as an A-module. Moreover, the element c is integral over R_{n-1}, and hence, according to Lemma 9.3, $R_{n-1}[c]$ is a finitely generated R_{n-1}-module. Hence it is also a finitely generated R-module. Since it is non-zero and $cR_{n-1}[c] \subset R_{n-1}[c]$, it follows from Lemma 9.3 that c is integral over R, i.e. $c \in S$. ∎

That the condition (iii) is satisfied is an immediate consequence of the above lemma and hence we see that R_K is actually a Dedekind ring. ∎

6.   The ring $\mathscr{Z}$ of rational integers has the property that every of its non-trivial epimorphic images, i.e. every quotient ring $\mathscr{Z}/N\mathscr{Z}$ (with $N \neq 0$), is finite. We shall show that the same is true of each of the rings R_K. (Here it is worth noting that it is not so in every Dedekind ring, for if e.g. R is a polynomial ring over any infinite field K, then in it every ideal is principal, and so it is a Dedekind ring. Here for any non-constant polynomial $f \in R$, the ring R/fR contains the isomorphic image of K, and hence is infinite.

Theorem 9.14. *If* R_K *is the ring of integers in some finite extension field of* $\mathscr{Q}$ *and* $I \subset R_K$ *is its any non-zero ideal, then the quotient ring* R_K/I *is finite.*

Moreover, if $N(I)$ *denotes the number of elements of* R_K/I, *then* $N(AB) = N(A) N(B)$.

Proof. By theorem 9.12 it suffices to prove the assertion for ideals which are powers of a prime ideal. Now observe that if $I = P$ is a prime ideal and $P \cap \mathscr{Z} = p\mathscr{Z}$, then the field R_K/P is an extension of the field $\mathscr{Z}/p\mathscr{Z}$. If $a \in R_K$ generates the field K over $\mathscr{Q}$, i.e. $K = \mathscr{Q}(a)$, and a' is the image of a in R_K/P, then $R_K/P = \mathscr{Z}/p\mathscr{Z}(a')$, and moreover, if $F(x) \in \mathscr{Z}[x]$ is the minimal polynomial for a over $\mathscr{Q}$, then the polynomial arising from $F(x)$, by replacing its coefficients by their residues (mod p) has a root a', so that the field R_K/P is a finite extension of the field $\mathscr{Z}/p\mathscr{Z}$ of p elements and is itself finite.

There remains the case where $I = P^n$ ($n \geqslant 2$) and the ideal P is prime. The following lemma enables us to reduce it to the case considered above:

Lemma 9.15. *If P is a non-zero prime ideal in a Dedekind ring R and n is an arbitrary natural number, then the additive groups of the rings R/P and P^n/P^{n+1} are isomorphic.*

Proof. Fix arbitrarily an element $a \in P^n \setminus P^{n+1}$ and consider the homomorphism $g: x \to ax$ of the additive group R into the additive group P^n. Since $g(P) \subset P^{n+1}$, g induces a homomorphism h of the additive group A of the ring R/P into the additive group B of the ring P^n/P^{n+1}. Let us show that h is an isomorphism. Indeed, if x lies in the kernel of h and y is a representative of the coset x in R, then $ay \in P^{n+1}$, so that $y \in P$ and we obtain $x = 0$, i.e. h is a monomorphism. To prove that it is an epimorphism, we choose any element $c \in B$ and let $c' \in P^n$ be any representative of c. From Lemma 9.11 it follows that there exists an $x \in R$ satisfying $ax \equiv c' \pmod{P^{n+1}}$, and then the coset $x \pmod P \in R/P$ satisfies $h(x \pmod P) = c$. ∎

From this lemma it follows that if R/P^n is a finite ring for some n, then the ring R/P^{n+1} is finite, for

$$(R/P^{n+1}) / (R/P^n) \simeq P^n/P^{n+1} \quad ,$$

and moreover, $N(P^{n+1}) = N(P^n) N(P)$. Easy induction leads to $N(P^n) = N(P)^n$, and using Theorem 9.12, we obtain the equality $N(AB) = N(A) N(B)$ for any ideals A, B. ∎

We call the number $N(A)$ the *norm of an ideal* A.

7. If R is an arbitrary commutative ring, then every ideal $I \subset R$ can be treated as an R-module. A natural question arises, when two ideals of this ring will be isomorphic as R-modules. The answer in the case of integral rings is not difficult and is contained in the following lemma:

Lemma 9.16. *For two ideals I_1, I_2 of an integral ring R to be isomorphic with each other as R-modules it is necessary and sufficient that there exists a non-zero element a of the field K of fractions of R for which $I_1 = aI_2$.*

Proof. If $I_1 = aI_2$ and a is non-zero, then the mapping given by $x \to ax$ gives a required isomorphism. Conversely, if $F: I_1 \to I_2$ is an isomorphism, then for any x, y of I_1 we have $xf(y) = f(xy) = yf(x)$, so that for $x \neq 0$, the ratio $a = f(x)x^{-1}$ does not depend on x and $f(x) = ax$ holds for all x in I_1 (since this equality holds also for $x = 0$). ∎

Corollary 1. *For two fractional ideals A_1, A_2 to be isomorphic as R-modules it is necessary and sufficient that there exists an element a such that $A_1 = aA_2$.*

Proof. Let $c_i \in R$ be chosen so that $c_i \neq 0$ and $c_i A_i = I_i \subset R$. Then A_i is isomorphic with I_i, so that we can apply the preceding lemma. ∎

Corollary 2. *If* I_1, I_2, J_1, J_2 *are given fractional ideals and* $I_1 \simeq J_1$, $I_2 \simeq J_2$, *then* $I_1 I_2 \simeq J_1 J_2$.

Proof. We have $J_i = a_i I_i$ $(i = 1, 2)$ with some $a_i \in K$, $a_i \neq 0$, and hence $J_1 J_2 = a_1 I_1 a_2 I_2 = a_1 a_2 I_1 I_2 \simeq I_1 I_2$. ∎

Now we divide all ideals $\neq 0$ of a ring R into classes, putting two ideals in the same class if they are isomorphic as R-modules. Since from Corollary 1 to Lemma 9.16 and the definition of a fractional ideal, it follows that for every fractional ideal of R, there exists an ideal of R isomorphic with it. We consider the corresponding division of the set of all non-zero fractional ideals.

Denote the set of these classes by Ω.

Theorem 9.15. *If* R *is a Dedekind ring, then it is possible to introduce the structure of a group into the set* Ω *in the following way: If* $X_1, X_2 \in \Omega$ *and* I_1, I_2 *are ideals of* R *satisfying* $I_i \in X_i$ $(i = 1, 2)$, *then the product* $X_1 X_2$ *is the class containing the product* $I_1 I_2$.

This group is commutative and its unit element is the class E *consisting of all principal ideals of* R.

Proof. The product defined above does not depend on the choice of ideals I_1, I_2 in classes X_1, X_2 in view of Corollary 2 to Lemma 9.16. Associativity and commutativity of the operation are obvious, and moreover for any class $X \in \Omega$, we have $EX = XE = X$ because $IR = RI = I$. It remains to prove the existence of the inverse element. Let I be an arbitrary ideal of R lying in a given class X. Then the ideal I^{-1} is a fractional ideal, but with a suitably chosen $c \in R$, cI^{-1} is an ideal in R. If it lies in the class Y then $XY = E$ in view of $I \cdot cI^{-1} = cR \in E$. ∎

If we denote by G the group of all fractional ideals and by P its subgroup consisting of all principal fractional ideals, i.e. ideals of the form cR, then the set Ω with the multiplication introduced above is clearly a group isomorphic to the quotient group G/P.

8.   Now let K be any finite extension of the rational field and R_K be its ring of integers. We may apply the theorem proved above to this ring and obtain in this way a group which is called the *ideal class group* of the field K and denoted by $H(K)$. As we shall see below, the structure of this group has an essential arithmetic meaning in the ring R_K and the investigation of these ideal class groups for various fields has been one of the main subjects of research in algebraic number theory recently.

It happens that the ideal class group of a field K is always a finite group. We prove this result by the method of A. Hurwitz:

Theorem 9.16. *The group* $H(K)$ *is finite.*

Proof. We begin with a simple lemma:

Lemma 9.17. *There exists a constant* $B(K)$ *depending solely on the field* K *with the property that for every element* a *of this field one can find an integer* x *of this field and a rational integer* r, $1 \leqslant r \leqslant B(K)$ *for which*

$$|N_K(ra - x)| < 1 \ .$$

Proof. Denote by n the degree of K and choose any integral basis $w_1, \ldots, w_n$ in K. Let $\varphi_1, \ldots, \varphi_n$ be embeddings of K into $\mathfrak{Z}$ and let $w_j^{(i)} = \varphi_i(w_j)$ $(i, j = 1, 2, \ldots, n)$. Finally let

$$C = \prod_{i=1}^{n} \sum_{j=1}^{n} |w_j^{(i)}|$$

and let k be a natural number greater than $C^{1/n}$. For $r = 0, 1, 2, \ldots, k^n$ we can choose a number $b_r \in R_K$ so that in the expression

$$ra - b_r = \sum_{j=1}^{n} a_{jr} w_j \qquad (a_{jr} \in \mathfrak{Q})$$

the coefficients a_{jr} lie in the interval $[0, 1)$. Now we divide the n-dimensional unit cube $[0, 1)^n$ into k^n cubes of side $1/k$. With each integer $0 \leqslant r \leqslant k^n$ we associate that one among them which contains the point $\langle a_{1r}, \ldots, a_{nr} \rangle$. Then there must exist two integers with which the same cube is associated. Their difference has the form

$$ma - b = \sum_{j=1}^{n} A_j w_j \ ,$$

where $0 \leqslant m \leqslant k^n$, $b \in R_K$, $A_j \in \mathfrak{Q}$ and $|A_j| < \dfrac{1}{k}$ $(j = 1, 2, \ldots, n)$. Moreover,

$$|N_K(ma - b)| = \left| \prod_{i=1}^{n} \sum_{j=1}^{n} A_j w_j^{(i)} \right| \leqslant \prod_{i=1}^{n} \sum_{j=1}^{n} |A_j| \, |w_j^{(i)}|$$

$$\leqslant \prod_{i=1}^{n} \frac{1}{k} \sum_{j=1}^{n} |w_j^{(i)}| = \frac{C}{k^n} < 1 \ .$$

Taking $B(K) = k^n$, we obtain the assertion of the lemma. ∎

In order to complete the proof of the theorem, we show now that for each ideal I of the ring R_K, one can find an ideal J, isomorphic with I as an R_K-module that contains the natural number $B(K)!$. This will be sufficient for us because, by Theorem 9.11, any given number $a \neq 0$ can be contained only in finitely many ideals of R_K.

Hence let I be any ideal in R_K and let a be an element of I with the smallest non-zero absolute value of the norm. If $b \in I$, then, applying the lemma proved above to the quotient b/a, we find a number $0 \leqslant r \leqslant B(K)$ and an element $c \in R_K$ such that

$$|N_K(r\frac{b}{a} - c)| < 1 \ ,$$

so that

$$|N_K(rb - ca)| < |N_K(a)|$$

Since $rb - ca \in I$, we must have, by the choice of a, $rb - ca = 0$ and

$$aR_K \,|\, rbR_K \,|\, B(K)! \, bR_K \,|\, B(K)! \, IR_K \quad,$$

and hence there exists an ideal J such that

$$B(K)! I = aJ \quad.$$

This ideal is isomorphic with I, and moreover, by

$$B(K)! \, a \in B(K)! I = aJ \quad,$$

we have $B(K)! \in J$, which, as we have seen above, is sufficient for the proof of the theorem. ∎

Remark. If $K = \mathcal{Q}(\sqrt{D})$ $(D \neq 1,\ \mu(D) \neq 0)$ is a quadratic field and $D \not\equiv 1 \pmod 4$, then obviously we have $C = (1 + |D|^{\frac{1}{2}})^2$, k is the smallest natural number greater than $1 + |D^{\frac{1}{2}}|$ and $B(K) = k^2$.

We call the number of elements of the group $H(K)$ the *class number* of the field K and denote it by $h(K)$. We shall show that the value of this number has an effect on the arithmetic properties of the ring R_K.

Let us recall fundamental definitions concerning the unique factorization domains. We say that in an integral domain R the theorem of unique factorization holds if the following two conditions are satisfied:

(i)   Every non-zero element $a \in R$ which is not a unit can be expressed in the form of a product of irreducible elements.

(ii)   If $a = b_1 \ldots b_m = c_1 \ldots c_n$ are two such representations, then $m = n$ and moreover, the elements c_i can be renumbered so that for every i, the elements b_i and c_i are associated that is, differ from each other by unit factors.

In the fundamental course of algebra it is proved that if in a ring R, every ideal is principal, and in particular, if in R the Euclid algorithm holds, then the theorem of unique factorization holds in R. Using this fact, we shall now prove the following theorem:

Theorem 9.17.   *For the theorem on unique factorization to hold in a ring R it is necessary and sufficient that $h(K) = 1$.*

Proof.   Sufficiency of the condition follows from the remark that $h(K) = 1$ is equivalent to the fact that every ideal of R_K is principal.

To prove its necessity let us suppose that in the ring R_K the theorem on unique factorization holds and that $h(K) \neq 1$. First we shall show that one can find infinitely many prime ideals that are not principal ideals. Indeed, suppose that $p_1, \ldots, p_M$ are all the prime ideals, non-principal in R_K. Here we have $M \geqslant 1$, for if all the prime ideals in our ring are principal, then by virtue of Theorem 9.8 all the ideals in it would be principal, contrary to the assumptions. Choose for each $i = 1, \ldots, M$ an element a_i lying in $p_i \setminus p_i^2$ and put $Q_i = p_1 \ldots p_M / p_i$. If now x_i is a solution of the system of congruences

$$x_i \equiv a_i \pmod{p_i^2} \quad ,$$

$$x_i \equiv 1 \pmod{Q_i} \quad ,$$

then clearly,

$$x_i R = p_i I_i \quad ,$$

where the ideal I_i is relatively prime to Q_i. If I_i were not divisible by any non-principal prime ideal, then it would be principal, and hence p_i would be principal. Hence we found a non-principal prime ideal, distinct from all the ideals $p_1, \ldots, p_M$.

From the remark that the class number $h(K)$ is, by the preceding theorem, finite, we can choose a class X from the group $H(K)$, different from the class E which contains infinitely many prime ideals. Let g be the smallest natural number such that $X^g = E$. Then the ideals $p_1 \ldots p_g$ and p_i^g (with $p_i \in X$) are principal ideals, and if a, a_i are their generators, then they are irreducible elements, and moreover, the element $a_1 \ldots a_g$ has two distinct decomposition into irreducible elements, viz. $a_1 \ldots a_g = a^g$, which contradicts our assumption. ∎

9.   We shall close this section by proving a theorem concerning the decomposition of ideals pR_K into the product of prime ideals, where p is any rational prime and K a given quadratic field. Then we shall apply this to the practical determination of the class number of a given quadratic field.

Theorem 9.18. *If K is a quadratic field and d is its discriminant, then for every rational prime p we have*

$$pR_K = \begin{cases} p^2 , & \text{if } p \mid d \quad , \\[2mm] P_1 P_2 , & \text{if } \left(\dfrac{d}{p} \right) = +1 , \\[2mm] p , & \text{if } \left(\dfrac{d}{p} \right) = -1 , \end{cases}$$

where P, P_1, P_2 are prime ideals in R_K and furthermore, $P_1 \neq P_2$.

Proof. We begin with a simple lemma:

Lemma 9.18. *If K is an extension of the field of rational numbers of degree n and $m \in \mathcal{Z}$ is different from zero, then $N(mR_K) = |m|^n$.*

Remark. This is a special case of a general result: If a is a non-zero element of R_K, then $N(aR_K) = |N_K(a)|$. However, we shall not prove this here. (See Problem 2 at the end of the section).

Proof. Let $w_1, \ldots, w_n$ be an integral basis of the field K. We shall show that each coset of $R_K / m R_K$ contains an element of the form

$$a_1 w_1 + \ldots + a_n w_n \qquad (a_i \in \mathcal{Z} , \quad 0 \leqslant a_i \leqslant m-1) \quad ,$$

and that distinct elements of this form lie in distinct cosets. Indeed, if

$$b = b_1 w_1 + \ldots + b_n w_n \qquad (b_i \in \mathbb{Z})$$

is any element of a given coset and a_i is a residue of $b \pmod m$ lying in the interval $(0, m-1]$, then

$$\xi = \sum_{i=1}^{n} (b_i - a_i) w_i \in m R_K$$

and

$$\sum_{i=1}^{n} a_i w_i = b - \xi \in b + m R_K \ .$$

If $\sum_{i=1}^{n} a_i' w_i$ is another element with this property, then

$$\sum_{i=1}^{n} (a_i - a_i') w_i \in m R_K \ ,$$

i.e.

$$\sum_{i=1}^{n} \frac{a_i - a_i'}{m} w_i \in R_K$$

and we obtain

$$a_i \equiv a_i' \pmod m \qquad (i = 1, 2, \ldots, n) \ ,$$

so that, by $0 < a_i$, $a_i' \leq m - 1$ we get $a_i = a_i'$. ∎

Corollary. *If K is an extension of the field of rational numbers of degree n and p is a prime number, then in the decomposition of the ideal pR_K into the product of prime ideals there appear at most n factors and the norm of each factor is a power of p.*

Proof. If $pR_K = P_1 \ldots P_t$, then

$$p^n = N_K(p) = N_K(pR_K) = N(P_1) \ldots N(P_t) \ ,$$

whence follows that for suitable $c_i \in \mathbb{Z}$, $c_i \geq 1$ $(i = 1, 2, \ldots, t)$, $N(P_i) = p^{c_i}$, and $c_1 + \ldots + c_t = n$, and so $t \leq n$. ∎

From the above corollary it follows that in the case of a quadratic field the following three possibilities arise:

a) $pR_K = P^2$, $N(P) = p$,

b) $pR_K = P_1 P_2$, $P_1 \neq P_2$, $N(P_1) = N(P_2) = p$,

c) $pR_K = P$, $N(P) = p^2$.

In order to examine which of those possibilities holds for a given prime number p we shall need a lemma which enables us to check, whether the ideal pR_K is a prime ideal.

Lemma 9.19. *Let* $K = \mathcal{Q}(D^{\frac{1}{2}})$ *be a quadratic field,* $D \neq 1$, $\mu(D) \neq 0$ *and* p *is a prime number. The ideal* pR_K *is a principal ideal in* R_K *iff for any* $a, b \in \mathcal{Z}$ *not both divisible by* p, *the following system of congruences is solvable:*

In the case $D \equiv 1 \,(\mathrm{mod}\, 4)$, *under the notation* $\delta = \dfrac{D-1}{4}$

$$ax + b\delta y \equiv 1 \,(\mathrm{mod}\, p) \quad,$$

(9.9)
$$bx + (a+b)y \equiv 0 \,(\mathrm{mod}\, p) \quad,$$

and in the case $D \equiv 1 \,(\mathrm{mod}\, 4)$

$$ax + bDy \equiv 1 \,(\mathrm{mod}\, p) \quad,$$

(9.10)
$$bx + ay \equiv 0 \,(\mathrm{mod}\, p) \quad.$$

If this system has no solution for some a, b *not both divisible by* p, *then the ideal* $pR_K + (a + bw)R_K$ *is a prime ideal dividing* pR_K. *Here*

$$
w = \begin{cases}
\sqrt{D} & \text{when } D \not\equiv 1 \,(\mathrm{mod}\, 4) \quad, \\[2mm]
\dfrac{1+\sqrt{D}}{2} & \text{when } D \equiv 1 \,(\mathrm{mod}\, 4) \quad.
\end{cases}
$$

Proof. By Theorem 9.13 every non-zero prime ideal in R_K is maximal and the ideal pR_K will be maximal iff for any element $c \notin pR_K$ we have

$$1 \in pR_K + cR_K \quad,$$

since, then, $pR_K + cR_K = R_K$. The elements $1, w$ form an integral basis of K. If $c = a + bw \notin pR_K$, then at least one of the elements a, b is not divisible by p and clearly, *vice versa*. Hence the ideal pR_K will be a maximal ideal iff for any such c with some $x, y, u, v \in \mathcal{Z}$ the equality

(9.11)
$$1 = (x + yw)(a + bw) + p(u + vw)$$

will hold, i.e.

$$1 = ax + pu + (ay + bx + pv)w + byw^2 \quad,$$

which, in view of $w^2 = D$ in the case $D \not\equiv 1 \,(\mathrm{mod}\, 4)$, and $w^2 = w + \delta$ in the case $D \equiv 1 \,(\mathrm{mod}\, 4)$, takes the form

$$1 = (ax + byD + pu) + (ay + bx + pv)w \qquad (D \not\equiv 1 \,(\mathrm{mod}\, 4)) \quad,$$

$$1 = (ax + by\delta + pu) + (ay + bx + by + pv)w \qquad (D \equiv 1 \,(\mathrm{mod}\, 4))$$

and it is not difficult to observe that it holds iff the system of congruences (9.9), (9.10) is satisfied.

Conversely, if this system has no solutions for some a, b not both divisible by p, then (9.11) cannot hold, so that the ideal $P = pR_K + (a + bw)R_K$ is different from R_K and contains pR_K. Its norm $N(P)$ divides $N(pR_K) = p^2$, so we must have $N(P) = p$ and we see that P is a prime ideal. ∎

Corollary. *If p is an odd prime number not dividing the discriminant $d = d(K)$, then the ideal pR_K is a prime ideal iff $\left(\dfrac{d}{p}\right) = -1$. If, on the contrary, $p = 2 \nmid d$, then the ideal pR_K is a prime ideal iff $d \not\equiv 1 \pmod 8$.*

Proof. We treat the congruences (9.9), (9.10) as systems of linear equations in the field of p elements. In the case $D \equiv 1 \pmod 4$ we are concerned with the system (9.9) with determinant $a^2 + ab - \delta b^2$. For $p \neq 2$, the equation $X^2 + Xb - \delta b^2 = 0$ has a solution in the field of p elements iff its discriminant $b^2(1 + 4\delta) = Db^2$ is a square, i.e. $\left(\dfrac{D}{p}\right) = 1$. Hence, if $\left(\dfrac{D}{p}\right) = -1$, then the determinant of the system (9.9) does not vanish $\pmod p$, and so the system is consistent and pR_K is a maximal ideal. In case $\left(\dfrac{D}{p}\right) = 1$ we denote by a a solution of the congruence $X^2 + X - \delta \equiv 0 \pmod p$. For a so chosen and $b = 1$ the determinant of (9.9) equals $0 \pmod p$ and it is easily seen that this system is inconsistent. Therefore pR_K is not a maximal ideal. If, on the contrary, $p = 2$, $d \equiv 1 \pmod 8$, then δ is even, the determinant of (9.9) is equal to $a^2 + ab \pmod 2$, and we see that for $a = b = 1$ the system (9.9) is inconsistent. On the other hand, for $d \equiv 5 \pmod 8$ we have $2 \nmid \delta$ and hence the determinant of (9.9) equals $a^2 + ab - b^2 \pmod 2$, i.e. it is congruent to zero $\pmod 2$ exclusively in the case when both the numbers a, b are even. Thus we see that for $D \equiv 1 \pmod 4$ the assertion of the theorem is true.

We now go on to the case $D \not\equiv 1 \pmod 4$. The argument here proceeds analogously although we shall consider the system (9.10) with determinant equal to $a^2 - Db^2$. Since $d = 4D$ and $p \nmid d$, p is an odd number. In the case $\left(\dfrac{d}{p}\right) = \left(\dfrac{D}{p}\right) = -1$, we have $p \nmid a^2 - Db^2$ for an arbitrary choice of the numbers a, b not both divisible by p and in this case pR_K is a prime ideal. If, on the contrary, $\left(\dfrac{d}{p}\right) = \left(\dfrac{D}{p}\right) = 1$, then we find an a satisfying the congruence $a^2 \equiv D \pmod p$, and then for this a and $b = 1$ the determinant of (9.10) is equal to zero $\pmod p$ and the system (9.10) is inconsistent. Hence pR_K is not a prime ideal. ∎

For the proof of the theorem it now suffices to show that in the case $p \mid d$ we have $pR_K = P^2$ and in the case $\left(\dfrac{d}{p}\right) = +1$ we have $pR_K = P_1 P_2$. First let p be a divisor of D. Then from $P \mid pR_K$ it follows that

$$P \mid pR_K \mid DR_K = D^{\frac{1}{2}} R_K D^{\frac{1}{2}} R_K \quad,$$

and therefore P must divide $D^{\frac{1}{2}} R_K$ and we see that P^2 divides DR_K. Since the ideals generated by distinct prime numbers cannot have common divisors, hence pR_K is divisible by P^2, but from the equality of norms of these ideals follows their coincidence and finally we obtain $pR_K = P^2$.

Now let p divide d, but $p \nmid D$. This can occur only in the case where $D \not\equiv 1 \,(\text{mod } 4)$ and $p = 2$. Note that the ideal $P = 2pR_K + (1 + D^{\frac{1}{2}})R_K$ contains pR_K, is different from it and $P \neq R_K$ because $1 \notin P$, for, from the equality

$$1 = (x + y\sqrt{D})(1 + \sqrt{D}) + 2(u + v\sqrt{D})$$

it would follow that $x + 2u + yD = 1$ and $x + y + 2v = 0$, and so $1 \equiv x + yD \equiv x + y \equiv 0 \,(\text{mod } 2)$. Hence $2R_K$ is divisible by P, and the norm of P must be 2. Let us show that $P^2 = 2R_K$. Indeed,

$$P^2 = 4R_K + 2(1 + D^{\frac{1}{2}})R_K + (1 + D + 2D^{\frac{1}{2}})R_K \subset 2R_K \quad,$$

since $1 + D \equiv 0 \,(\text{mod } 2)$ and in view of $N(P^2) = N(2R_K) = 4$ we obtain $2R_K = P^2$.

Now let $\left(\dfrac{d}{p}\right) = 1$. From the Corollary to Lemma 9.19 it follows that pR_K is not a prime ideal, and so we must have $pR_K = P_1 P_2$ and it is only necessary to show that $P_1 \neq P_2$. From the proof of the quoted Corollary an explicit form of the ideal P_1 results, namely

$$P_1 = \begin{cases} pR_K + (a + w)R_K & \text{if} \quad D \equiv 1\,(\text{mod } 4),\ p \neq 2,\ a^2 + a - \dfrac{D-1}{4} \equiv 0\,(\text{mod } p) \quad, \\[2mm] 2R_K + (1 + w)R_K & \text{if} \quad D \equiv 1\,(\text{mod } 4),\ p = 2 \quad, \\[2mm] pR_K + (a + w)R_K & \text{if} \quad D \not\equiv 1\,(\text{mod } 4),\ a^2 \equiv D\,(\text{mod } p) \quad, \end{cases}$$

where $w = \sqrt{D}$ when $D \not\equiv 1\,(\text{mod } 4)$ and $w = \dfrac{1 + \sqrt{D}}{2}$ when $D \equiv 1\,(\text{mod } 4)$.

It remains to check that $P_1^2 \neq pR_K$, but this easily follows from the observation that in the respective cases we have

$$P_1^2 = \begin{cases} p^2 R_K + p(a + w)R_K + \left(a^2 + \dfrac{D-1}{4} + (2a+1)w\right) R_K \quad, \\[2mm] 4R_K + 2(1 + w)R_K + \left(1 + 3w + \dfrac{D-1}{4}\right) R_K \quad, \\[2mm] p^2 R_K + p(a + w)R_K + (a^2 + D + 2aw)R_K \quad, \end{cases}$$

and the numbers

$$a^2 + \frac{D-1}{4} + (2a+1)w, \qquad 1 + 3w + \frac{D-1}{4}, \qquad a^2 + D + 2aw$$

lie in the corresponding ideals P_1^2, not belonging, however, to pR_K.　■

10.　We shall apply the theorem obtained above to the determination of the ideal class group $H(K)$ of some quadratic fields. The method given by us can be applied in principle for any quadratic field, but the necessary calculations are rather cumbersome already for fields with small discriminants.

From the proof of Theorem 9.16 it follows that in each ideal class one can find an ideal dividing the ideal $B(K)!R_K$, where $B(K)$ is defined in Lemma 9.17. Note that every such ideal is a product of prime ideals dividing $B(K)!R_K$, and hence the group $H(K)$ is generated by classes containing such prime ideals. Every one of such ideals contains some prime number p not exceeding $B(K)$, since R_K/P is a field, and so $N(P)$ is a power of a prime p. Moreover, we have $p^k = N(P) \in P$ because the additive group R_K/P contains $N(P)$ elements, and so for any $a \in R_K$ the image $N(P)a$ in R_K/P lies in P, and this in particular, holds for $a = 1$ implying $p \in P$. Finally we see that it is necessary to determine classes containing prime ideals dividing pR_K, where p ranges over prime numbers not greater than $B(K)$.

Let us begin with the field $\mathscr{Q}(2^{\frac{1}{2}})$. Here $D = 2$, $d = 8$ and from the proof of Lemma 9.17 we easily obtain $B(K) = 9$. Hence we need to consider the primes $2, 3, 5, 7$. In view of $(\frac{2}{3}) = -1$, $(\frac{2}{5}) = -1$, $(\frac{2}{7}) = 1$ the preceding theorem gives the decompositions

$$2R_K = P_2^2, \qquad 3R_K = P_3, \qquad 5R_K = P_5, \qquad 7R_K = P_7 Q_7.$$

The ideals P_3 and P_5 are clearly principal, and so is the ideal P_2 since it is equal to $2^{\frac{1}{2}}R_K$ and there remains to check whether the ideal P_7 is principal or not, i.e. whether there exists in R_K an element with norm ± 7. But the equation $x^2 - 2y^2 = 7$ has a solution $x = 3$, $y = 1$ and we find that P_7 is a principal ideal, and hence $h(K) = 1$.

Now consider the field $\mathscr{Q}\big((-5)^{\frac{1}{2}}\big)$. Here $D = -5$, $d = -20$, $B(K) = 16$ and we have to check the prime numbers $p = 2, 3, 5, 7, 11, 13$. Considering the Legendre symbols, we can easily find decompositions

$$2R_K = P_2^2, \qquad 3R_K = P_3 Q_3, \qquad 5R_K = P_5^2,$$

$$7R_K = P_7 Q_7, \qquad 11R_K = P_{11}, \qquad 13R_K = P_{13}.$$

The ideal P_5 is equal to $(-5)^{\frac{1}{2}}R_K$, and so it is principal, but the ideals P_2, P_3, Q_3, P_7 and Q_7 are not because the equation

$$x^2 + 5y^2 = p$$

has no solution for $p = 2, 3$, and 7. We shall show that all these ideals belong to the same class.

From the proof of Lemma 9.19 it follows that we may put

$$P_2 = 2R_K + (1 + \sqrt{-5})R_K , \qquad P_3 = 3R_K + (1 + \sqrt{-5})R_K .$$

We shall show that the product $P_2 P_3$ is a principal ideal generated by $1 + (-5)^{\frac{1}{2}}$.

Indeed, the number $1 + \sqrt{5}$ lies in $P_2 P_3 = P_2 \cap P_3$ and

$$P_2 P_3 = 6R_K + 2(1 + \sqrt{-5})R_K + 3(1 + \sqrt{-5})R_K + (1 + \sqrt{-5})^2 R_K$$

and all the numbers on the right-hand side lie in $(1 + \sqrt{-5})R_K$.

If X is the class in which P_2 lies, then $X^2 = E$, the unit class and if Y is the class in which P_3 lies, then $XY = E$, so that $X = Y$ and moreover Q_3 also belongs to X. Now consider P_7. We may assume that

$$P_7 = 7R_K + (3 + \sqrt{-5})R_K .$$

The number $3 + \sqrt{-5}$ lies in $P_2 P_7 = P_2 \cap P_7$ and moreover,

$$P_2 P_7 = 14R_K + 2(3 + \sqrt{-5})R_K + 7(1 + \sqrt{-5})R_K + (1 + \sqrt{-5})(3 + \sqrt{-5})R_K ,$$

but the numbers 14, $2(3 + \sqrt{-5})$, $7(1 + \sqrt{-5})$, $(1 + \sqrt{-5})(3 + \sqrt{-5})$ are divisible by $3 + \sqrt{-5}$, and this gives the equality

$$P_2 P_7 = (3 + \sqrt{-5})R_K .$$

Hence, if A is the class in which P_7 lies, then $XA = E$, so $A = X$ and we see that $H(K) = \{E, X\}$, i.e. $h(K) = 2$.

The method presented in the above examples of finding the class number can be improved by using the following results that will not be proved in this book:

(i) If $a \in R_K$, then $N(aR_K) = |N_K(a)|$.

(ii) *Minkowski's theorem: In each ideal class in R_K there is an ideal with norm not exceeding.*

$$\frac{n^n}{n!} \left(\frac{4}{\pi}\right)^{r_2} |d(K)|^{\frac{1}{2}} ,$$

where n is the degree of the field K and $2r_2$ is the number of embeddings of the field K into the field of complex numbers whose images are not contained in the field of real numbers.

The use of Eq. (ii) leads to a remarkable decrease in the number of primes that are to be examined. Thus, for example, from this it follows immediately that the quadratic fields with discriminants -3, -4, -7, -8, 5, 8, 12, 13 have class number equal to 1. The usefulness of (i) becomes clear in checking whether two given ideals lie in the same class. In the example of the field $\mathcal{Q}\left((-5)^{\frac{1}{2}}\right)$ considered above we had to show that the ideal $P_2 P_7$ is principal and we did this by making use of an explicit form of the ideals considered. Theorem (i) enables us to omit it due to the following reasoning:

The unique ideals with norm 14 are $P_2 P_7$ and $P_2 Q_7$ and it suffices to find two principal ideals with norm 14 to be able to assert that $P_2 P_7$ is principal. But this amounts to finding two numbers generating distinct ideals with norm 14. Since $N(x + y(-5)^{\frac{1}{2}}) = x^2 + 5y^2$, it is sufficient to consider the numbers $3 + (-5)^{\frac{1}{2}}$ and $-3 + (-5)^{\frac{1}{2}}$ which generate distinct ideals because neither of them is divisible by the other.

The reader who is interested in algebraic number theory can study it by using the book of Z. I. Borević and I. R. Šafarević [1], or the author's book "Elementary and analytic theory of algebraic numbers", Warsaw (1974).

Exercises

1.   Prove that if I is an ideal in R_K and $w_1, \ldots, w_n$ is an integral basis of K, then there exist elements $v_1, \ldots, v_n \in I$ such that every element of I can be uniquely expressed in the form

$$m_1 v_1 + \ldots + m_n v_n \qquad (m_k \in \mathscr{Z}, \quad k = 1, 2, \ldots, n),$$

where

$$v_k = \sum_{i=1}^{k} \alpha_{ik} w_k \qquad (\alpha_{ik} \in \mathscr{Z}, \quad k = 1, 2, \ldots, n).$$

Furthermore, show that $N(I) = \displaystyle\prod_{k=1}^{n} |\alpha_{kk}|$.

2.   Show that if $a \neq 0$, $a \in R_K$, then

$$N(aR_K) = |N_K(a)|.$$

3 (L. Carlitz [1]). Show that if $h(K) \leqslant 2$, then all the decompositions into irreducible factors of a given integer a of the field K contains the same number of factors. Under the assumption that in every ideal class there are infinitely many prime ideals, show that the converse implication also holds.

§3. *p*-adic numbers

1.   In this section we shall study another generalization of integers, the *p*-adic integers. We shall begin by giving the definition and fundamental properties of *p*-adic numbers, and then we shall introduce *p*-adic integers and show the relations between the equations in the *p*-adic field and the congruences in the ring $\mathscr{Z}$.

Let p be any prime. We know that every non-zero rational number w can be written in the form

$$w = \frac{p^k a}{b},$$

where $a, b \in \mathscr{Z}$, $(a, b) = 1$, p does not divide ab and $k \in \mathscr{Z}$. For such a rational number we put

$$v_p(w) = p^{-k} \ .$$

We call the function obtained in this way the *p-adic valuation*. Additionally we put $v_p(0) = 0$. From Theorem 1.7 it follows without difficulty that such a valuation satisfies the following conditions:

$$(9.12) \qquad v_p(xy) = v_p(x) \, v_p(y)$$

and

$$(9.13) \qquad v_p(x + y) \leqslant \max\{v_p(x), v_p(y)\} \leqslant v_p(x) + v_p(y) \ .$$

From the condition (9.13) it follows that, putting $d(x, y) = v_p(x - y)$ we obtain a metric in $\mathscr{Q}$, so that this field can be treated as a metric space. We may consider, in particular, the limits of sequences — a sequence a_n of rational numbers will be convergent to a rational number a if $\lim v_p(a - a_n) = 0$. It is not difficult to observe that this space is not complete, i.e. there exist fundamental sequences (satisfying the Cauchy condition) which are not convergent. This follows from the remark that no complete metric space with a non-discrete topology can be countable and our topology is not discrete, for, e.g. the sequence $p, p^2, \ldots, p^n, \ldots$ is not stationary but converges to 0.

We shall call the completion of $\mathscr{Q}$ in the metric induced by the p-adic valuation the *space of p-adic numbers* Q_p. We shall call the elements of this space the *p-adic numbers*. It is known from topology that we obtain this space by considering the family of all fundamental sequences, introducing in it the equivalence relation

$$\{a_n\} \sim \{b_n\} \qquad \text{iff} \qquad \lim_{n \to \infty} (a_n - b_n) = 0$$

and dividing it by this relation. For us the construction itself is not really important and we formulate the needed topological facts in the following lemma:

Lemma 9.20. *For each prime p there exists a unique metric space $\langle Q_p ; d \rangle$ with the following properties:*

  (i) Q_p *is a complete space containing $\mathscr{Q}$,*
  (ii) *for $x, y \in \mathscr{Q}$ we have $d(x, y) = v_p(x - y)$,*
  (iii) $\mathscr{Q}$ *is a dense subspace of Q_p.* ∎

The following lemma enables us to extend the valuation v_p to Q_p:

Lemma 9.21. *If a is a non-zero element of Q_p and the sequence $c_1, c_2, \ldots$ of rational numbers is convergent to a, then the sequence $v_p(c_n)$ is constant from some place on and its limit does not depend on the choice of the sequence $\{c_n\}$, but solely on a. If moreover a is a rational number, then $\lim v_p(c_n) = v_p(a)$.*

Proof. For natural m, n we have

$$v_p(c_m) = v_p(c_n + c_m - c_n) \leq v_p(c_n) + v_p(c_m - c_n) ,$$

i.e.

$$v_p(c_m) - v_p(c_n) \leq v_p(c_m - c_n) .$$

Changing the role of m and n we obtain

$$v_p(c_n) - v_p(c_m) \leq v_p(c_n - c_m) = v_p(c_m - c_n) ,$$

and so

$$|v_p(c_n) - v_p(c_m)| \leq v_p(c_m - c_n) = d(c_m, c_n) .$$

Since the sequence $\{c_n\}$ is fundamental, this inequality shows that the sequence $\{v_p(c_n)\}$ is fundamental in R, hence it is convergent. It cannot be convergent to zero, for $a \neq 0$, and so, for suitable M, $\epsilon > 0$ we must have $0 < \epsilon \leq v_p(c_n) \leq M$ for sufficiently large n. But in the interval $[\epsilon, M]$ there are only finitely many numbers of the form p^{-k}, so that our sequence $v_p(c_n)$ must be a constant from some place on. If $a = \lim c_n = \lim d_n$ $(c_n, d_n \in \mathcal{Q})$, then the sequence $c_1, d_1, c_2, d_2, \ldots$ is also convergent to a and from the preceding argument we obtain $\lim v_p(c_n) = \lim v_p(d_n)$. The last part of the assertion now becomes *clear*, because for a rational a we may take $c_n = a$ for $n = 1, 2, \ldots$ ∎

For any p-adic number a we may now define the value of the function v_p by the formula

$$v_p(a) = \begin{cases} 0 & \text{if } a = 0 , \\ \\ \lim v_p(c_n) & \text{if } a \neq 0 , \ \lim c_n = a, \ c_n \in \mathcal{Q} . \end{cases}$$

The last part of the preceding lemma shows that the function thus defined is an extension of the p-adic valuation to the space Q_p. We shall now show that in Q_p one can define arithmetical operations in a natural way to obtain a field.

Theorem 9.19. *If $a, b \in Q_p$, $a = \lim a_n$, $b = \lim b_n$, a_n, $b_n \in \mathcal{Q}$, then the sequences $a_n + b_n$, $a_n - b_n$, $a_n b_n$ and in the case b, $b_n \neq 0$, also a_n/b_n, are all convergent and their limits depend solely on a and b, and not on the choice of the sequences a_n, b_n. If we denote these limits by $a + b$, $a - b$, ab and a/b, respectively, then the space Q_p forms a field with respect to the operations thus defined.*

Proof. Since the sequences $v_p(a_n)$, $v_p(b_n)$ are bounded, and under the assumption that b and b_n do not vanish, the sequence $1/v_p(b_n)$ is also bounded, it follows from the following inequalities that the sequences $a_n + b_n$, $a_n - b_n$, $a_n b_n$, a_n/b_n are Cauchy sequences, and hence are convergent:

$$v_p\left((a_n \pm b_n) - (a_m \pm b_m)\right) \leqslant v_p(a_n - a_m) + v_p(b_n - b_m) \ , \qquad .$$

$$v_p(a_n b_n - a_m b_m) \leqslant v_p(a_n) \, v_p(b_n - b_m) + v_p(b_m) \, v_p(a_n - a_m) \ ,$$

$$v_p\left(\frac{a_n}{b_n}\right) - \left(\frac{a_m}{b_m}\right) \leqslant \frac{v_p(a_n) \, v_p(b_m - b_n) + v_p(b_n) \, v_p(a_n - a_m)}{v_p(b_n) \, v_p(b_m)} \ .$$

Since this is true for any sequences a_n, b_n with limits a, b, the limits of the sequences $a_n + b_n$, $a_n - b_n$, $a_n b_n$, a_n/b_n actually depend solely upon a and b. Checking the validity of axioms of a field now presents no difficulties. ∎

It happens that the function v_p defined above has in Q_p the same formal properties as its restriction to Q.

Lemma 9.22. *For $x, y \in Q_p$ we have*
(i) $v_p(x + y) \leqslant \max\{v_p(x), \ v_p(y)\}$,
(ii) $v_p(xy) = v_p(x) \, v_p(y)$,
(iii) *if $d(x, y)$ is the metric in Q_p appearing in Lemma 9.20, then $d(x, y) = v_p(x - y)$,*
(iv) *if $\lim x_n = x$, then $\lim v_p(x_n) = v_p(x)$.*

Proof. In the cases (i) and (ii) we select sequences a_n, b_n of rational numbers converging respectively to x and y. Then

$$v_p(a_n + b_n) \leqslant \max\{v_p(a_n), \ v_p(b_n)\}$$

and

$$v_p(a_n b_n) = v_p(a_n) \, v_p(b_n)$$

and it remains to pass to the limit.

To prove (iii) we note that in the above notation we have

$$d(x, y) \leqslant d(x, a_n) + d(a_n, b_n) + d(b_n, y)$$

and

$$d(a_n, b_n) \leqslant d(a_n, x) + d(x, y) + d(y, b_n) \ ,$$

and so

$$d(x, y) \leqslant \lim\inf d(a_n, b_n)$$

and

$$d(x, y) \geqslant \lim\sup d(a_n, b_n) \ ,$$

which gives by Lemma 9.20 (ii)

$$d(x, y) = \lim d(a_n, b_n) = \lim v_p(a_n - b_n) = v_p(x - y) \ .$$

Now (iv) follows easily owing to

$$\lim v_p(x_n) = \lim d(x_n, 0) = d(x, 0) = v_p(x) . \qquad \blacksquare$$

Corollary. *If* $F(x_1, \ldots, x_n)$ *is a polynomial in* n *variables with coefficients in* Q_p, *then* F *is a continuous function from* Q_p *into* Q_p.

Proof. It suffices to prove the assertion for polynomials $x_1 + x_2$ and $x_1 x_2$, but this follows immediately from (i) and (ii). $\blacksquare$

Before we turn to the definition of p-adic integers, we shall prove a simple but interesting lemma concerning the convergence of sequences and series formed of p-adic numbers:

Lemma 9.23.

(i) *For a sequence* $\{a_n\}$ *of* p-*adic numbers to be convergent it is necessary and sufficient that*

$$\lim_{n \to \infty} (a_n - a_{n+1}) = 0 .$$

(ii) *For a series* $\displaystyle\sum_{n=1}^{\infty} a_n$ *with* p-*adic number terms to be convergent it is necessary and sufficient that*

$$\lim_{n \to \infty} a_n = 0 .$$

Proof. Clearly it is sufficient to prove (i), and, since Q_p is complete, it suffices to check that the condition $(a_n - a_{n+1}) \to 0$ is equivalent to $\displaystyle\lim_{\substack{m \to \infty \\ n \to \infty}} (a_m - a_n) = 0$. But this is easy since for $m > n$ we have

$$v_p(a_m - a_n) = v_p(a_m - a_{m-1} + a_{m-1} - \ldots + \ldots - a_n)$$

$$\leqslant \max_{1+n \leqslant j \leqslant m} v_p(a_j - a_{j-1}) \to 0 . \qquad \blacksquare$$

Corollary. *In the field* Q_p *we have the equality*

$$\sum_{n=0}^{\infty} p^n = \frac{1}{1-p} .$$

Proof. The series on the left-hand side is convergent by the lemma because $v_p(p^n) = p^{-n}$ tends to zero, and if S is its sum, then

$$(1-p)S = \sum_{n=0}^{\infty} p^n - \sum_{n=0}^{\infty} p^{1+n} = 1 . \qquad \blacksquare$$

2.   It is not difficult to check that the closure of the ring $\mathscr{Z}$ of rational integers in the field Q_p is a ring. We shall call it the *ring of p-adic integers* and denote it by Z_p, and call its elements the *p-adic integers*. It is clear that every rational number is a p-adic integer, but there exist non-integral rational numbers that are p-adic integers. As an example we may take the number $1/(1-p)$, which lies in Z_p by the Corollary to the preceding lemma, and does not lie in $\mathscr{Z}$ if $p \neq 2$. The following theorem describes all rationals lying in Z_p:

Theorem 9.20. *For a p-adic number a to belong to Z_p it is necessary and sufficient that $v_p(a) \leqslant 1$. In particular, a rational number w is a p-adic integer iff it can be expressed in the shape $w = A/B$, with A, B lying in $\mathscr{Z}$ and B not divisible by p.*

Proof. If a lies in Z_p and $\{a_n\}$ is a sequence of rational integers converging to a, then

$$v_p(a) = \lim v_p(a_n) \leqslant 1 \quad ,$$

because the inequality $v_p(x) \leqslant 1$ for $x \in \mathscr{Z}$ follows immediately from the definition of v_p. Now suppose that the condition $v_p(a) \leqslant 1$ is satisfied. Then by Lemma 9.22 (iv) we can find a sequence $\{a_n\}$ of rational numbers converging to a and satisfying $v_p(a) < \frac{3}{2}$, but the possible values of the function v_p greater than 1 are $p, p^2, \ldots$, hence $v_p(a_n) \leqslant 1$. This means that each of the numbers a_n can be written in the form $a_n = b_n/c_n$, where $b_n, c_n \in \mathscr{Z}$, $p \nmid c_n$. Now let x_n be a solution of the congruence

$$c_n x_n \equiv b_n \,(\mathrm{mod}\, p^n)$$

that exists in view of Theorem 1.22. We show that the sequence x_n converges to a. Indeed, we have

$$v_p(x_n - a) \leqslant \max \;\; v_p(x_n - a_n), \; v_p(a_n - a) \quad ,$$

but

$$v_p(x_n - a_n) = v_p\left(x_n - \frac{b_n}{c_n}\right) = v_p(c_n^{-1})\,v_p(c_n x_n - b_n) = v_p(c_n x_n - b_n) \leqslant p^{-n} \to 0$$

and

$$\lim_{n \to \infty} v_p(a_n - a) = 0 \quad ,$$

which gives

$$\lim_{n \to \infty} v_p(x_n - a) = 0 \qquad \text{and} \qquad \lim_{n \to \infty} x_n = a \;\; .$$

Hence a lies in Z_p. The second part of the theorem is an immediate consequence of the first.   ∎

Corollary 1. *For a rational number w to lie in Z_p for every p it is necessary and sufficient that $w \in \mathscr{Z}$.*

Corollary 2. *For a rational integer a to be divisible by p in the ring Z_p it is necessary and sufficient that $p \mid a$ in $\mathscr{Z}$.*

Corollary 3. *If F is a polynomial with coefficients in Z_p and the elements $a, b \in Z_p$ satisfy the condition*

$$v_p(a - b) \leqslant p^{-n} ,$$

then

$$v_p\left(F(a) - F(b)\right) \leqslant p^{-n} .$$

Proof. If $F(x) = \sum_{k=0}^{m} a_k x^k$, then

$$F(a) - F(b) = \sum_{k=0}^{m} a_k (a^k - b^k) = (a - b) \sum_{k=1}^{m} a_k (a^{k-1} + a^{k-2} b + \ldots + b^{k-1})$$

$$= (a - b)c ,$$

where $c \in Z_p$ and

$$v_p\left(F(a) - F(b)\right) = v_p(a - b) \, v_p(c) \leqslant v_p(a - b) \leqslant p^{-n} ,$$

because

$$v_p(c) \leqslant 1 . \quad \blacksquare$$

The following theorem describes the multiplicative structure of the ring Z_p .

Theorem 9.21.

(i)   The set $U_p = \{a \in Q_p : v_p(a) = 1\}$ is the set of all invertible elements of the ring Z_p .

(ii)   The set $P = \{a \in Q_p : v_p(a) < 1\}$ is a maximal ideal in Z_p and every non-zero ideal of this ring is of the form $I = p^n Z_p = P^n$ for some $n \in \mathscr{N}_0$, (where $P^0 = Z_p$) and $P^n = \{a \in Q_p : v_p(a) \leqslant p^{-n}\}$.

The embedding of $\mathscr{Z}$ into Z_p induces an isomorphism of $\mathscr{Z}/p^n \mathscr{Z}$ onto Z_p/P^n, and so each coset $Z_p \pmod{P^n}$ contains rational integers.

Proof.

(i)   Since $v_p(a^{-1}) = v_p(a)^{-1}$, invertible elements of the ring Z_p are precisely those elements a which satisfy the conditions $v_p(a) \leqslant 1$ and $v_p(a)^{-1} \leqslant 1$ at the same time, but this is possible only when $v_p(a) = 1$.

(ii)   Owing to Lemma 9.22 we know that the set P is an ideal in Z_p . Its complement $Z_p \backslash P$ coincides with the set of invertible elements, and so it is a maximal ideal. Now let I be any non-zero ideal of Z_p, let $p^{-k} = \max_{x \in I} v_p(x)$ and let a be an element of I such that $v_p(a) = p^{-k}$. First, let us show that I coincides with ideal

$$P_k = \{x : v_p(x) \leqslant p^{-k}\} .$$

If y is any non-zero element of P_k, then $v_p(y) = p^{-n}$, where $n \geqslant k$, and then $v_p(ya^{-1}) = p^{-n+k} \leqslant 1$, i.e. $ya^{-1} \in Z_p$ and $y \in aZ_p \subset I$, $P_k \subset I$. If the strict inclusion holds here and $y \in I \backslash P^k$, then $v_p(y) > p^{-k}$, contrary to the choice of the element a.

The next step is to show that $P_k = P^k$. If $k = 0$, then this equality follows from Theorem 9.20. Suppose it is valid for all integers less than n and let us show that $P_n = P^n$. To do this we note that

$$P^n = P^{n-1} P = P_{n-1} P = \left\{ \sum_{i=1}^{s} a_i b_i : v_p(a_i) \leqslant p^{-n+1}, \; v_p(b_i) \leqslant p^{-1} \right\},$$

so that for $x \in P^n$ we have $v_p(x) \leqslant p^{-n}$, i.e. $P^n \subset P_n$.

On the other hand, if $x \in P_n$, i.e. $v_p(x) \leqslant p^{-n}$, then $v_p(xp^{-1}) = p^{-n+1}$, and this gives $xp^{-1} \in P_{n-1} = P^{n-1}$ and we see that $x \in pP^{n-1} \subset PP^{n-1} = P^n$, for $p \in P$.

To prove the equality $P^k = p^k Z_p$, we observe that the inclusion $p^k Z_p \subset P^k$ follows from $p \in P$, and the reverse inclusion is a consequence of $v_p(p^k) = p^{-k}$, i.e. $p^k \in P_k = P^k$.

Since $p^n \mathcal{L} \subset p^n Z_p = P^n$, the embedding $\mathcal{L}$ into Z_p induces an embedding of $\mathcal{L}/p^n \mathcal{L}$ into Z_p/P^n and there remains to show that it is surjective. For this purpose we take an arbitrary element $a \in Z_p$ and choose a rational integer b such that $v_p(a - b) \leqslant p^{-n}$. Then $a = b + c$, where $c \in P^n$ and we see that $a + P^n = b + P^n$, i.e. the images of a and b in Z_p/P^n coincide. ∎

Corollary 1. *The ring Z_p is a principal ideal ring, and hence it is a Dedekind domain and a ring with unique factorization.* ∎

Corollary 2. *We may write every non-zero element $a \in Q_p$ uniquely in the form $a = up^k$, with $k \in \mathcal{L}$, $u \in U_p$. Moreover, $v_p(a) = p^{-k}$ and the element a lies in Z_p iff $k \in \mathcal{L}_0$.*

Proof. If $v_p(a) = p^{-k}$, then $u = ap^{-k}$ lies in U_p and $a = up^k$. If $a = up^k = u_1 p^t$, with $u, u_1 \in U_p$, $k, t \in \mathcal{L}_0$, then $v_p(a) = p^{-k} = p^{-t}$, whence $k = t$ and $u = u_1$. ∎

3.   It happens that p-adic numbers can be expressed as a sum of a particularly simply formed series with rational terms. We shall now prove this result. It is worth noting that at the beginning of the development of the theory, p-adic numbers were defined precisely as such formal series.

Theorem 9.22. *Every non-zero p-adic number a can be expressed uniquely as a sum of the series of the form*

(9.14) ·
$$\sum_{n=N}^{\infty} a_n p^n,$$

where $a_n \in \mathscr{Z}$, $0 \leqslant a_n \leqslant p-1$ and $a_N \neq 0$. Here N is determined by the condition $v_p(a) = p^{-N}$. The number a is integral iff $N \geqslant 0$, and $a \in U_p$ iff $N = 0$.

Proof. We begin with the case where a is a p-adic integer. Let a_n be a sequence of rational integers converging to a. Then for a fixed N and a sufficiently large m the congruence

(9.15) $$a_m \equiv a_{m+1} \equiv a_{m+2} \equiv \ldots \pmod{p^N}$$

holds.

Suppose that this holds for $m \geqslant M_0(N)$, where we may assume that $M_0(N+1) > M_0(N)$.

For a fixed N we may write

$$a_{M_0(N)} = \sum_{k=0}^{N-1} a_k^{(N)} p^k + p^N \xi_N \qquad (a_k^{(N)}, \, \xi_N \in \mathscr{Z}) \, ,$$

where $0 \leqslant a_k^{(N)} \leqslant p-1$ are uniquely determined. Note that for $N > r$ the numbers $a_r^{(N)}$ do not depend on N. Indeed, if $N_1 > N > r$, then

$$a_{M_0(N_1)} = \sum_{k=0}^{N-1} a_k^{(N_1)} p^k + p^{N_1} \xi_{N_1} \qquad (a_k^{(N_1)}, \, \xi_{N_1} \in \mathscr{Z}, \, 0 \leqslant a_k^{(N_1)} \leqslant p-1)$$

and by $M_0(N_1) > M_0(N)$ and (9.15) we have

$$a_{M_0(N)} \equiv a_{M_0(N_1)} \pmod{p^N} \, ,$$

and consequently

$$\sum_{k=0}^{N-1} a_k^{(N)} p^k \equiv \sum_{k=0}^{N-1} a_k^{(N_1)} p^k \pmod{p^N} \, .$$

Since the numbers on both sides of this congruence are positive and less than p^N, they must be identical. Making use of the uniqueness of p-adic expansion of natural numbers, we get $a_r^{(N)} = a_r^{(N_1)}$.

Therefore we see that the sequence $a_k^{(N)}$ is, for a fixed k, constant from some place on. Let a_k' be its limit. Then the series

$$\sum_{k=0}^{\infty} a_k' p^k$$

is convergent in view of $v_p(a_k' p^k) = p^{-k} \to 0$, and for $S_N = \sum_{k=0}^{N-1} a_k' p^k$ we obtain the equality

$$S_N = \sum_{k=0}^{N-1} a_k^{(N)} p^k = a_{M_0(N)} - p^N \xi_N \, ,$$

which gives

$$v_p(a_{M_0(N)} - S_N) = v_p(p^N \xi_N) \leqslant p^{-N} \to 0 ,$$

so that

$$a = \lim_{n \to \infty} a_n = \lim_{N \to \infty} a_{M_0(N)} = \lim_{N \to \infty} S_N = \sum_{k=0}^{\infty} a'_k p^k .$$

If now b is any non-zero element of Q_p, then by Corollary 2 to Theorem 9.21 we may write $b = b_0 p^n$, where $n \in \mathcal{Z}$ and $b_0 \in U_p$. Applying to the element b_0 that part of the theorem already proved, we may write

$$b_0 = \sum_{k=0}^{\infty} a_k p^k \qquad (a_k \in \mathcal{Z}, \ 0 \leqslant a_k \leqslant p-1) ,$$

and then

$$b = \sum_{k=-n}^{\infty} a_{k-n} p^k$$

which is the desired representation.

Now note that if a is the sum of the series (9.14), then $v_p(a) = p^{-N}$. Indeed, the series

$$\sum_{k=N}^{\infty} a_k p^{k-N}$$

is convergent to an element $b \in Z_p$ and $a = p^N b$, hence $v_p(a) \leqslant p^{-N}$, and if we had $v_p(a) < p^{-N}$, then $b \in P = pZ_p$, which would contradict

$$b = a_N + p \sum_{k=N+1}^{\infty} a_k p^{k-N-1} \in a_N + pZ_p \neq pZ_p .$$

It remains to show the uniqueness of the expansion (9.14) since the remaining parts of the assertion are consequences of Theorem 9.21. If a p-adic number $b \neq 0$ had two such expansions, then with N, satisfying $v_p(b) = p^{-N}$ we would have for suitable $a_k, \ b_k \in \mathcal{Z}, \ 0 \leqslant a_k, \ b_k \leqslant p-1$ the equality

$$\sum_{k=N}^{\infty} a_k p^k = \sum_{k=N}^{\infty} b_k p^k ,$$

and there would exist an index k for which $a_k \neq b_k$. Let r be the smallest such index. Then

$$\sum_{k=r}^{\infty} a_k p^k = \sum_{k=r}^{\infty} b_k p^k ,$$

so that

$$a_r + p \sum_{k=1+r}^{\infty} a_k p^{k-r} = b_r + p \sum_{k=1+r}^{\infty} b_k p^{k-r}$$

and $a_r - b_r \in pZ_p$. From Corollary 2 to Theorem 9.17 now follows $a - b \in p\mathcal{Z}$, but this is possible only when $|a_r - b_r| < p$. ∎

4.   We shall now give fundamental topological properties of the field of p-adic numbers.

Theorem 9.23.

(i)   The ideals P^n $(n = 0, 1, \ldots)$ are compact and open sets in Q_p.

(ii)   The ideal P^n is the closure of $p^n \mathcal{Z}$ for $n = 0, 1, \ldots$

(iii) If a is any p-adic number, then the family $\{a + P^n\}$ forms a basis of neighbourhoods of a.

Proof.

(i)   Openness of P^n follows from the equality

$$P^n = \{x \in Q_p : v_p(x) < p^{1-n}\} \quad .$$

To prove compactness it is enough to observe that the mapping f of the product D of a countable number of sets $\{0, 1, \ldots, p-1\}$ with p elements into P^n given by

$$f(\langle a_k \rangle_{k=0}^{\infty}) = \sum_{k=0}^{\infty} a_k p^{n+k}$$

is surjective by virtue of the preceding theorem and is continuous in the product topology. Since D is a compact set, P^n must also be a compact set as a continuous image of a compact set.

(ii)   This part is a consequence of Theorem 9.22, and (iii) follows from the remark that $a + P^n = \{x \in Q_p : v_p(x-a) < p^{1-n}\}$ is an open disk with radius p^{1-n}. ∎

Corollary. If $F(X_1, \ldots, X_k)$ is a polynomial in k variables with coefficients in $\mathcal{Z}$, then the equation

(9.16) $F(x_1, \ldots, x_k) = 0$

has a solution in Z_p iff for any n the congruence

(9.17) $F(x_1, \ldots, x_k) \equiv 0 \pmod{p^n}$

has a solution in $\mathcal{Z}$.

Proof. Necessity of the given condition follows from Corollary to Lemma 9.22 and the fact that $\mathcal{Z}$ is dense in Z_p. To prove sufficiency we denote by $x_1^{(n)}, \ldots, x_k^{(n)}$ an arbitrary solution of the congruence (9.17). For each i the sequence $x_i^{(n)}$ contains, because of the compactness of Z_p, some convergent subsequence, so that choosing a suitable subsequence we may confine ourselves to the case where all the sequences $x_i^{(n)}$ are convergent. Let

$$x_i = \lim_{n \to \infty} x_i^{(n)} \quad .$$

Since $F(x_1^{(n)}, \ldots, x_n^{(n)}) \in P^n \subset P^{n-1} \subset \ldots$ and F is a continuous function,

$$F(x_1, \ldots, x_n) \in \bigcap_{n=1}^{\infty} P^n = \{0\} \quad . \quad \blacksquare$$

As an example, consider the equation $X^2 = a$ in the field Q_p with an odd prime p, where a is a non-zero rational integer. We write it in the form $a = p^k A$, where $p \nmid A$. According to Theorem 1:23 the congruence $x^2 \equiv a \pmod{p^n}$ has a solution for any n iff 2 divides k and $\left(\dfrac{A}{p}\right) = 1$, and hence by the preceding corollary this is a necessary and sufficient condition for the solvability of the equation $X^2 = a$ in the field of p-adic numbers for $p \neq 2$.

5.   We shall close this section by proving the so-called *Hensel's lemma* presenting a simple sufficient condition for the existence of a solution of an algebraic equation in the field of p-adic numbers:

Theorem 9.24. *Let* $F(X) = \displaystyle\sum_{k=0}^{N} a_k X^k$ *be a polynomial with p-adic integer coefficients and let* $F'(X)$ *be its derivative, i.e.* $F'(X) = \displaystyle\sum_{k=1}^{N} k a_k X^{k-1}$. *Suppose that for some* $a \in Z_p$ *and* $\delta \in \mathscr{L}$ *the following conditions are satisfied:*

(i) $F'(a) \in P^{\delta} \setminus P^{1+\delta}$,
(ii) $F(a) \in P^{1+2\delta}$.

Then there exists a p-adic integer b *which is a root of the polynomial* $F(X)$ *and moreover,* $b \in a + P^{1+\delta}$.

Proof. Our proof imitates Newton's method. Let $a_0 = a$ and

(9.18) $a_{n+1} = a_n - F(a_n)/F'(a_n)$

provided $F'(a_n) \neq 0$. From the next lemma it follows that this condition is always satisfied.

Lemma 9.24. *If the sequence* a_n *is defined by (9.18), then*

(i)   *for* $n = 0, 1, \ldots$ *we have* $a_n \in Z_p$, *and* $v_p(F'(a_n)) = p^{-\delta}$,
(ii)   *for* $n = 0, 1, \ldots$ *we have* $F(a_n) \in P^{n+1+2\delta}$,
(iii) *for* $n = 1, 2, \ldots$ *we have* $v_p(a_n - a_{n-1}) \leq p^{-n-\delta}$.

Proof. For $n = 0$ all the assertions of the lemma are true (part (iii) being vacuously true). Therefore we suppose that (i), (ii), (iii) hold for some $n \geqslant 0$. By (iii) (and for $n = 0$ trivially) we have

$$v_p(a_n - a_0) \leqslant p^{-1 - \delta} \ .$$

Corollary 3 to Theorem 9.20 now gives us

$$v_p(F'(a_n) - F'(a_0)) \leqslant p^{-1 - \delta} \ ,$$

but from the assumption it follows that $F'(a_0) = u_0 \, p^\delta$ with $u_0 \in U_p$, so that

$$F'(a_n) = u_n \, p^\delta \neq 0 \ ,$$

where $u_n \in U_p$.

Hence we find that

$$\frac{F(a_n)}{F'(a_n)} = \frac{F(a_n)}{u_n \, p^\delta} \in P^{n+1+\delta} \subset Z_p \ ,$$

and hence $a_{n+1} \in Z_p$ and $a_{n+1} - a_n \in P^{n+1+\delta}$, which proves the validity of (i) and (iii) for $n + 1$. To prove (ii) we observe that there exists a polynomial $G(X)$ with p-adic integer coefficients such that

$$F(X) = F(a_n) + F'(a_n)(X - a_n) + G(X)(X - a_n)^2 \ .$$

Indeed, an easy calculation shows that this is true for a monomial $F(X) = cX^j$, and in the general case it suffices to express F as a sum of such monomials.

Substituting in the above equality $X = a_{n+1}$, we obtain

$$F(a_{n+1}) = F(a_n) + F'(a_n)(a_{n+1} - a_n) + G(a_{n+1})(a_{n+1} - a_n)^2$$

$$= \frac{F(a_n)^2}{F'(a_{n+1})^2} \, G(a_{n+1}) \in P^{2\delta + 2 + n} \ . \qquad \blacksquare$$

From Lemma 9.23 (i) and part (iii) of Lemma 9.24 we obtain the convergence of the sequence a_n. If b is its limit, then

$$v_p(F'(b)) = \lim_{n \to \infty} v_p(F'(a_n)) = p^{-\delta} \ ,$$

hence $F'(b) \neq 0$, and using (9.18), we obtain

$$b = b - F(b)/F'(b) \ ,$$

i.e. $F(b) = 0$. Furthermore, $v_p(a - b) = \lim v_p(a - a_n) \leqslant p^{-1 - \delta}$. $\blacksquare$

Corollary. *The equation* $X^{p-1} = 1$ *has* $p - 1$ *distinct solutions in* Z_p *which lie in distinct cosets of* Z_p/P.

Proof. It is sufficient to apply the theorem just proved to the polynomial $F(X) = X^{p-1} - 1$ with $a = 1, 2, \ldots, p - 1$ and $\delta = 0$. ∎

Exercises

1.   Prove that *if* $v(x)$ *is a function defined in* $\mathscr{Q}$ *with non-negative real values which satisfies* (9.12) *and* (9.13) *and does not vanish identically, then there exist a prime number* p *and a positive constant* a *such that* $v(x) = v_p(x)^a$. (This is *Ostrowski's theorem*).

2.   Let U_p^n denotes the set $1 + P^n \subset Z_p$ for $n = 1, 2, \ldots$
   (i) Show that U_p^n forms a group under multiplication.
   (ii) Prove that every element of U_p may be expressed in the form au, where a is a solution of the equation $X^{p-1} = 1$ and u lies in U_p^1.
   (iii) Show that if $m > 1$ and $a \in U_p^{m+1}$, then the equation $X^p = a$ has a root lying in U_p^m. Prove that in the case $p \neq 2$ the assertion holds already for $m = 1$.

3.   The p-adic exponential function is defined by the formula

$$\exp x = \sum_{n=0}^{\infty} x^n/n!$$

wherever this series converges. Prove that this holds for $x \in P$ ($p \neq 2$), or $x \in P^2$ ($p = 2$). Prove that $\exp x$ is a continuous function.

4.   Prove that for $p \neq 2$ the function $\exp x$ gives a one to one mapping of P onto U_p^1.

5.   The p-adic logarithm is defined by the formula

$$\log(1 + x) = \sum_{n=1}^{\infty} (-1)^{n+1} x^n/n$$

wherever the series is convergent. Find the region of convergence of this series.

6.   Show that the equality $\log \exp x = x$ holds whenever the function $\exp x$ is defined.

7.   Prove that a p-adic number of the form (9.14) is a rational number iff the sequence a_n is constant from some place on.

8.   Find a necessary and sufficient condition for the solvability of the equation $X^n = a$ in the field Q_p in the case when p does not divide n.

9.   Show that if a polynomial with coefficients in Z_p and with leading coefficient equal to 1 has a root in Q_p, then this root must belong to Z_p.

Suggestions for further reading

The reader who is interested in number theory may widen his knowledge by consulting the following books:

For Chapters I and II: H. Halberstam, K. F. Roth [1], G. H. Hardy, E. M. Wright [1], J. L. Mordell [2], W. Sierpiński [3], [4].

For Chapter III: W. J. Ellison [1], A. G. Postnikov [1], K. Prachar [1].

For Chapter IV: H. Halberstam, K. F. Roth [1], H. L. Montgomery [1], K. Prachar [1].

For Chapter V: J. W. S. Cassels [2], H. G. Eggleston [1], C. G. Lekkerkerker [1].

For Chapter VI: H. Halberstam, K. F. Roth [1], H. H. Ostmann [1].

For Chapter VII: I. P. Kubilius [2], A. G. Postnikov [1].

For Chapter VIII: J. W. S. Cassels [1], A. J. Hinčin [1], L. Kuipers, H. Niederreiter [1].

For Chapter IX: Z. I. Borević, I. R. Šafarević [1], W. Narkiewicz [2], A. Weil [2].

LITERATURE CITED

Baker, A.

[1] *Linear forms in the logarithms of algebraic numbers*, Mathematika 13 (1966), pp. 204-216, 14 (1967), pp. 102-107, 15 (1968), pp. 204-216.

[2] *Contributions to the theory of diophantine equations, I. On the representation of integers by binary forms*, Philos. Trans. Roy. Soc. Ser. A 263 (1967/68), pp. 173-191.

Bell, E. T.

[1] *Factorizability of numerical functions*, Bull. Amer. Math. Soc. 37 (1931), pp. 251-253.

Bellman, R.

[1] *A note on the divergence of a series*, Amer. Math. Monthly 50 (1943), pp. 318-319.

Billingsley, P.

[1] *On the central limit theorem for the prime divisor function*, Amer. Math. Monthly 76 (1969), pp. 132-139.

Birch, B. J.

[1] *Multiplicative functions with non-decreasing normal order*, J. London Math. Soc. 42 (1967), pp. 149-151.

Blankinship, W. A.

[1] *A new version of the Euclidean algorithm*, Amer. Math. Monthly 70 (1963), pp. 742-745.

Bochner, S.

[1] *Lectures on Fourier integrals*, Ann. of Math. Studies, No. 42, translation from the original, Princeton Univ. Press, Princeton (1959).

Bohman, J.

[1] *Some computational results regarding the prime numbers below 2 000 000*, Nordisk Tidskr. Informationsbehandling (BIT) 13 (1973), pp. 242-244.

Bokowski, J., Hadwiger, H., Wills, J. M.

[1] *Eine Ungleichung zwischen Volumen, Oberfläche und Gitterpunktanzahl konvexer Körper im n-dimensionalen euklidischen Raum*, Math. Zeitschr. 127 (1972), pp. 363-364.

Bombieri, E.

[1] *On the large sieve*, Mathematika 12 (1965), pp. 201-225.

Borević, Z. I., Shafarević, I. R.

[1] *Number Theory*, Pure Appl. Math. Vol. 20, Academic Press, New York, London (1966).

Bovey, J.D.

[1] $\Gamma^*(8)$, Acta Arith. 25 (1974), pp. 145-150.

Bradley, G. H.

[1] *Algorithm and bound for the greatest common divisor of n integers*, Comm. ACM 13 (1970), pp. 433-436.

Brauer, A., Reynolds, T. L.

[1] *On a theorem of Aubry-Thue*, Canad. J. Math. 3 (1951), pp. 367-374.

Buck, R. C.

[1] *Prime-representing functions*, Amer. Math. Monthly 53 (1946), p. 256.

Burgess, D. A.

[1] *On character sums and primitive roots*, Proc. London Math. Soc. (3) 12 (1962), pp. 179-192.

Carlitz, L.

[1] *A characterization of algebraic number fields with class number two*, Proc. Amer. Math. Soc. 11 (1960), pp. 391-392.

Cassels, J. W. S.

[1] *Introduction to diophantine approximations*, Cambridge Math. Tracts No. 45, Cambridge Univ. Press, Cambridge (1957); reprint: Hafner Pub. Co., New York (1972).

[2] *An introduction to the geometry of numbers*, Grundl. Math. Wiss. Bd. 99, Springer Verlag, Berlin-Heidelberg-New York (1959).

Chen, J. J.

[1] *On Waring's problem for nth powers*, Acta Math. Sinica 8 (1958), pp. 253-257,

Chinese Math. Acta 8 (1967), pp. 849-853.

[2] *Waring's problem for g(5)*, Sci. Sinica 13 (1964), pp. 1547-1568.

[3] *On the representation of a large even integer as the sum of a prime and the product of at most two primes*, Kexue Tongbao 17 (1966), pp. 385-386, Sci. Sinica 16 (1973), 157-176.

Chowla, S., Shimura, G.

[1] *On the representation of zero by a linear combination of k-th powers*, Norske Vid. Selsk. Forh. (Trondheim) 36 (1960), pp. 169-176.

Cohen, E.

[1] *The number of unitary divisors of an integer*, Amer. Math. Monthly 67 (1960), pp. 879-880.

[2] *Arithmetical functions associated with unitary divisors of an integer*, Math. Zeitschr. 74 (1960), pp. 66-80.

Davenport, H.

[1] *On a generalization of Euler's function $\phi(n)$*, J. London Math. Soc. 7 (1932), pp. 290-296; Collected works, Vol. 4, pp. 1827-1833, Academic Press, London-New York-San Francisco (1977).

Davenport, H., Lewis, D. J.

[1] *Homogeneous additive equations*, Proc. Roy. Soc. London Ser. A 274 (1963), pp. 433-460, Collected works, Vol. 3, pp. 1313-1330.

Davison, T. M. K.

[1] *On arithmetic convolutions*, Canad. Math. Bull. 9 (1966), pp. 287-296.

Delange, H.

[1] *Généralisation du theorème de Ikehara*, Ann. Sci. École Norm. Sup. (3) 71 (1954), pp. 213-242.

[2] *Sur les fonctions arithmétiques multiplicatives*, Ann. Sci. École Norm. Sup. (3) 78 (1961), pp. 273-304.

Dickson, L. E.

[1] *Solution of Waring's problem*, Amer. J. Math. 58 (1936), pp. 530-535.

Dodson, M.

[1] *Homogeneous additive congruences*, Philos. Trans. Roy. Soc. London Ser. A 261 (1967), pp. 163-210.

Eggleston, H. G.

[1] *Convexity*, Cambridge Math. Tracts No. 47, Cambridge Univ. Press, Cambridge (1958).

Elliott, P. D. T. A.
[1]   *A conjecture of Kátai,* Acta Arith. 26 (1974), pp. 11-20.

Ellison, W. J.
[1]   *Waring's problem,* Amer. Math. Monthly 78 (1971), pp. 10-36.
[2]   *Les nombres premiers,* Hermann, Paris 1975 (in collaboration with M. Mendès, France).

Erdös, P.
[1]   *On the normal number of prime factors of p − 1 and some related problems concerning Euler's φ-function,* Quart. J. Math. Oxford Ser. 6 (1935), pp. 205-213.
[2]   *Über die Reihe Σ 1/p,* Mathematika 7 (1938), pp. 1-2.
[3]   *On the density of some sequences of numbers,* III. J. London Math. Soc. 13 (1938), pp. 119-127.
[4]   *On the distribution of additive functions,* Ann. of Math. 47 (1946), pp. 1-20.

Erdös, P., Fuchs, W. H. J.
[1]   *On a problem of additive number theory,* J. London Math. Soc. 31 (1956), pp. 67-73.

Erdös, P., Kac, M.
[1]   *The Gaussian law of errors in the theory of additive number theoretic functions,* Amer. J. Math. 62 (1940), pp. 738-742.

Erdös, P., Wintner, A.
[1]   *Additive arithmetical functions and statistical independence,* Amer. J. Math. 61 (1939), pp. 713-721.

Fel'dman, N. I.
[1]   *An effective refinement of the exponent in Liouville's theorem* (Russian). Izv. Akad. Nauk SSSR Ser. Mat. 35 (1971), pp. 973-990; translation in Math. USSR Izv. 5 (1971), pp. 985-1002.

Fel'dman, N. I., Shidlovskii, A. B.
[1]   *The development and present state of the theory of transcendental numbers* (Russian). Uspehi Mat. Nauk 22 (1967), No. 3 (135), pp. 3-81; translation in Russian Math. Surveys 22, No. 3 (1967), pp. 1-79.

Feller, W.
[1]   *An introduction to probability theory and its applications,* I (3rd ed.) 1968, II (3rd ed.) 1971, Wiley & Sons Inc., New York.

Fredman, M. L.
[1]   *Arithmetical convolution products and generalizations,* Duke Math. J. 37 (1970), pp. 231-242.

Gallagher, P. X.

[1]   *The large sieve*, Mathematika 14 (1967), pp. 14-20.

Gerst, O., Brillhart, J.

[1]   *On the prime divisors of polynomials*, Amer. Math. Monthly 78 (1971), pp. 250-266.

Hadamard, J.

[1]   *Sur la distribution des źeros de la fonction ξ(s) et ses conséquences arithmétiques*, Bull. Soc. Math. France 24 (1896), pp. 199-200.

Halberstam, H., Roth, K. F.

[1]   *Sequences*, I, Oxford Univ. Press, Oxford (1966).

Hardy, G. H.

[1]   *On Dirichlet's divisor problem*, Proc. London Math. Soc. (2) 15 (1916), pp. 1-25; Collected papers, Vol. II, Oxford Univ. Press, Oxford (1967), pp. 268-293.

Hardy, G. H., Littlewood, J. E.

[1]   *Partitio Numerorum V; A further contribution to the study of Goldbach's problem*, Proc. London Math. Soc. (2) 22 (1924), pp. 46-50; Collected papers, Vol. I, Oxford Univ. Press, Oxford (1966), pp. 632-642.

Hardy, G. H., Ramanujan, S. S.

[1]   *The normal number of prime factors of a number n*, Quart. J. of Pure and Appl. Math. 48 (1917), pp. 76-92; Collected papers of Srinivasa Ramanujan, Chelsea Pub. Co., New York (1962), pp. 262-275.

[2]   *Asymptotic formulae in combinatory analysis*, Proc. London Math. Soc. (2) 17 (1918), pp. 75-115; Collected papers of Srinivasa Ramanujan, Chelsea Pub. Co., New York (1962), pp. 276-309.

Hardy, G. H., Wright, E. M.

[1]   *An introduction to the theory of numbers*, 5th ed. Oxford Univ. Press, London (1979).

Heath-Brown, D. R., Iwaniec, H.

[1]   *On the difference between consecutive prime numbers*, Invent. Math. 55 (1979), pp. 49-69.

Hilbert, D.

[1]   *Beweis für die Darstellbarkeit der ganzen Zahlen durch eine feste Anzahl n-ter Potenzen* (Waringsche Problem), Math. Ann. 67 (1909), pp. 281-300; Gesammelte Abhandlungen, Bd. I, Chelsea Pub. Co., New York (1965).

Hinćin, A. Ja
[1]   *Cepnye drobi,* Gosudarstv. Izdat. Tehn.-Teor. Lit., Moscow-Leningrad (1961);
      translation: *Continued fractions,* The Univ. of Chicago Press, Chicago, Ill.-
      London (1964).

Hooley, C.
[1]   *On Artin's conjecture,* J. Reine Angew. Math. 225 (1967), pp. 209-220.

Hua, L. K.
[1]   *On an exponential sum,* J. Chinese Math. Soc. 2 (1940), pp. 301-312.
[2]   *Abschätzungen von Exponentialsummen und ihre Anwendung in der
      Zahlentheorie,* Enzyklopädie Math. Wiss. 12, Heft 13, I, B. G. Teubner
      Verlargesellschaft, Leipzig (1959).
[3]   *Introduction to Number Theory,* Springer Verlag, Berlin-Heidelburg-New York
      (1982). [(Translation of the Chinese edition, "Su lung taoyen", Science Press,
      Beijing (1964).]

Huxley, M. N.
[1]   *Small differences between consecutive primes,* Mathematika 20 (1973), pp. 229-
      232.

Ingham, A. E.
[1]   *On two classical lattice point problems,* Proc. Cambridge Philos. Soc. 36 (1940),
      pp. 131-138.

Járnik, V.
[1]   *O mříżových bodech v rovině,* Rozprawy Ćeskoslovenske Akad. Véd Řada Mat.
      Přirod. Véd 33 (1924), p. 36.

Kátai, I.
[1]   *A remark on additive arithmetic functions,* Ann. Univ. Sci. Budapest. Eötvös
      Sect. Math. 10 (1967), pp. 81-83.
[2]   *A remark on number-theoretical functions,* Acta Arith. 14 (1967), pp. 409-415.
[3]   *Characterization of additive functions by its local behavior,* Ann. Univ. Sci.
      Budapest. Eötvös Sect. Math. 12 (1969), pp. 35-37.
[4]   *On sets characterizing number-theoretical functions,* II, Acta Arith. 16 (1969),
      pp. 1-4.
[5]   *On a problem of P. Erdös,* J. Number Theory 2 (1970), pp. 1-6

Kolesnik, G. A.
[1]   *On the estimation of certain trigonometric sums,* Acta Arith. 25 (1973), pp. 7-30.

Kubilius, I. P.
[1]   *On the distribution of the values of additive arithmetic functions,* (Russian),
      Dokl. Akad. Nauk SSSR 100 (1955), pp. 623-626.

[2] *Probabilistic methods in the theory of numbers*, Transl. of Math. Monographs, Vol. 11, Amer. Math. Soc., Providence, R. I. (1964).

Kuipers, L., Niederreiter, H.

[1] *Uniform distribution of sequences*, John Wiley and Sons, Inc., New York-London (1974).

Lambek, J.

[1] *Arithmetical functions and distributivity*, Amer. Math. Monthly 73 (1966), pp. 969-973.

Landau, E.

[1] *Ueber die zahlentheoretische Funktion* $\phi(n)$ *und ihre Bedeutung zum Goldbachschen Satz*, Nachr. Akad. Wiss. Göttingen Math.-Phys. Kl. (1900), pp. 181-184.

[2] *Uber den Verlauf der zahlentheoretischer Funktion* $\phi(n)$, Arch. Math. Phys. 5 (1903), pp. 86-91.

Lang, S.

[1] *Transcendental numbers and diophantine approximations*, Bull. Amer. Math. Soc. 77 (1971), pp. 635-677.

Lehmer, D. H.

[1] *On a conjecture of Ramanujan*, J. London Math. Soc. 11 (1936), pp. 114-118.

Lekkerkerker, C. G.

[1] *Geometry of numbers*, Bibliotheca Mathematica, Vol. 8, Wolters-Noordhoff Pub., Groningen; North-Holland Pub. Co., Amsterdam-London (1969).

Linnik, Ju. V.

[1] *The large sieve*, C. R. (Dokl.) Acad. Sci. USSR (N.S.) 30, (1941), pp. 292-294; Izbrannye Trudy (Selected works), Vol. II, Nauka, Leningrad (1979), pp. 293-296.

Mahler, K.

[1] *On the fractional parts of the powers of a rational number*, II, Mathematika 4 (1957), pp. 122-124.

Mann, H. B.

[1] *A proof of the fundamental theorem on the density of sums of sets of positive integers*, Ann. of Math. 43 (1942), pp. 523-527.

Mauclaire, J. L.

[1] *Sur la régularité des fonctions additives*, Enseign. Math. (2), 18 (1972), pp. 169-174.

McCarthy, P. J.

[1] *On a certain family of arithmetic functions*, Amer. Math. Monthly **65** (1959), pp. 586-590.

Montgomery, H. L.

[1] *Topics in multiplicative number theory*, Lect. Notes Math. Vol. **227**, Springer Verlag, Berlin-Heidelberg-New York (1971).

Montgomery, H. L., Vaughan, R. C.

[1] *The large sieve*, Mathematika **20** (1973), pp. 119-134.

Mordell, J. L.

[1] *On a sum analogous to Gauss's sum*, Quart. J. Math. Oxford Ser. **3** (1932), pp. 161-167.

[2] *Diophantine equations*, Pure Appl. Math. Vol. **30**, Academic Press, London-New York (1969).

Moser, L.

[1] *On the series* $\Sigma \ 1/p$, Amer. Math. Monthly **65** (1958), pp. 104-105.

Moser, L., Lambek, J.

[1] *On monotone multiplicative functions*, Proc. Amer. Math. Soc. **4** (1953), pp. 544-545.

Nagell, T.

[1] *Sur un théorème d'Axel Thue*, Ark. Mat. **1** (1951), pp. 489-496.

Narkiewicz, W.

[1] *On a class of arithmetical convolutions*, Colloq. Math. **10** (1963), pp. 81-94.

[2] *Elementary and analytic theory of algebraic numbers*, Monographie Matematyczne, Vol. **52**, PWN (Polish Scientific Publishers), Warsaw (1974).

Niven, I.

[1] *An unsolved case of the Waring problem*, Amer. J. Math. **66** (1944), pp. 137-143.

Nosarzewska, M.

[1] *Evaluation de la différence entre l'aire d'une région plane convexe et le nombre des points aux coordonnées entières couverts par elle*, Colloq. Math. **1** (1948), pp. 305-311.

Ostmann, H.-H.

[1] *Additive Zahlentheorie*, I, II, Ergebnisse Math., Bd. **7, 11**, Springer Verlag, Berlin-Heidelberg-New York (1968).

Ostrowski, A.
[1] *Über einige Lösungen der Funktionalgleichung* $\phi(x)\phi(y) = \phi(xy)$, Acta Math. 41 (1918), pp. 271-284.

Page, A.
[1] *On the number of primes in an arithmetic progression*, Proc. London Math. Soc. (2), 39 (1935), pp. 116-141.

Pillai, S.
[1] *On Waring's problem.* $g(6) = 73$, Proc. Indian Acad. Sci. Sect. A 12 (1940), pp. 30-40.

Pollard, J. M.
[1] *An algorithm for testing the primality of any integer*, Bull. London Math. Soc. 3 (1971), pp. 337-340.

Postnikov, A. G.
[1] *Introduction to analytic number theory*, Nauka, Moscow (1971).

Prachar, K.
[1] *Primzahlverteilung*, Grundl. Math. Wiss. 91, Springer Verlag, Berlin-Heidelberg-New York (1957).

Rédei, L.
[1] *Endlich-projektivgeometrisches Analogon des Minkowskischen Fundamentalsatzes*, Acta Math. 84 (1950), pp. 155-158.

Remak, R.
[1] *Vereinfachung eines Blichfeldtschen Beweises aus der Geometrie der Zahlen*, Math. Zeitschr. 26 (1927), pp. 694-699.

Rényi, A.
[1] *On a theorem of P. Erdös and its applications to information theory*, Mathematika 1 (1959), pp. 341-344.
[2] *A new proof of a theorem of Delange*, Pub. Math. Debrecen 12 (1965), pp. 323-329.
[3] *Probability theory*, North-Holland Pub. Co., Amsterdam-London (1970).

Riemann, B.
[1] *Ueber die Anzahl der Primzahlen unter einer gegebener Grösse*, Monatsberichte Akad. Berlin, November (1859); Collected works, Dover Pub. Inc., New York (1953), pp. 145-153.

Rubugunday, R.
[1] *On g(k) in Waring's problem*, J. Indian Math. Soc. 6 (1942), pp. 192-198.

Ryavec, C.
[1] *On additive functions*, Michigan Math. J. 16 (1969), pp. 321-329.

Saltykov, A. I.
[1] *On Euler's function*, Vestnik Moskov. Univ. Ser. I Mat. Meh. 15 (1960), No. 6, pp. 34-50.

Schinzel, A.
[1] *Integer points on conics*, Prace Matematyczne 16 (1972), pp. 133-135; Errata, ibid. 17 (1973), p. 305.

Schmidt, E.
[1] *Zum Hilbertschen Beweise des Waringschen Theorems*, Math. Ann. 74 (1913), pp. 271-274.

Schmidt, W. M.
[1] *Simultaneous approximation to algebraic numbers by rationals*, Acta Math. 125 (1970), pp. 189-201.

Schur, I.
[1] *Über die Existenz unendlich vieler Primzahlen in einiger speziellen arithmetischen Progressionen*, Sitz. Berichte Berliner Math. Gesellschaft 11 (1912), pp. 40-50; Gesammelte Abhandlungen Bd. II pp. 1-11, Springer Verlag, Berlin-Heidelberg-New York (1973).

Serre, J. P.
[1] *A course in arithmetic*, 2nd ed., Springer Verlag, Berlin-Heidelberg-New York (1979).

Shapiro, H. N.
[1] *On the number of primes less than or equal x*, Proc. Amer. Math. Soc. 1 (1950), pp. 346-348.
[2] *Distribution functions of additive arithmetic functions*, Proc. Nat. Acad. Sci. U.S.A. 42 (1956), pp. 426-430.
[3] *On the convolution ring of arithmetical functions*, Comm. Pure Appl. Math. 25 (1972), pp. 287-336.

Siegel, C. L.
[1] *Über die Classenzahl quadratischer Zahlkorper*, Acta Arith. 1 (1936), pp. 83-86; Gesammelte Abhandlungen, Bd. I, pp. 406-409, Springer Verlag, Berlin-Heidelberg-New York (1966).

Sierpiński, W.
[1] *Sur un probleme du calcul des fonctions asymptotiques*, Prace Mat.-Fiz. 17 (1906), pp. 77-118 (Polish); Oeuvres Choisies, Tom I, pp. 73-108.

[2] *Sur une formule donnant tous les nombers premiers*, C. R. Acad. Sci. Paris Ser. A-B **235** (1952), pp. 1078-1079.

[3] *Teoria liczb* [*Théorie des nombers*], 3-ème éd. Monografje Matematyczne (Monographies Mathématiques, Tom 19), PWN (Polish Scientific Publishers), Warszawa-Wrocław (1950).
Teoria liczb, Czesc II (*Théorie des nombers II*], Monografje Matematyczne (Monographies Mathématiques, Tom 38), PWN (Polish Scientific Publishers), Warszawa (1959).

[4] *Elementary theory of numbers*, Monografje Matematyczne (Monographies Mathematiques), Tom 42, PWN (Polish Scientific Publishers), Warszawa (1964).

Slowinski, D.

[1] *Searching for the 27th Mersenne prime*, J. Recreational Math. **11** (1978/79), pp. 258-261.

Steinhaus, H.

[1] *Sur un théoréme de M. V. Járnik*, Colloq. Math. 1 (1948), pp. 1-5.

Stemmler, R. M.

[1] *The ideal Waring theorem for exponents 401-200 000*, Math. Comp. **18** (1964), pp. 144-146.

Stephens, P. J.

[1] *An average result for Artin's conjecture*, Mathematika **16** (1969), pp. 178-188.

Subbarao, M. V.

[1] *Arithmetic functions and distributivity*, Amer. Math. Monthly **75** (1968), pp. 984-988.

[2] *On some arithmetic convolutions*, in *"The theory of arithmetic functions"*, Lect. Notes Math. **251** (1972), pp. 247-271, Springer Verlag, Berlin-Heidelberg-New York.

Thomas, H. E. Jr.

[1] *Waring's problem for twenty-two biquadrates*, Trans. Amer. Math. Soc. **193** (1974), pp. 427-430.

Tietäväinen, A.

[1] *On a problem of Chowla and Shimura*, J. Number Theory **3** (1971), pp. 247-252.

Titchmarsh, E. C.

[1] *A divisor problem*, Rend. Circ. Mat. Palermo **54** (1930), pp. 414-429, corrigenda, ibid. **57** (1933), pp. 478-479.

Tuckerman, B.
[1]   The 24th Mersenne prime, Proc. Nat. Acad. Sci. U.S.A. **68** (1971), pp. 2319-2320.

Turán, P.
[1]   On a theorem of Hardy and Ramanujan, J. London Math. Soc. **9** (1934), pp. 274-276.
[2]   Über einige Verallgemeinerungen eines Satzes von Hardy und Ramanujan, ibid. **11** (1936), pp. 125-133.
[3]   Results in number theory in the Soviet Union (Hungarian) Math. Lápok, (1950), pp. 243-266.

de la Vallée-Poussin, C.
[1]   Recherches analytiques sur la théorie des nombres premiers, Ann. Soc. Sci. Bruxelles **202** (1896), pp. 183-256, 281-297.

van der Waerden, B. L.
[1]   Algebra (in two vols.), Fred. Ungar Pub. Co., New York (1970).

Vinogradov, I. M.
[1]   On an upper bound for $G(n)$, (Russian). Izv. Akad. Nauk SSSR, Ser. Mat. **23** (1959), pp. 637-642.
[2]   The method of trigonometrical sums in the theory of numbers, (Russian), 2nd revised and enlarged ed., Nauka, Moscow (1980); translation of the 1st ed., Statistical Pub. Soc., Calcutta (1975).

Waldschmidt, M.
[1]   Nombres Transcendantes, Lect. Notes Math. Vol. 402, Springer Verlag, Berlin-Heidelberg-New York (1974).

Walfisz, A.
[1]   Gitterpunkte in mehrdimensionalen Kugeln, PWN (Polish Scientific Publishers), Warszawa (1957).
[2]   Weylsche Exponentialsummen in der neueren Zahlentheorie, VEB Deutscher Verlag der Wissenschaften, Berlin (1963).

Weber, H.
[1]   Lehrbuch der Algebra, II (3 vols., 3rd ed.) Chelsea Pub. Co. New York (1961).

Weil, A.
[1]   Sur les courbes algébriques et les variétés qui s'en déduisent, Pub. Inst. Math. Strasbourg No. 7, Hermann & Cie, Éditeurs, Paris 1948; also appears in Courbes algébriques et variétés abéliennes, Hermann, Paris (1971).
[2]   Basic Number Theory, Grundl. Math. Wiss. Bd. 144, Springer Verlag, Berlin-Heidelberg-New York, 3rd ed. (1974).

Weinstock, R.

[1] *Greatest common divisor of several integers and an associated linear diophantine equation*, Amer. Math. Monthly 67 (1960), pp. 664-667.

Weyl, H.

[1] *Über die Gleichverteilung von Zahlen mod Eins*, Math. Ann. 77 (1916), pp. 313-352; Gesammelte Abhandlungen, Bd. I, pp. 563-599, Springer Verlag, Berlin-Heidelberg-New York (1968).

Wieferich, A.

[1] *Beweis des Satzes, dass sich eine jede ganze Zahl als Summen von höchsten neun Kuben darstellen lässt*, Math. Ann. 66 (1909), pp. 95-101.

Williams, H. C., Zarnke, C. R.

[1] *Some prime numbers of the forms $2A3^n + 1$ and $2A3^n - 1$*, Math. Comp. 26 (1972), pp. 995-998.

Wirsing, E.

[1] *Characterization of the logarithm as an additive function*, Proc. Sympos. Pure Math. 20 (1969), pp. 375-381, Amer. Math. Soc., Providence R. I.

INDEX

OTHER TITLES IN MATHEMATICS FROM WORLD SCIENTIFIC